THE ULTIMATE
ENGAA COLLECTION

UniAdmissions

Published by *RAR Medical Services Limited*
www.uniadmissions.co.uk
info@uniadmissions.co.uk
Tel: +44 (0) 208 068 0438

ABOUT THE AUTHORS

Madhivanan graduated with a **1st class degree in Engineering** from the University of Cambridge coming in the top 10% of his year and specialising in bioengineering, instrumentation and control and mechanical engineering. In his final year, he completed a project on the automated diagnosis of otitis media and was awarded the 3rd Year Computer-based Project Prize.

He has won gold medals in the UK Mathematics and Chemistry Olympiads. Madhi has worked with UniAdmissions since 2016 and has successfully tutored many students into Oxbridge. In his spare time, Madhivanan enjoys badminton, chess and travelling.

Peter is one of our Oxbridge admissions tutors at *UniAdmissions*. He is currently in his 4th year of studying Physics at St. Catherine's College, Oxford. He has achieved a distinction in each of his first three years and has been awarded an ATV Scholarship.

Next year he is starting a PhD at Imperial College, focussing on plasma atmospheres around comets. Outside of academia, Peter enjoys playing rugby and escaping into the countryside to climb some rocks.

Rohan is the **Director of Operations** at UniAdmissions and is responsible for its technical and commercial arms. He graduated from Gonville and Caius College, Cambridge and is a fully qualified doctor. Over the last five years, he has tutored hundreds of successful Oxbridge and Medical applicants. He has also authored ten books on admissions tests and interviews.

Rohan has taught physiology to undergraduates and interviewed medical school applicants for Cambridge. He has published research on bone physiology and writes education articles for the Independent and Huffington Post. In his spare time, Rohan enjoys playing the piano and table tennis.

THE ULTIMATE ENGAA COLLECTION

MADHIVANAN ELANGO

PETER STEPHENSON

ROHAN AGARWAL

UniAdmissions

Contents

The Ultimate ENGAA Collection 9

The Basics 9

General Advice 11

Section 1: Physics 15

Physics Questions 19

Advanced Physics Questions 38

Section 1: Maths 45

Maths Questions 47

Advanced Maths Questions 61

Section 2 67

Questions 69

Answer Key 77

Section 1: Worked Answers 78

Section 2: Worked Solutions 126

ENGAA Past Paper Worked Solutions 134

Specimen 136

2016 154

2017 166

2018 176

2019 194

2020 223

2021 249

ENGAA Practice Papers 275

Revision Timetable 280

Getting the most out of Mock Papers 281

Things to have done before using this book 283

Revision Checklist 289

ANSWERS 336

Final Advice 377

Your Free Book 381

PREFACE

First of all, I'd like to thank you for trusting UniAdmissions to help you with your university application.

It's a tough and confusing process to approach in many ways. As time has passed, the application has gotten more competitive and convoluted with things like the introduction of Admissions Tests, and an extremely high calibre of students applying, making differentiating yourself even tougher.

When I was applying to Cambridge, there just weren't any good resources that I felt I could trust my future with. Unsurprisingly, many of my peers felt the exact same way, hence, UniAdmissions was born! We started in 2013 and have had an amazing response from the thousands of students we support with their applications each year.

The purpose of this book is to help you prepare for your upcoming admissions test, and ultimately, help you gain your place to study at Oxford or Cambridge. I sincerely believe this book will help you prepare to the best of its ability, however, words on a page can only take you so far.

Since UniAdmissions' inception in 2013, we've made it our mission to improve our student's Oxbridge acceptance rate year on year. We've been quite successful in doing that; while the average Oxbridge acceptance rate stands at around 18%, UniAdmissions' success rate is consistently at 61% - triple the national average. Everyone and I else behind the scenes at UniAdmissions are always looking at how we can get this success rate to be even higher.

Consider the circumstance – we deal exclusively with applications to Oxford and Cambridge. Both are world-renowned institutes of education, and both garner the brightest and best pupils from around to world to their gates. In order to set yourself apart from this exceptional group of prospective applicants, you must showcase yourself to be truly remarkable.

Over the years, we've made great headway to cracking the Oxbridge formula – as evidenced with our success rates above which we're extremely proud of.

In the past, we've offered our students books, physical courses, tuition, mock interviews, online resources and much more. All of these forms of support do make their mark on your application, however, what we've discovered is that a holistic approach works much better.

This is why, nowadays, we have shifted our focus at UniAdmissions completely towards our Oxbridge Programmes. The Programme is a structured syllabus that covers everything you need to know and practice in order to get your dream offer.

The support we provide splits broadly into four key categories; one-to-one teaching, intensive courses, materials and enrichment supervisions. Each category represents a different style of learning which has its own positives and negatives.

We believe in offering the very best support we can provide to each student who entrusts a portion of their application to us, as you have with this book.

With that said, I hope you enjoy this book and I wish you the very best of luck with your application!

– Dr Rohan Agarwal

HOW TO USE THIS BOOK

Congratulations on taking the first step to your ENGAA preparation! Introduced in 2016, the ENGAA is a difficult exam and you'll need to prepare thoroughly to ensure you get that dream university place.

The *Ultimate ENGAA Collection* is the most comprehensive ENGAA book available – it is the culmination of three top-selling ENGAA books:

- *The Ultimate ENGAA Guide*
- *ENGAA Past Paper Solutions*
- *ENGAA Practice Papers*

Whilst it might be tempting to dive straight in with mock papers, this is not a sound strategy. Instead, you should approach the ENGAA in the three steps shown below. Firstly, start off by understanding the structure, syllabus and theory behind the test. Once you are satisfied with this, move onto doing the 250 practice questions found in *The Ultimate ENGAA Guide* (not timed!). Then, once you feel ready for a challenge, do each past paper under timed conditions. Starting with the 2016 paper, work chronologically through the past papers and carefully check your solutions against the model answers given in *ENGAA Past Paper Worked Solutions*. Finally, once you have exhausted these, go through the two ENGAA Mock Papers found in *ENGAA Practice Papers* – these are a final boost to your preparation.

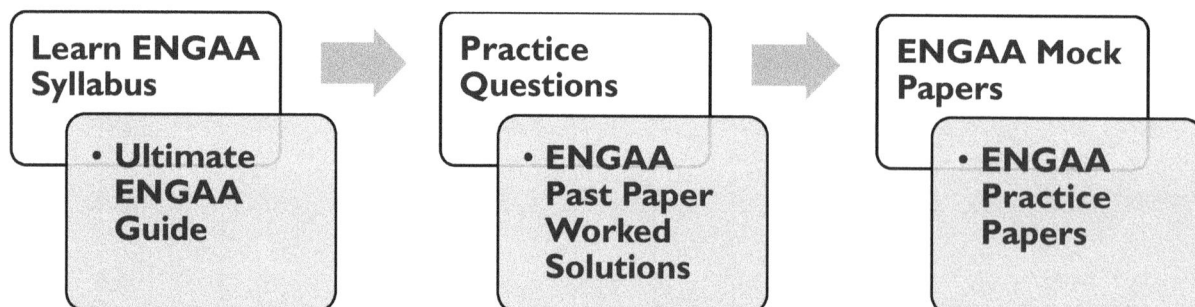

Learn ENGAA Syllabus		Practice Questions		ENGAA Mock Papers
• Ultimate ENGAA Guide	→	• ENGAA Past Paper Worked Solutions	→	• ENGAA Practice Papers

As you have probably realised by now, there are well over 500 questions for you to tackle - this preparation is clearly not intended to be completed within a single week. From our experience, the best students will prepare anywhere between four to eight weeks (although there are some notable exceptions!).

Remember, a methodical approach and regular practise are vital in obtaining a high ENGAA score. Do not fall into the trap that "you can't prepare for the ENGAA"– this could not be further from the truth. With knowledge of the test, some useful time-saving techniques, and plenty of practice, you can dramatically boost your score.

Work hard, persevere through any challenges, and do yourself justice. Good luck!

THE ULTIMATE ENGAA COLLECTION

THE BASICS

What is the ENGAA?

The Engineering Admissions Assessment (ENGAA) is a two-hour written exam taken by prospective Cambridge Engineering applicants.

What does the ENGAA consist of?

Section	Timing	SKILLS TESTED	Questions	Calculator
ONE	60 Minutes	IA: Maths and Physics IB: Advanced Maths & Advanced Physics	20 MCQs 20 MCQs	Not Allowed
TWO	60 Minutes	Advanced Physics	20 MCQs	Not Allowed

Why is the ENGAA used?

Cambridge Engineering applicants tend to be a bright bunch and usually have excellent grades, with many achieving over 90% in all their A level subjects. This means that competition is fierce – the ENGAA, therefore, is required for universities to differentiate between these top candidates.

When do I sit ENGAA?

The ENGAA normally takes place in the final week of October or the first week of November. As the date varies from year to year, it is best to confirm this on the department's site for prospective students:

https://www.undergraduate.study.cam.ac.uk/courses/engineering#entry-requirements

Can I resit the ENGAA?

No, you can only sit the ENGAA once per admissions cycle.

Where do I sit the ENGAA?

You can usually sit the ENGAA at your school or college (ask your exams officer for more information). Alternatively, if your school is not registered, or you are not attending school, you can sit the ENGAA at an authorised test centre.

Do I have to resit the ENGAA if I reapply?

Yes - you cannot use your score from any previous attempts.

How is the ENGAA scored?

In both sections, each question carries one mark and there is no negative marking.

How is the ENGAA used?

The individual weightings of the ENGAA components vary between Cambridge colleges, so it is important that you email the admissions office, of your chosen college, to understand how your marks will be used. In general, the university will interview a high proportion of realistic applicants, so the ENGAA score is not vital for making the interview shortlist. However, it can play a huge role in the final decision after your interview.

GENERAL ADVICE

Start Early

It is much easier to prepare if you practise in shorter, more frequent sessions. Start your preparation well in advance - ideally by mid-September but at the latest by early October. This will give you plenty of time to complete as many papers as required to help you adequately prepare and will avoid last-minute panicking and cramming which is a less effective way to learn. In general, an early start will give you the opportunity to identify the complex issues and work at your own pace.

Prioritise

Some questions in the ENGAA can be time-consuming and complex; given the intense time pressure, you need to be aware of your limits. It is essential that you do not get stuck on very difficult questions. If a question looks particularly complicated or involved, mark it for review and move on. You do not want to be caught 5 questions short at the end because you took 3 minutes to solve a multi-step physics question. If a question is taking too long, choose a sensible answer and move on. Remember that each question carries equal weighting and, therefore, you should adjust your timing accordingly. With practice and discipline, you can become increasingly efficient and learn to maximise your performance.

Positive Marking

There are no penalties for incorrect answers in the ENGAA; you will gain one mark for each right answer and miss a mark for each wrong or unanswered question. This provides you with the luxury of resorting to a guess if you run behind on time or are unable to identify the correct answer. Since each question provides you with 4 to 6 possible answers, you have a 16-25% chance of guessing correctly. Therefore, if you are completely unsure or short on time, then it is best to make an educated guess and move on.

Before simply 'guessing', you should try to eliminate a couple of answers to increase your chances of getting the question correct. For example, if a question has 5 options and you manage to eliminate 2 options, your chances of answering the question correctly increase from 20% to 33%!

Avoid losing easy marks on other questions because of poor exam technique; if you do not manage to finish the exam, take the last 10 seconds to guess the remaining questions to at least give yourself a chance of obtaining more marks.

Practice

This is the best way of familiarising yourself with the style of questions and the timing for this section. Without practise, you are unlikely to be familiar with the style and variety of questions you will encounter in the real examination. Therefore, you want to be comfortable with approaching the style of questions, by adequately practising before you sit the test.

By practising questions, you are less likely to panic during your real examination, which will increase your likelihood of performing well. Initially, work through the questions at your own pace, and spend time carefully reading the questions and looking at any additional data. As you approach the exam date, **make sure you practice the questions under exam conditions.**

Past Papers

The format of the ENGAA has changed for 2019, so past papers do not perfectly reflect the exam you will be sitting this year. Specimen papers are freely available online at **www.uniadmissions.co.uk/ENGAA**. Once you have worked your way through the questions in this book, you are highly advised to attempt them.

Repeat Questions

When checking through answers, pay attention to questions you have answered incorrectly or left blank. If there is a worked answer, follow through the solution carefully, ensuring that you understand the reasoning behind every step, and then repeat the question to check that you can do it independently. If only the answer is given, have another look at the question and attempt to work backwards using the provided answer. This is the best way to learn from your mistakes, and means you are less likely to make similar mistakes when it comes to the test.

The same procedure applies to questions for which you made an educated guess – even if your guess was correct. When working through this book, **make sure you highlight any questions you are unsure of** to indicate you need to spend more time looking over them once marked.

Calculators

You are not permitted to use calculators – thus, it is essential that you have strong numerical skills. For instance, you should be able to rapidly convert between percentages, decimals and fractions. You will seldom get questions that would require a calculator for an exact answer – in this case, you would be expected to arrive at a sensible estimate.

Consider for example:

Estimate 3.962 × 2.322:

3.962 is approximately 4 and 2.323 is approximately 2.33 = 7/3.

Thus, $3.962 \times 2.322 \approx 4 \times \frac{7}{3} = \frac{28}{3} = 9.33$.

Since you will rarely be asked to perform difficult calculations, you can use this as an indicator of whether you are tackling a question correctly. For example, when solving a physics question in section 1, you may end up dividing 8,079 by 357; this should raise alarm bells as calculations in the ENGAA are rarely this difficult.

A word on timing...

"If you had all day to do your ENGAA, you would get 100%. But you don't."

Whilst this is not completely true, it illustrates a very important point. Once you have practiced and know how to approach the style of questions, the clock is your biggest enemy. This seemingly obvious statement has one very important consequence. **The way to improve your ENGAA score is to improve your speed.** There is no magic bullet. But there are a great number of techniques that, with practice, will give you significant time gains, allowing you to answer more questions and score higher marks.

Timing is tight throughout the ENGAA – **mastering timing is the first key to success.** Some candidates choose to work as quickly as possible to allow time at the end to check back, but this is generally not the best method. ENGAA questions can contain a lot of information – each time you start answering a question, it takes time to get familiar with the instructions and information. By splitting the question into two sessions (the first run-through and the return-to-check) you double the amount of time you spend on familiarising yourself with the data, which costs valuable time.

In addition, candidates who do check back may spend 2–3 minutes doing so and yet not make any actual changes. Whilst this can be reassuring, it is a false reassurance as it is unlikely to have a significant effect on your actual score. Therefore, it is usually best to pace yourself very steadily, aiming to spend the same amount of time on each question and finish the final question of the section just as time runs out. This reduces the time spent on re-familiarising with questions and maximises the time spent on the first attempt, gaining more marks.

It is essential that you do not get stuck with the hardest questions – there will undoubtedly be some. In the time spent answering only one of these, you may miss out on answering three easier questions. If a question is taking too long, choose a sensible answer and move on. Never see this as giving up or in any way failing, rather it is the smart way to approach a test with a tight time limit.

With practice and discipline, you can get very good at this and learn to maximise your efficiency. t is not about being a hero and aiming for full marks – this is almost impossible and very much unnecessary (even for Cambridge!). It is about optimising your exam performance and gaining the maximum possible number of marks within the given timeframe.

> *Top tip!* Ensure that you take a watch that can show you the time, in seconds, into the exam. This will allow you have a much more accurate idea of the time you are spending on a question. In general, if you have spent >120 seconds on a section one question, move on regardless of how close you think you are to solving it.

Use the Options:

Some questions may try to overload you with information. When presented with large tables and data, it is essential you look at the answer options so you can focus on the main task. This can allow you to reach the correct answer a lot quicker. Consider the example below:

The table below shows the results of a study investigating antibiotic resistance in staphylococcus populations. A single staphylococcus bacterium is chosen at random from a similar population. Resistance to any one antibiotic is independent of resistance to others.

Antibiotic	Number of Bacteria tested	Number of Resistant Bacteria
Benzyl-penicillin	1011	98
Chloramphenicol	109	1200
Metronidazole	108	256
Erythromycin	105	2

Calculate the probability that the bacterium selected will be resistant to all four drugs.

1 in 10^6	1 in 10^{20}	1 in 10^{30}
1 in 10^{12}	1 in 10^{25}	1 in 10^{35}

Studying the options first makes it clear that there is **no need to calculate exact values** – you only need the powers of 10. This makes your life a lot easier. If you had not noticed this, you might have spent well over 90 seconds trying to calculate the exact value when it was not required for the answer.

In other cases, you may be able to use the options to arrive at the solution quicker, by process of elimination. Consider the example below:

A region is defined by the two inequalities: $x - y^2 > 1 \ and \ xy > 1$. Which of the following points is in the defined region?

(10, 3)	(-10, 3)	(-10, -3)
(10, 2)	(-10, 2)	

Whilst it's possible to solve this question both algebraically and graphically by manipulating the identities, by far **the quickest way is to actually use the options**. Note that options C, D and E violate the second inequality, narrowing down the answers to either A or B. For A: $10 - 3^2 = 1$, which is on the boundary of the defined region and not actually within it. Thus, the answer is B (as $10 - 2^2 = 6$, which satisfies the first inequality).

In general, it pays dividends to inspect the options briefly, to determine whether you can eliminate certain answers. Get into this habit early – it may feel unnatural at first, but it is guaranteed to save you time in the long run.

Keywords

If you are stuck on a question, pay attention to the options that contain key modifiers like "**always**", "**only**", "**all**" as examiners like using them to test if there are any gaps in your knowledge. For example, the statement "arteries carry oxygenated blood" would normally be true; however, "All arteries carry oxygenated blood" would be false because the pulmonary artery carries deoxygenated blood.

SECTION 1: PHYSICS

Section 1 tests your aptitude in Mathematics and Physics. The questions are, approximately, split evenly between GCSE level and A level. You must answer 40 questions in 60 minutes - therefore, this is a very time-pressured section.

The questions can be quite difficult, and it is easy to get bogged down. The challenging nature of the questions, coupled with the intense time pressure of having to do one question every 90 seconds, makes this a difficult section.

Gaps in Knowledge

In addition to GCSE level physics and maths, you are also expected to have a firm command of A-level topics. This can be a problem if you have not studied these topics at school yet. A summary of the specification is provided later in the book, but you are highly advised to go through the official ENGAA Specification and ensure that you have covered all examinable topics. An electronic copy of this can be obtained from **www.uniadmissions.co.uk/ENGAA**.

The questions in this book will help highlight any gaps in your knowledge or areas of weakness that you may have. Upon discovering these, make sure you take some time to revise these topics before carrying on – there is little to be gained by attempting questions with huge gaps in your knowledge.

Maths

Mathematical aptitude is extremely important for the ENGAA; many students find that improving their numerical and algebraic skills usually results in notable improvements in their section 1 and 2 scores. Maths pervades the ENGAA – so, if you find yourself consistently running out of time in practice papers, spending a few hours on brushing up your basic maths skills may do wonders for you.

Physics

The syllabus of assumed knowledge, for the physics and advanced physics questions, can be found within

the official specification here:

https://www.undergraduate.study.cam.ac.uk/files/publications/engaa_specification_2020.pdf

Multi-Step Questions

Most ENGAA physics questions require two step calculations. Consider the following example:

A metal ball is released from the roof of a 20-metre building. Assuming air resistance is negligible, calculate the velocity at which the ball hits the ground. You are given that $g = 10$ ms^{-2}.

- 5 ms^{-1}
- 15 ms^{-1}
- 25 ms^{-1}
- 10 ms^{-1}
- 20 ms^{-1}

When the ball hits the ground, all of its gravitational potential energy has been converted into kinetic energy.

$$E_p = E_k \ \Rightarrow \ \text{mg}\Delta\text{h} = \frac{\text{mv}^2}{2}$$

$$\therefore v = \sqrt{2gh} = \sqrt{2 \times 10 \times 20} = \sqrt{400} = 20 \, ms^{-1}$$

In this example, you were required to not only recall two equations, but apply and rearrange them to find a numerical answer – all in under 60 seconds. Note that, if you were comfortable with basic Newtonian mechanics, you could have solved this using a single SUVAT equation:

$$v^2 = u^2 + 2as \ \Rightarrow \ v = \sqrt{2 \times 10 \times 20} = 20 \, ms^{-1}$$

SI Units

Remember that, in order to get the correct answer, you must always work in SI units! Do your calculations in terms of metres (not centimetres) and kilograms (not grams), and so on.

Top tip! Knowing SI units is extremely useful because they allow you to **'work out' equations** if you ever forget them. For example, the units for density are kg/m³. Since *kg* is the SI unit for mass, and *m³* is represented by volume, the equation for density must be Density = Mass/Volume.

This can also work the other way; for example, we know that the unit for Pressure is Pascal (Pa). But based on the fact that Pressure = Force/Area, a Pascal must be equivalent to N/m².

Formulas you MUST know:

Equations of Motion (where s = displacement, u = initial velocity, v = final velocity, a = acceleration and t = time):

- $s = ut + \frac{1}{2}at^2$
- $v = u + at$
- $a = \frac{v-u}{t}$
- $v^2 = u^2 + 2as$

Equations Relating to Force:

- Force = Mass x Acceleration (for **constant mass**).
- Force = Rate of change of momentum.
- Pressure = Force / Area.
- Moment of a Force = Force × Distance from pivot.
- Work done = Force x Displacement in direction of force.

Mechanics & Motion:

- Conservation of momentum: $\Delta p = \Delta mv = 0$ (when there is **no net external force**).
- Force = Rate of change of momentum $= \frac{\Delta p}{\Delta t} = \frac{\Delta mv}{\Delta t}$.
- Angular velocity = Rate of change of angular displacement $= \omega = \frac{v}{r} = 2\pi f$ (not explicitly in the ENGAA syllabus but useful for some questions).
- Stress = Force / Area.
- Strain = Extension / Original length.
- Young's modulus = Stress / Strain.

Equations relating to Energy:

- Kinetic Energy $= \frac{1}{2}mv^2$ (for an object with mass **m** and velocity **v**).
- Gravitational Potential Energy $= mgh$ (for an object with mass **m** at height **h** above ground).
- Energy Efficiency = (Useful energy / Total energy) × 100%.

Equations relating to Power:

- Power = Work done / Time.
- Power = Energy transferred / Time.
- Power = Force × Velocity (for **constant force** and velocity **in direction** of the force)

Factor	Text	Symbol
10^{12}	Tera	T
10^9	Giga	G
10^6	Mega	M
10^3	Kilo	k
10^2	Hecto	h
10^{-1}	Deci	d
10^{-2}	Centi	c
10^{-3}	Milli	m
10^{-6}	Micro	μ
10^{-9}	Nano	n
10^{-12}	Pico	p

Magnetic Fields:

- Magnetic force on a **straight, current-carrying wire** in field: $F = BIL$.
- Magnetic flux linkage = Magnetic Flux x Number of coils $= \emptyset N = BAN$.
- Magnitude of induced EMF: $\varepsilon = N\frac{\Delta\emptyset}{\Delta t}$.

Electrical Equations:

- Charge (Q) = Current × Time = It.
- Voltage (V) = Current × Resistance = IR.
- Power = IV = I²R = V²/R.
- Electromotive Force (EMF) = $\varepsilon = \frac{E}{Q} = I(R + r)$.
- Resistivity = $\rho = \frac{RA}{l}$.
- Resistors in series: $R_T = \sum_{i=1}^{n} R_i$ = **Sum** of all resistances.
- Resistors in parallel: $\frac{1}{R_T} = \sum_{i=1}^{n} \frac{1}{R_i}$ = **Sum** of **reciprocals** of all resistances.

For objects in equilibrium:

- Sum of Clockwise moments = Sum of Anti-clockwise moments.
- Sum of all Resultant Forces = 0.

Waves:

- Speed = Frequency × Wavelength = $c = f\lambda$ (only for **electromagnetic** waves).
- Time Period = 1 / Frequency.
- Snell's law: $n_1 sin\theta_1 = n_2 sin\theta_2$ (where n_1, n_2 are the refractive indices, θ_1 = angle of incidence and θ_2 = angle of refraction).

Radioactivity:

- Decay: $N = N_0 e^{-\lambda t}$ (where N_0 = Initial population size, λ = Decay constant).
- Half-life: $T_{1/2} = \frac{ln2}{\lambda}$ (time taken for the population to half).
- Activity: $A = \lambda N$.
- Energy: $E = mc^2$ (where **m** = mass, **c** = speed of light).

Other:

- Weight = Mass × g.
- Density = Mass / Volume.
- Momentum = Mass × Velocity.
- **g = 9.81 ms-2** (unless otherwise stated).

PHYSICS QUESTIONS

Question 1:
Which of the following statements is **FALSE**?

A. Electromagnetic waves can cause substances to heat up.
B. X-rays and gamma rays can knock electrons out of their orbits.
C. Loud sounds can make solid objects vibrate.
D. Wave power can be used to generate electricity.
E. As a wave propagates outwards, the source of the wave loses energy.
F. The amplitude of a wave determines its mass.

Question 2:
A spacecraft is analysing a newly discovered exoplanet. A rock of unknown mass falls, from rest, on the planet from a height of 30 m. Given that $g = 5.4$ ms^{-2} on the planet, calculate the speed of the rock when it hits the ground and the time it took to fall.

	Speed (ms^{-1})	Time (s)
A	18	3.3
B	18	3.1
C	12	3.3
D	10	3.7
E	9	2.3
F	1	0.3

Question 3:
A canoe floating on the sea rises and falls 7 times in 49 seconds. The waves pass it at a speed of 5 ms^{-1}. How long are the waves?

A. 12 m
B. 22 m
C. 25 m
D. 35 m
E. 57 m
F. 75 m

Question 4:
Miss Orrell lifts her 37.5 kg bike vertically for a distance of 1.3 m in 5 s. The acceleration of free fall is 10 ms^{-2}. What is the average power that she develops?

A. 9.8 W
B. 12.9 W
C. 57.9 W
D. 79.5 W
E. 97.5W
F. 100.0 W

Question 5:

A truck accelerates at 5.6 ms⁻², from rest, for 8 seconds. Calculate the final speed and the distance travelled in 8 seconds.

	Final Speed (ms⁻¹)	Distance (m)
A	40.8	119.2
B	40.8	129.6
C	42.8	179.2
D	44.1	139.2
E	44.1	179.7
F	44.2	129.2
G	44.8	179.2
H	44.8	199.7

Question 6:

Which of the following statements is true when a sky diver jumps out of a plane?

A. The sky diver will accelerate until the air resistance is greater than their weight.
B. The sky diver will accelerate until the air resistance is less than their weight.
C. The sky diver will accelerate until the air resistance equals their weight.
D. The sky diver will accelerate until the air resistance equals their weight squared.
E. The sky diver will travel at a constant velocity after leaving the plane.

Question 7:

A 100 g apple falls on Isaac's head from a height of 20 m. Calculate the apple's momentum before the point of impact. Take $g = 10$ ms⁻².

A. 0.2 kgms⁻¹   C. 1 kgms⁻¹   E. 10 kgms⁻¹
B. 0.5 kgms⁻¹   D. 2 kgms⁻¹   F. 20 kgms⁻¹

Question 8:

Which of the following properties do all electromagnetic waves all have in common?

1. They can travel through a vacuum.
2. They can be reflected.
3. They have the same wavelength.
4. They have the same amount of energy.
5. They can be polarised.

A. 1, 2 and 3 only   D. 3 and 4 only
B. 1, 2, 3 and 4 only   E. 1, 2 and 5 only
C. 4 and 5 only   F. 1 and 5 only

Question 9:

A battery with an internal resistance of 0.8 Ω and an EMF of 36 V is used to power a drill with resistance 1 Ω. What is the current in the circuit when the drill is connected to the power supply?

A. 5 A   C. 15 A   E. 25 A
B. 10 A   D. 20 A   F. 30 A

Question 10:

Officer Bailey throws a 20 g dart at a speed of 100 ms^{-1}. It strikes the dartboard and is brought to rest in 10 milliseconds. Calculate the average force exerted on the dart by the dartboard.

A. 0.2 N   C. 20 N   E. 2,000 N
B. 2.0 N   D. 200 N   F. 20,000 N

Question 11:

Professor Huang lifts a 50 kg bag through a distance of 0.7 m in 3 s. What average power does she develop to 3 significant figures? Take g = 10 ms^{-2}.

A. 112 W   C. 114 W   E. 116 W
B. 113 W   D. 115 W   F. 117 W

Question 12:

An electric scooter is travelling at a constant speed of 30 ms^{-1}. Its constant speed is maintained by a driving force of 300 N in the direction of motion, which works against a frictional force. Given that the engine runs at 200 V, calculate the current in the scooter, given that the engine is running at 100% efficiency.

A. 4.5 A   C. 450 A   E. 45,000 A
B. 45 A   D. 4,500 A   F. More information needed.

Question 13:

Which of the following statements about the physical definition of work are correct?

1. Work done $= \frac{\text{Force}}{\text{distance}}$.
2. The unit of work is equivalent to kg ms^{-2}.
3. Work is defined as a force multiplied by displacement of the body in the direction of the force.

A. Only 1   D. 1 and 2
B. Only 2   E. 2 and 3
C. Only 3   F. 1 and 3

Question 14:

Which of the following statements about kinetic energy are correct?

1. It is defined as $E_k = \frac{mv^2}{2}$, where m = mass and v = velocity.
2. The unit of kinetic energy is equivalent to Pa $\times$ m³.
3. Kinetic energy is equal to the minimum amount of energy needed to decelerate the body in question from its current speed to rest.

A. Only 1
B. Only 2
C. Only 3
D. 1 and 2
E. 2 and 3
F. 1, 2 and 3

Question 15:

In relation to radiation, which of the following statements is **FALSE**?

A. Radiation is the emission of energy in the form of waves or particles.
B. Radiation can be either ionizing or non-ionizing.
C. Gamma radiation has very high energy.
D. Alpha radiation is of lower energy than beta radiation.
E. X-rays are an example of particle radiation.

Question 16:

In relation to the physical definition of half-life, which of the following statements are correct?

1. In radioactive decay, the half-life is independent of atom type and isotope.
2. Half-life is defined as the time required for exactly half of the population to decay.
3. A mass with a constant half-life will exhibit exponential decay.

A. Only 1
B. Only 2
C. Only 3
D. 1 and 2
E. 2 and 3
F. 1 and 3

Question 17:

In relation to nuclear fusion, which of the following statements is **FALSE**?

A. Nuclear fusion is initiated by the absorption of neutrons.
B. Nuclear fusion describes the fusion of hydrogen atoms to form helium atoms.
C. Nuclear fusion releases great amounts of energy.
D. Nuclear fusion requires high activation temperatures.
E. Nuclear fusion is used in stars.

Question 18:

In relation to nuclear fission, which of the following statements is correct?

A. Nuclear fission is the basis of many nuclear weapons.
B. Nuclear fission is triggered by the shooting of neutrons at unstable atoms.
C. Nuclear fission can trigger chain reactions.
D. Nuclear fission commonly results in the emission of ionizing radiation.
E. All of the above.

Question 19:

Two identical resistors (R_a and R_b) are connected in a series circuit. Which of the following statements are true?

1. The current through both resistors is the same.
2. The voltage across both resistors is the same.
3. The voltage across the two resistors is given by Ohm's Law.

A. Only 1	E. 2 and 3
B. Only 2	F. 1 and 3
C. Only 3	G. 1, 2 and 3
D. 1 and 2	H. None of the statements are true.

Question 20:

The Sun is 8 light-minutes away from the Earth. Estimate the circumference of the Earth's orbit around the Sun. Assume that the Earth is in a circular orbit around the Sun. You may take the speed of light to be 3×10^8 ms^{-1}.

A. 10^{24} m	C. 10^{18} m	E. 10^{12} m
B. 10^{21} m	D. 10^{15} m	F. 10^9 m

Question 21:

Which of the following statements are true?

1. Speed is the same as velocity.
2. The standard (SI) unit for speed is ms^{-2}.
3. Velocity = Distance / Time.

A. Only 1	E. 2 and 3
B. Only 2	F. 1 and 3
C. Only 3	G. 1, 2 and 3
D. 1 and 2	H. None of the statements are true.

Question 22:

Which of the following statements best defines Ohm's Law?

A. The current through an insulator between two points is indirectly proportional to the potential difference across the two points.

B. The current through an insulator between two points is directly proportional to the potential difference across the two points.

C. The current through a conductor between two points is inversely proportional to the potential difference across the two points.

D. The current through a conductor between two points is proportional to the square of the potential difference across the two points.

E. The current through a conductor between two points is directly proportional to the potential difference across the two points.

Question 23:

Which of the following statements regarding Newton's Second Law are always correct?

1. For objects at rest, the resultant force acting upon them must be zero.
2. Force = Mass × Acceleration.
3. Force = Rate of change of Momentum.

A. Only 1

B. Only 2

C. Only 3

D. 1 and 2

E. 2 and 3

F. 1 and 3

G. 1, 2 and 3

Question 24:

Which of the following equations concerning electrical circuits are correct?

1. $\text{Charge} = \dfrac{\text{Voltage x time}}{\text{Resistance}}$

2. $\text{Charge} = \dfrac{\text{Power x time}}{\text{Voltage}}$

3. $\text{Charge} = \dfrac{\text{Current x time}}{\text{Resistance}}$

A. Only 1

B. Only 2

C. Only 3

D. 1 and 2

E. 2 and 3

F. 1 and 3

G. 1, 2 and 3

H. None of the statements are true.

24

Question 25:

An elevator has a mass of 1,600 kg and is carrying passengers that have a combined mass of 200 kg. A constant frictional force of 4,000 N acts upon the elevator. What force must the motor provide for the elevator to move with an upward acceleration of 1 ms^{-2}? You may assume that $g = 10$ ms^{-2}.

A. 1,190 N   C. 18,000 N   E. 23,800 N
B. 11,900 N   D. 22,000 N

Question 26:

A 1,000 kg car accelerates from rest at 5 ms^{-2} for 10 seconds. Then, a constant braking force is applied to bring it to rest within 20 seconds. What distance has the car travelled in total?

A. 125 m   C. 650 m   E. 1,200 m
B. 250 m   D. 750 m   F. More information needed.

Question 27:

An electric heater is connected to 120 V mains by a copper wire which has a resistance of 8 ohms. What is the power of the heater?

A. 90 W   E. 9,000W
B. 180 W   F. 18,000 W
C. 900 W   G. More information needed.
D. 1800 W

Question 28:

In a particle accelerator, electron pulses are accelerated through a potential difference of 40 MV and emerge with an energy of 40 MeV (1 MeV = 1.60 x 10^{-13} J). Each pulse contains 5,000 electrons. Assuming that the electrons have zero energy prior to being accelerated, what is the power delivered by the electron beam?

A. 1 kW   C. 100 kW   E. 10,000 kW
B. 10 kW   D. 1,000 kW   F. More information needed.

Question 29:

Which of the following statements regarding heat processes is correct?

A. When an object is in equilibrium with its surroundings, there is no energy transferred to or from the object and so its temperature remains constant.
B. When an object is in equilibrium with its surroundings, it radiates and absorbs energy at the same rate and so its temperature remains constant.
C. Radiation is faster than convection but slower than conduction.
D. Radiation is faster than conduction but slower than convection.
E. None of the above.

Question 30:

A 6 kg block is pulled from rest along a horizontal, frictionless surface by a constant, horizontal force of 12 N. Calculate the speed of the block after it has moved 300 cm.

A. $2\sqrt{3}\,\text{ms}^{-1}$

B. $4\sqrt{3}\,\text{ms}^{-1}$

C. $4\sqrt{3}\,\text{ms}^{-1}$

D. $12\,\text{ms}^{-1}$

E. $\sqrt{\frac{3}{2}}\,\text{ms}^{-1}$

Question 31:

A 100 V heater heats 1.5 litres of pure water from 10°C to 50°C in 50 minutes. Given that 1 kg of pure water requires 4,000 J to raise its temperature by 1∘C, calculate the resistance of the heater.

A. 12.5 ohms

B. 25 ohms

C. 125 ohms

D. 250 ohms

E. 500 ohms

F. 850 ohms

Question 32:

Which of the following statements are true?

1. Nuclear fission is the basis of nuclear energy.
2. Following fission, the resulting atoms are a different element to the original type.
3. Nuclear fission often results in the production of free neutrons and photons.

A. Only 1

B. Only 2

C. Only 3

D. 1 and 2

E. 2 and 3

F. 1 and 3

G. 1, 2 and 3

H. None of the statements are true.

Question 33:

Which of the following statements are correct? You may assume that $g = 10\,\text{ms}^{-2}$.

- Gravitational potential energy is defined as $\Delta E_p = m \times g \times \Delta h$.
- Gravitational potential energy is a measure of the work done against gravity.
- A reservoir situated 1 km above ground level with 10^6 litres of water has a potential energy of 1 Giga Joule, in reference to ground level.

A. Only 1

B. Only 2

C. Only 3

D. 1 and 2

E. 2 and 3

F. 1 and 3

G. 1, 2 and 3

H. None of the statements are true.

Question 34:

Which of the following statements are correct in relation to Newton's Third Law?

1. For every action, there is an equal and opposite reaction.
2. According to Newton's third law, there are no isolated forces.
3. Rockets cannot accelerate in deep space because there is no matter to generate an equal and opposite force.

A. Only 1

B. Only 2

C. Only 3

D. 1 and 2

E. 2 and 3

F. 1 and 3

Question 35:

Which of the following statements regarding electrostatic charge are correct?

1. Positively charged objects have gained electrons.
2. The amount of electrical charge flow in a circuit, over a known period of time, can be calculated if the voltage and resistance of the circuit are known.
3. Objects can be charged by friction.

A. Only 1
B. Only 2
C. Only 3
D. 1 and 2

E. 2 and 3
F. 1 and 3
G. 1, 2 and 3

Question 36:

Which of the following statements regarding gravity is true?

A. The gravitational force between two objects is independent of their mass.
B. Each planet in the solar system exerts a gravitational force on the Earth.
C. For satellites in a geostationary orbit, acceleration due to gravity is equal and opposite to the lift from engines.
D. Two objects that are dropped from the Eiffel tower will always land on the ground at the same time if they have the same mass.
E. All of the above.
F. None of the above.

Question 37:

Which of the following best defines an electrical conductor?

A. Conductors are usually made from metals and they conduct electrical charge in multiple directions.
B. Conductors are usually made from non-metals and they conduct electrical charge in multiple directions.
C. Conductors are usually made from metals and they conduct electrical charge in one fixed direction.
D. Conductors are usually made from non-metals and they conduct electrical charge in one fixed direction.
E. Conductors allow the passage of electrical charge with zero resistance because they contain freely mobile charged particles.
F. Conductors allow the passage of electrical charge with maximal resistance because they contain charged particles that are fixed and static.

Question 38:

An 800 kg compact car delivers 20% of its power output to its wheels. If the car has a mileage of 30 miles/gallon and travels at a speed of 60 miles/hour, how much power is delivered to the wheels? 1 gallon of petrol contains 9×10^8 J.

A. 10 kW
B. 20 kW
C. 40 kW
D. 50 kW
E. 100 kW

Question 39:

Which of the following statements regarding beta radiation are true?

1. After a beta particle is emitted, the atomic mass number remains unchanged.
2. Beta radiation can penetrate paper but not aluminium foil.
3. A moving beta particle is deflected in both electric and magnetic fields.

A. 1 only	C. 1 and 3	E. 2 and 3
B. 2 only	D. 1 and 2	F. 1, 2 and 3

Question 40:

A car with a weight of 15 kN is travelling at a speed of 15 ms⁻¹. It then crashes into a wall and is brought to rest in 10 milliseconds. Calculate the average braking force exerted on the car by the wall. Take $g = 10$ ms⁻².

A. 1.25×10^4N	C. 1.25×10^6N	E. 2.25×10^5N
B. 1.25×10^5N	D. 2.25×10^4N	F. 2.25×10^6N

Question 41:

Which of the following statements are correct?

1. Electrical insulators are usually metals, such as copper or aluminium.
2. The flow of charge through electrical insulators is extremely low.
3. Electrical insulators can be charged by rubbing them together.

A. Only 1	D. 1 and 2	G. 1, 2 and 3
B. Only 2	E. 2 and 3	
C. Only 3	F. 1 and 3	

The following information is needed for Questions 42 and 43:

The graph below represents a car's motion. At t = 0, the car's displacement is zero.

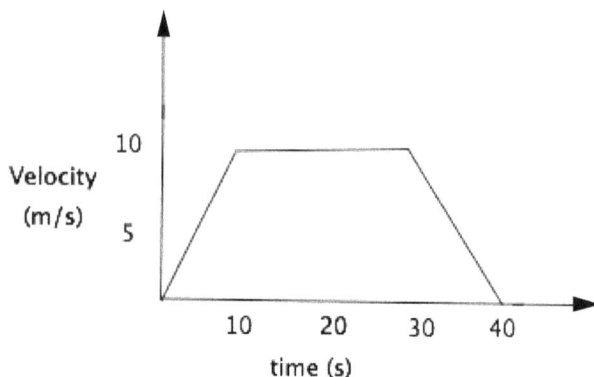

Question 42:

Which of the following statements are **incorrect**?

1. The car is reversing after t = 30 seconds.
2. The car moves with constant acceleration from t = 0 to t = 10 seconds.
3. The car moves with constant speed from t = 10 to t = 30 seconds.

A. Only 1
B. Only 2
C. Only 3
D. 1 and 2

E. 2 and 3
F. 1 and 3
G. 1, 2 and 3

Question 43:

Calculate the total distance travelled by the car during its motion.

A. 200 m
B. 300 m
C. 350 m

D. 400 m
E. 500 m
F. More information needed.

Question 44:

A 1,000 kg rocket is launched and reaches a constant velocity in 30 seconds. Suddenly, a strong gust of wind acts on the rocket for 5 seconds with a force of 10,000 N in its direction of motion. Assuming that the rocket's engines provide a constant force throughout the launch, what is the resulting change in velocity?

A. 0.5 ms^{-1}
B. 5 ms^{-1}
C. 50 ms^{-1}

D. 500 ms^{-1}
E. 5000 ms^{-1}
F. More information needed.

Question 45:

A 0.5 tonne crane lifts a car, with a weight of 0.01 tonnes, by 100 cm vertically in 5,000 milliseconds. Calculate the average power developed by the crane. Take g = 10 ms^{-2}.

A. 0.2 W
B. 2 W
C. 5 W

D. 20 W
E. 50 W
F. More information needed

Question 46:

A 20 V battery is connected to a circuit consisting of a 1 Ω and 2 Ω resistor in parallel. Calculate the total current of the circuit.

A. 6.67 A
B. 8 A

C. 10 A
D. 12 A

E. 20 A
F. 30 A

Question 47:

Which of the following statements is correct?

A. The speed of light changes when it enters water.
B. The speed of light changes when it leaves water.
C. The direction of light changes when it enters water.
D. The direction of light changes when it leaves water.
E. All of the above.
F. None of the above.

Question 48:

In a parallel circuit, a 60 V battery is connected to two branches. Branch *A* contains 6 identical 5 Ω resistors and branch *B* contains 2 identical 10 Ω resistors.

Calculate the current in branches *A* and *B*.

	I_A (A)	I_B (A)
A	0	6
B	6	0
C	2	3
D	3	2
E	3	3
F	1	5
G	5	1

Question 49:

Calculate the voltage of an electrical circuit that has a power output of 50,000,000,000 nW and a current of 0.000000004 GA.

A. 0.0125 GV
B. 0.0125 MV
C. 0.0125 kV
D. 0.0125 V

E. 0.0125 mV
F. 0.0125 μV
G. 0.0125 nV

Question 50:

Which of the following statements regarding radioactive decay is correct?

A. Radioactive decay of an individual atom is highly predictable.
B. An unstable element will continue to decay until it reaches a stable nuclear configuration.
C. All forms of radioactive decay release gamma rays.
D. All forms of radioactive decay release X-rays.
E. An atom's nuclear charge is unchanged after it undergoes alpha decay.
F. None of the above.

Question 51:

A circuit contains three identical resistors of unknown resistance connected in series with a 15 V battery. The power output of the circuit is 60 W.

Calculate the overall resistance of the circuit when two further identical resistors are added to it in series.

A. 0.125 Ω   C. Ω   E. 18.75 Ω
B. 1.25 Ω   D. 6.25 Ω   F. More information needed.

Question 52:

A 5,000 kg tractor's engine uses 1 litre of fuel to move 0.1 km. 1 ml of the fuel contains 20 kJ of energy.

Calculate the engine's efficiency. Take $g = 10$ ms^{-2}.

A. 2.5 %   C. 38 %   E. 75 %
B. 25 %   D. 50 %   F. More information needed.

Question 53:

Which of the following statements are correct?

1. Electromagnetic induction occurs when a current-carrying wire moves relative to a magnet.
2. Electromagnetic induction occurs when a magnetic field changes.
3. An electrical current is generated when a coil rotates in a magnetic field.

A. Only 1   D. 1 and 2   G. 1, 2 and 3
B. Only 2   E. 2 and 3
C. Only 3   F. 1 and 3

Question 54:

Which of the following statements are correct regarding parallel circuits?

1. The current flowing through a branch is dependent on the branch's resistance.
2. The total current flowing into the branches is equal to the total current flowing out of the branches.
3. An ammeter will always give the same reading regardless of its location in the circuit.

A. Only 1   D. 1 and 2   G. All of the above
B. Only 2   E. 2 and 3
C. Only 3   F. 1 and 3

Question 55:

Which of the following statements regarding series circuits are true?

1. The overall resistance of a circuit is given by the sum of all resistors in the circuit.
2. Conventional electrical current moves from the positive terminal to the negative terminal.
3. Electrons move from the positive terminal to the negative terminal.

A. Only 1

B. Only 2

C. Only 3

D. 1 and 2

E. 2 and 3

F. 1 and 3

Question 56:

The graphs below show current vs. voltage plots for 4 different electrical components.

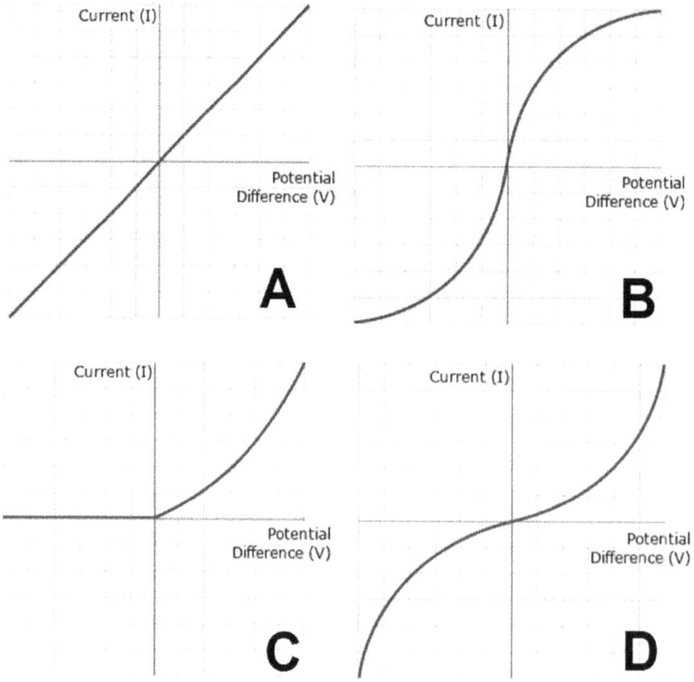

Which of the following graphs represents a resistor at constant temperature, and which a filament lamp?

	Fixed Resistor	Filament Lamp
A	A	B
B	A	C
C	A	D
D	C	A
E	C	C
F	C	D

Question 57:

Which of the following statements are true about vectors?

A. Vectors can be added or subtracted.

B. All vector quantities have a defined magnitude.

C. All vector quantities have a defined direction.

D. Displacement is an example of a vector quantity.

E. All of the above.

F. None of the above.

Question 58:

The acceleration due to gravity on the Earth is six times greater than that on the moon. Dr Tyson records the weight of a rock as 250 N on the moon.

Calculate the rock's density given that it has a volume of 250 cm³. Take g_{Earth} = 10 ms⁻².

A. 0.2 kg/cm³	C. 0.6 kg/cm³	E. 0.8 kg/cm³
B. 0.5 kg/cm³	D. 0.7 kg/cm³	F. More information needed.

Question 59:

A radioactive element X_{78}^{225} undergoes alpha decay. What is the atomic mass and atomic number after 5 alpha particles have been released?

	Mass Number	Atomic Number
A	200	56
B	200	58
C	205	64
D	205	68
E	215	58
F	215	73
G	225	78
H	225	83

Question 60:

A 20 A current passes through a circuit with resistance of 10 Ω. The circuit is connected to a transformer that contains a primary coil with 5 turns and a secondary coil with 10 turns. Calculate the potential difference exiting the transformer. You may use the following formula:

$$\frac{\text{Voltage in secondary coil}}{\text{Voltage in primary coil}} = \frac{\text{Turns in secondary coil}}{\text{Turns in primary coil}}$$

A. 100 V	D. 500 V	G. 5,000 V
B. 200 V	E. 2,000 V	
C. 400 V	F. 4,000 V	

Question 61:

A metal sphere of unknown mass is dropped from an altitude of 1 km and reaches terminal velocity 300 m before it hits the ground. Given that resistive forces do a total of 10 kJ of work for the last 100 m before the ball hits the ground, calculate the mass of the ball. Take $g = 10ms^{-2}$.

A. 1 kg
B. 2 kg
C. 5 kg
D. 10 kg
E. 20 kg
F. More information needed.

Question 62:

Which of the following statements is true about the electromagnetic spectrum?

A. The wavelength of ultraviolet radiation is shorter than that of x-rays.
B. For waves in the electromagnetic spectrum, wavelength is directly proportional to frequency.
C. Most electromagnetic waves can be stopped with a thin layer of aluminium.
D. Waves in the electromagnetic spectrum travel at the speed of sound.
E. Humans are able to visualise the majority of the electromagnetic spectrum.
F. None of the above.

Question 63:

In relation to the Doppler Effect, which of the following statements are true?

1. If an object emitting a wave moves towards the sensor, the wavelength increases and frequency decreases.
2. An object that originally emitted a wave of a wavelength of 20 mm, followed by a second reading delivering a wavelength of 15 mm, is moving towards the sensor.
3. The faster the object is moving away from the sensor, the greater the increase in frequency.

A. Only 1
B. Only 2
C. Only 3
D. 1 and 2
E. 1 and 3
F. 2 and 3
G. 1, 2 and 3
H. None of the above statements are true.

Question 64:

A 5 g bullet travels at 1 km/s and hits a brick wall. It penetrates 50 cm, before being brought to rest, 100 ms after impact. Calculate the average braking force exerted by the wall on the bullet.

A. 50 N
B. 500 N
C. 5,000 N
D. 50,000 N
E. 500,000 N
F. More information needed.

Question 65:

Polonium (Po) is a highly radioactive element that has no known stable isotope. Po^{210} undergoes radioactive decay to Pb^{206} and Y. Calculate the number of protons in 10 moles of Y. [Avogadro's Constant = 6×10^{23}].

A. 0

C. 1.2×10^{25}

E. 2.4×10^{25}

B. 1.2×10^{24}

D. 2.4×10^{24}

F. More information needed

Question 66:

Dr Sale measures a spike of 16,000 Bq from a nuclear rod composed of an unknown material. 300 days later, he visits and can no longer detect a reading higher than 1,000 Bq from the rod, even though the sample has not been disturbed.

What is the longest possible half-life of the nuclear rod?

A. 25 days

C. 75 days

E. 150 days

B. 50 days

D. 100 days

F. More information needed

Question 67:

A radioactive element Y_{89}^{200} undergoes a series of alpha and gamma decays. What are the number of protons and neutrons in the element after the emission of 3 alpha particles and 2 gamma waves?

	Protons	Neutrons
A	79	101
B	83	105
C	83	115
D	89	111
E	89	105
F	93	111
G	93	105
H	109	111

Question 68:

Most symphony orchestras tune to 'standard pitch' (frequency = 440 Hz). When they are tuning, sound directly from the orchestra reaches audience members that are 500 m away in 1.5 seconds.

Estimate the wavelength of 'standard pitch'.

A. 0.05 m

C. 0.75 m

E. 15 m

B. 0.5 m

D. 1.5 m

F. More information needed

Question 69:

A 1-kg cylindrical artillery shell with a radius of 50 mm is fired at a speed of 200 ms^{-1}. It strikes an armour-plated wall and is brought to rest in 500 μs.

Estimate the average pressure exerted on the entire shell, by the wall, at the time of impact.

A. 5×10^6 Pa   C. 5×10^8 Pa   E. 5×10^{10} Pa

B. 5×10^7 Pa   D. 5×10^9 Pa   F. More information needed

Question 70:

A 1,000 W display fountain launches 120 litres of water straight up every minute. Given that the fountain is 10% efficient, calculate the maximum possible height that the stream of water could reach.

Assume that there is negligible air resistance and $g = 10$ ms^{-2}.

A. 1 m   C. 10 m   E. 50 m

B. 5 m   D. 20 m   F. More information needed.

Question 71

In relation to transformers, which of the following statements are true?

1. Step-up transformers produce a greater voltage leaving the transformer, compared to the entering voltage.
2. In step-down transformers, the number of turns in the primary coil is smaller than in the secondary coil.
3. For transformers that are 100% efficient: $Ip \times Vp = Is \times Vs$.

A. Only 1   E. 1 and 3

B. Only 2   F. 2 and 3

C. Only 3   G. 1, 2 and 3

D. 1 and 2   H. None of the above.

Question 72:

The half-life of Carbon-14 is 5,730 years. A bone is found that contains 6.25% of the amount of C^{14} that would be found in a modern one. How old is the bone?

A. 11,460 years   C. 22,920 years   E. 34,380 years

B. 17,190 years   D. 28,650 years   F. 40,110 years

Question 73:

A wave has a velocity of 2,000 mms^{-1} and a wavelength of 250 cm. What is its frequency in MHz?

A. 8×10^{-3} MHz   C. 8×10^{-5} MHz   E. 8×10^{-7} MHz

B. 8×10^{-4} MHz   D. 8×10^{-6} MHz   F. 8×10^{-8} MHz

Question 74:

A radioactive element has a half-life of 25 days. After 350 days, it has a count rate of 50. What was its original count rate?

A. 102,400	D. 409,600	G. 3,276,800
B. 162,240	E. 819,200	
C. 204,800	F. 1,638,400	

Question 75:

Which of the following units is **NOT** equivalent to a Volt (V)?

A. $A\Omega$	C. $Nms^{-1}A^{-1}$	E. JC^{-1}
B. WA^{-1}	D. NmC	F. $JA^{-1}s^{-1}$

ADVANCED PHYSICS QUESTIONS

Question 76:

A ball is swung in a vertical circle from a string, of negligible mass, as shown in the diagram.

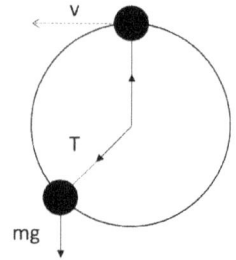

What is the minimum speed at the top of the arc for it to continue in a circular path? You may use the formula $a = \frac{v^2}{r}$, where a is the acceleration of a body undergoing circular motion with velocity v and radius r. The acceleration is directed towards the centre of the circle.

A. 0   C. $2r^2$ E. $\sqrt{gr}$

B. mgr   D. mg

Question 77:

A person pulls on a rope, at 60° to the horizontal, to exert a force on a mass m, as shown. What is the power needed to move the mass up the 30° incline at a constant velocity, v, given a friction force F?

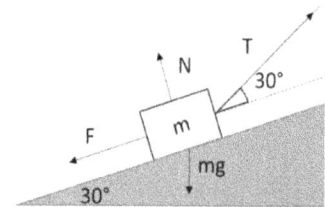

A. $\left(mg + \frac{F}{2}\right)v$ C. $\left(\frac{mg}{2}\right)v$ E. $\left(\frac{mg}{2} + F\right)v$

B. $\frac{mg}{\sqrt{2}} - F$ D. $\sqrt{2}Fv$

Question 78:

What is the maximum speed of a point mass, m, which is suspended from a pendulum of length l, and released from an angle θ?

A. $2gl(1 - \cos(\theta))$ C. $\sqrt{2gl(1 - \cos(\theta))}$ E. $\sqrt{2gl(1 - \cos^2(\theta))}$

B. $2gl(1 - \sin(\theta))$ D. $\sqrt{2gl(1 - \sin(\theta))}$

Question 79:

What would happen to V_{out} if the light intensity upon the circuit below is increased?

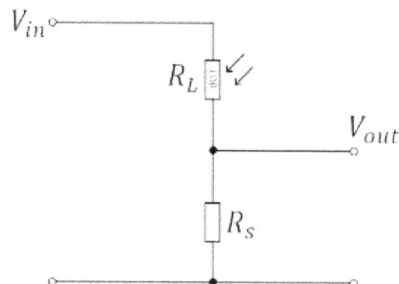

V_{in}

R_L

V_{out}

R_S

A. Go up   C. Stay the same   E. Increase infinitely

B. Go down   D. Decrease to zero

Question 80:

In the diagram below, the first ball is three times the mass of the other balls. If this is an elastic collision (no loss of kinetic energy), how many of the other balls move and at which velocity after the collision?

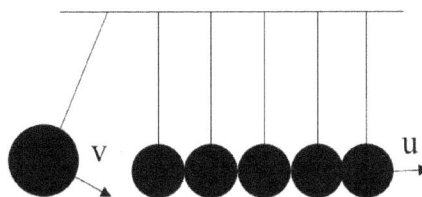

	Number of balls	Velocity
A	1	$3v$
B	1	$v/3$
C	3	$v/3$
D	3	v
E	3	$\sqrt{v}$

Question 81:

A ball is thrown into the air at a velocity $5ms^{-1}$ directly upwards and is caught on its descent. Assuming g = $10m/s^2$ and negligible air resistance, what is the displacement of the ball and the distance travelled by the ball after being caught?

A. 0 m, 2.5 m　　　　C. 5 m, 5 m　　　　E. 0 m, 2.5 m

B. 5 m, 0 m　　　　D. 2.5 m, 0 m　　　　F. 0 m, 10 m

Question 82:

The mechanism below is used to weigh two ridged uniform blocks of dimensions 2 metres by 1 metre, and mass 20 kg, on a lever connected to a weighing scale by a cable. What will the reading on the scale be?

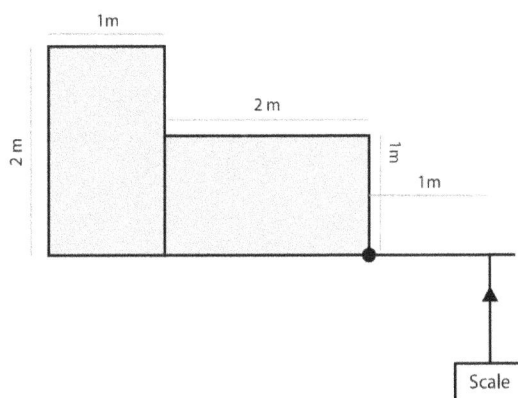

A. 35kg　　　　B. 40kg　　　　C. 70kg　　　　D. 80kg

Question 83:

A ball, of mass *m,* is dropped from 3 m above the ground and rebounds to a maximum height of 1 m. How much kinetic energy does it have just before hitting the ground and at the top of its bounce, and what is the maximum speed the ball reaches?

	E_k at Bottom	E_k at Top	Max Speed
A	0	30m	$2\sqrt{15}$
B	0	30m	30
C	30m	0	$2\sqrt{15}$
D	30m	0	60
E	60m	0	60

Question 84:

Which of the following beams could not be in equilibirum, regardless of the magnitudes of the forces (assuming they are not zero)? The arrows in the diagrams below represent forces.

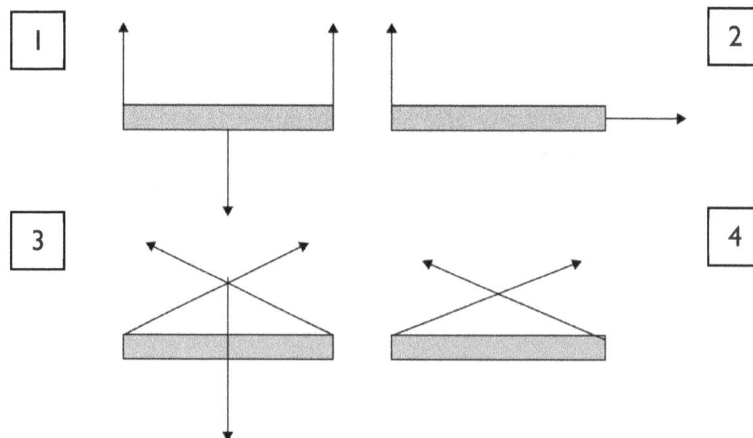

A. 1 and 2   C. 2 and 4   E. Only 4

B. 1 and 3   D. Only 1   F. 3 and 4

Question 85:

What is the stopping distance of a car, moving at a velocity *v,* if its braking force is half its weight?

A. v^2 C. $2mv$ E. $\sqrt{mg}$

B. $\dfrac{v^2}{g}$ D. $\dfrac{v^2}{2}$

Question 86:

The amplitude of a wave is damped from an initial amplitude of 200 to 25 over 12 seconds. How many seconds did it take to reach half its original amplitude? Assume that the wave undergoes exponential decay.

A. 1
B. 2
C. 3
D. 4
E. 6

Question 87:

A block is sliding at a speed v with an acceleration a along a rough ground. Which of the following expressions represents the power dissipated, assuming the friction coeffiecient is μ and its mass is m?

A. amg
B. μmgv
C. μamg
D. μmv

Question 88:

Study the diagram provided. The two waves can represent:

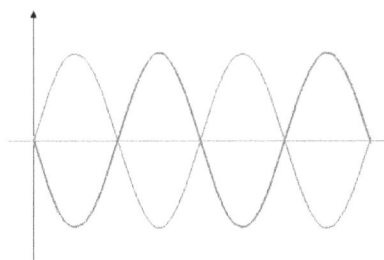

A. A standing wave with both ends fixed.
B. The 4th harmonic.
C. Destructive interference.
D. A reflection from a plane surface.
E. All of the above.

Question 89:

Radioactive element ^a_bX undergoes beta decay, in which a neutron is changed into a proton, and the product of this decay emits an alpha particle to become ^c_dY. What are the atomic number and atomic mass?

	c	d
A	a - 4	b + 1
B	a - 3	b - 2
C	a - 4	b - 1
D	a - 5	b
E	a - 1	b - 4

Question 90:

Consider a spherical shell of radius r and thickness t, where r >> t (r is significantly greater than t). If the inside is pressurised to p, above atmospheric pressure, what is the stress in the walls of the sphere?

A. pr/2t
B. 2rtπ
C. $4\pi r^3/3$
D. pr/t

Question 91:

For the arrangement of springs below, what is the spring constant of the whole system, assuming each spring has spring constant k?

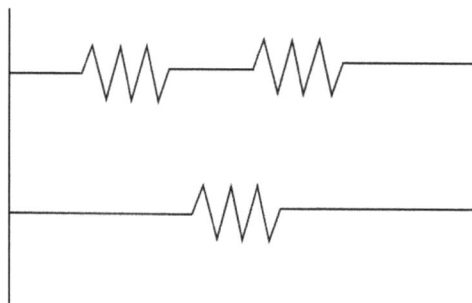

A. $k/2$ B. k C. $3k/2$ D. $2k^2$ E. $2k/3$

Question 92:

Consider the arrangement below showing a car dragging a trailer using a connection of stiffness 100000 N/m. At an instant they are accelerating at 2 m/s^2. Assuming the car weighs 20 Mg and the trailer weighs 10 Mg, what is the energy stored in the spring?

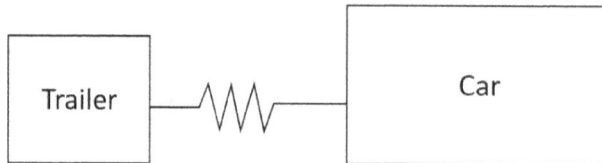

A. 2000J   B. 2J   C. 2000kJ   D. 8000J

Question 93:

A blue LED has a power output of 10 W and is connected to a source with an EMF of 5V. Estimate the number of electrons passing through the LED in 10 seconds and the energy provided by the battery to each individual electron.

	Number of electrons	Energy of an electron (J)
A	1.2×10^{19}	2×10^{-19}
B	2×10^{19}	8×10^{-19}
C	1.25×10^{20}	8×10^{-19}
D	1.5×10^{20}	12×10^{-19}
E	3×10^{25}	4×10^{-19}

Question 94:

A flowerpot hangs on the end of a rod protruding at right angles from a wall, held up by string attached two thirds of the way along from the wall. What must the tension in the string be if the rod is weightless and the system is at equilibrium?

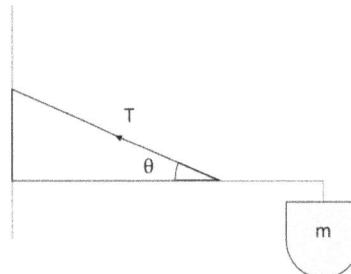

A. $mg \sin \theta$

B. $\dfrac{3mg}{2\sin \theta}$

C. $\dfrac{3mg}{2\cos \theta}$

D. $\dfrac{2mg}{3\sin \theta}$

E. $\dfrac{2mg}{3\cos \theta}$

Question 95:

Consider two signals: a 5V signal at 30kHz and a 10V signal at 50kHz. What is the time period of the combined signal?

A. 6.67 µs

B. 150 µs

C. 6.67 ms

D. 80 ms

E. 80 µs

Question 96:

For the graph of a material below estimate E, the Young's Modulus, and estimate the strain energy at point x, the elastic limit.

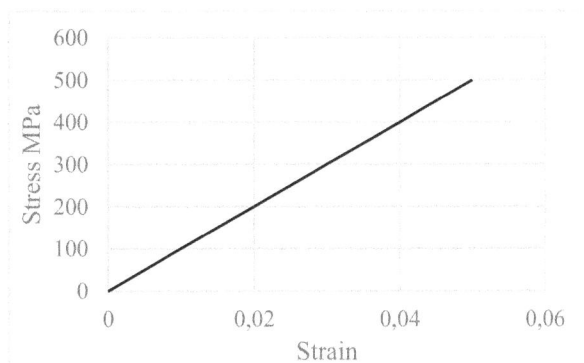

A. 210 GPa and 10000 kJ.

B. 10 GPa and 12.5 MJ.

C. 210 GPa and 6.125 MJ.

D. 10 GPa and 10000 kJ.

Question 97:

A pool cue gives a ball of mass 100g an impulse of 0.27 Ns. What is the velocity of the ball after the impact?

A. 27 m/s

B. 0.027 ms

C. 0.27 m/s

D. 2.7 m/s

E. 3.6 m/s

Question 98:

A ball is thrown at velocity v. Assuming negligible air resistance, and that the ball forms a projectile path, what is the optimal angle to throw the ball to get the furthest distance?

A. 10   C. 45   E. 80
B. 30   D. 60

Question 99:

An electric car travels at constant speed v, with a motor of efficiency 90% and consumes electrical power P. What is the work done by air resistance in a distance of 1 km?

A. 900v C. 0.9P/v E. 90P/v
B. 900P/v D. 900v/P

Question 100:

A block of mass m is on a smooth wedge. The wedge accelerates at a in the horizontal direction. What acceleration a is required from the wedge so that the block stays still? The wedge makes an angle θ to the horizontal.

A. mg C. $g\cos\theta$ E. m^2g
B. $g\tan\theta$ D. $mg\sin\theta$

SECTION 1: MATHS

The syllabus for the maths and advanced maths section of the ENGAA can be found within the official specification. A link is provided here:

https://www.undergraduate.study.cam.ac.uk/files/publications/engaa_specification2019.pdf.

Core Formulas:

2D Shapes			3D Shapes		
	Area			Surface Area	Volume
Circle	πr^2		**Cuboid**	Σ of 6 faces	Length × Width × Height
Parallelogram	Base × Vertical height		**Cylinder**	$2\pi r^2 + 2\pi rl$	$\pi r^2 \times h$
Trapezium	0.5 × Vertical height × (a + b)		**Cone**	$\pi r^2 + \pi rl$	$\pi r^2 \times (h/3)$
Triangle	0.5 × Base × Height		**Sphere**	$4\pi r^2$	$(4/3)\pi r^3$

Even good students who are studying maths at A2 can struggle with certain ENGAA maths topics because they're usually glossed over at school. These include:

Quadratic Formula

The solutions for a quadratic equation in the form $ax^2 + bx + c = 0$ are given by: $x = \frac{-b \pm \sqrt{b^2 - 4ac}}{2a}$.

Remember that you can also use the discriminant to quickly determine whether a quadratic equation has any solutions:

$b^2 - 4ac < 0 \Rightarrow$ No solutions

$b^2 - 4ac = 0 \Rightarrow$ 1 repeated solution

$b^2 - 4ac > 0 \Rightarrow$ 2 distinct solutions

Completing the Square

If a quadratic equation cannot be factorised easily, and is in the format $ax^2 + bx + c = 0$, then you can rearrange it into the form $a\left(x + \frac{b}{2a}\right)^2 \left[c - \frac{b^2}{4a}\right] = 0$

This looks more complicated than it is – remember that in the ENGAA, you're extremely unlikely to get quadratic equations where $a > 1$ or an equation which doesn't have any easy factors. This gives you an easier equation: $\left(x + \frac{b}{2}\right)^2 + \left[c - \frac{b^2}{4}\right] = 0$ and is best understood with an example.

Consider: $x^2 + 6x + 10 = 0$.

This equation cannot be factorised easily but note that: $x^2 + 6x - 10 = (x + 3)^2 - 19 = 0$.

Therefore, $x = -3 \pm \sqrt{19}$. Completing the square is an important skill – make sure you're comfortable with it.

Difference between 2 Squares

If you are asked to simplify expressions and find that there are no common factors, but the expression does involve square numbers, you might be able to factorise it by using the 'difference between two squares' trick.

For example, $x^2 - 25$ can also be expressed as$(x + 5)(x - 5)$.

MATHS QUESTIONS

Question 101:

Robert has a box of building blocks. The box contains 8 yellow blocks and 12 red blocks. He picks three blocks from the box and stacks them up high. Calculate the probability that he stacks two red building blocks and one yellow building block, in **any** order.

A. $\frac{8}{20}$

B. $\frac{44}{95}$

C. $\frac{11}{18}$

D. $\frac{8}{19}$

E. $\frac{12}{20}$

F. $\frac{35}{60}$

Question 102:

Solve $\frac{3x+5}{5} + \frac{2x-2}{3} = 18$.

A. 12.11

B. 13.49

C. 13.95

D. 14.2

E. 19

F. 265

Question 103:

Solve $3x^2 + 11x - 20 = 0$.

A. 0.75 and $-\frac{4}{3}$

B. -0.75 and $\frac{4}{3}$

C. -5 and $\frac{4}{3}$

D. 5 and $\frac{4}{3}$

E. 12 only

F. -12 only

Question 104:

Express $\frac{5}{x+2} + \frac{3}{x-4}$ as a single fraction.

A. $\frac{15x-120}{(x+2)(x-4)}$

B. $\frac{8x-26}{(x+2)(x-4)}$

C. $\frac{8x-14}{(x+2)(x-4)}$

D. $\frac{15}{8x}$

E. 24

F. $\frac{8x-14}{x^2-8}$

Question 105:

The value of p is directly proportional to the cube root of q. When p = 12, q = 27. Find the value of q when p = 24.

A. 32

B. 64

C. 124

D. 128

E. 216

F. 1728

Question 106:

Write 72^2 as a product of its prime factors.

A. $2^6 \times 3^4$

B. $2^6 \times 3^5$

C. $2^4 \times 3^4$

D. 2×3^3

E. $2^6 \times 3$

F. $2^3 \times 3^2$

Question 107:

Calculate: $\frac{2.302 \times 10^5 + 2.302 \times 10^2}{1.151 \times 10^{10}}$.

A. 0.0000202   C. 0.00002002   E. 0.000002002
B. 0.00020002   D. 0.00000002   F. 0.000002002

Question 108:

Given that $y^2 + \textbf{a}y + \textbf{b} = (y + 2)^2 - 5$, find the values of **a** and **b**.

	a	b
A	-1	4
B	1	9
C	-1	-9
D	-9	1
E	4	-1
F	4	1

Question 109:

Express $\frac{4}{5} + \frac{m-2n}{m+4n}$ as a single fraction in its simplest form.

A. $\frac{6m+6n}{5(m+4n)}$ C. $\frac{20m+6n}{5(m+4n)}$ E. $\frac{3(3m+2n)}{5(m+4n)}$

B. $\frac{9m+26n}{5(m+4n)}$ D. $\frac{3m+9n}{5(m+4n)}$ F. $\frac{6m+6n}{3(m+4n)}$

Question 110:

A is inversely proportional to the square root of B. When A = 4, B = 25.

Calculate the value of A when B = 16.

A. 0.8   C. 5   E. 10
B. 4   D. 6   F. 20

Question 111:

S, T, U and V are points on the circumference of a circle, as shown in the diagram, and O is the centre of the circle.

Given that angle SVU = 89°, calculate the size of the smaller angle SOU.

A. 89°
B. 91°
C. 102°
D. 178°
E. 182°
F. 212°

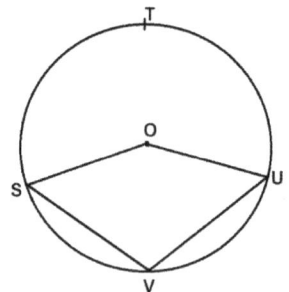

Question 112:

Open cylinder A has a surface area of 8π cm^2 and a volume of 2π cm^3. Open cylinder B is an enlargement of A and has a surface area of 32π cm^2. Calculate the volume of cylinder B.

A. 2π cm^3 C. 10π cm^3 E. 16π cm^3

B. 8π cm^3 D. 14π cm^3 F. 32π cm^3

Question 113:

Express $\frac{8}{x(3-x)} - \frac{6}{x}$ in its simplest form.

A. $\frac{3x-10}{x(3-x)}$ C. $\frac{6x-10}{x(3-2x)}$ E. $\frac{6x-10}{x(3-x)}$

B. $\frac{3x+10}{x(3-x)}$ D. $\frac{6x-10}{x(3+2x)}$ F. $\frac{6x+10}{x(3-x)}$

Question 114:

A bag contains 10 balls, where 9 are white and 1 is black. What is the probability that the black ball is drawn in the tenth and final draw if the drawn balls are not replaced?

A. 0 C. $\frac{1}{100}$ E. $\frac{1}{362,880}$

B. $\frac{1}{10}$ D. $\frac{1}{10^{10}}$

Question 115:

Gambit has an ordinary deck of 52 cards. What is the probability of Gambit drawing 2 Kings (without replacement)?

A. 0 C. $\frac{1}{221}$ E. None of the above.

B. $\frac{1}{169}$ D. $\frac{4}{663}$

Question 116:

There are two identical unfair dice, where the probability that the dice gets a 6 is twice as high as the probability of any other outcome, which are all equally likely. What is the probability that rolling both dice will give a total will be 12?

A. 0 C. $\frac{1}{9}$ E. None of the above.

B. $\frac{4}{49}$ D. $\frac{2}{7}$

Question 117:

A roulette wheel consists of 36 numbered spots and 1 zero spot (i.e. 37 spots in total).

What is the probability that the ball will stop in a spot either divisible by 3 or 2?

A. 0 C. $\frac{25}{36}$ E. $\frac{24}{37}$

B. $\frac{25}{37}$ D. $\frac{18}{37}$

Question 118:

An unbiased coin is flipped 4 times. What is the probability of obtaining 2 heads and 2 tails?

A. $\frac{1}{16}$

B. $\frac{3}{16}$

C. $\frac{3}{8}$

D. $\frac{9}{16}$

E. None of the above.

Question 119:

Shivun rolls two fair dice. What is the probability that he obtains a total of 5, 6 or 7?

A. $\frac{9}{36}$

B. $\frac{7}{12}$

C. $\frac{1}{6}$

D. $\frac{5}{12}$

E. None of the above.

Question 120:

Dr Savary has a bag that contains x red balls, y blue balls and z green balls. He pulls out a ball, replaces it, and then pulls out another. What is the probability that he picks one red ball and one green ball?

A. $\frac{2(x+y)}{x+y+z}$

B. $\frac{xz}{(x+y+z)^2}$

C. $\frac{2xz}{(x+y+z)^2}$

D. $\frac{(x+z)}{(x+y+z)^2}$

E. $\frac{4xz}{(x+y+z)^4}$

F. More information needed.

Question 121:

Mr Kilbane has a bag that contains x red balls, y blue balls and z green balls. He pulls out a ball, does **NOT** replace it, and then pulls out another. What is the probability that he picks one red ball and one blue ball?

A. $\frac{2xy}{(x+y+z)^2}$

B. $\frac{2xy}{(x+y+z)(x+y+z-1)}$

C. $\frac{2xy}{(x+y+z)^2}$

D. $\frac{xy}{(x+y+z)(x+y+z-1)}$

E. $\frac{4xy}{(x+y+z-1)^2}$

F. More information needed.

Question 122:

There are two tennis players. The first player wins the point with probability p, and the second player wins the point with probability 1- p. The first player to score four points wins the game, unless the score is 4 - 3. At this point, the first player to get two points ahead wins.

What is the probability that the first player wins in exactly 5 rounds?

A. $4p^4(1\text{-}p)$

B. $p^4(1\text{-}p)$

C. $4p(1\text{-}p)$

D. $4p(1\text{-}p)^4$

E. $4p^5(1\text{-}p)$

F. More information needed.

Question 123:

Solve the equation $\frac{4x+7}{2} + 9x + 10 = 7$.

A. $\frac{22}{13}$

B. $-\frac{22}{13}$

C. $\frac{10}{13}$

D. $-\frac{10}{13}$

E. $\frac{13}{22}$

F. $-\frac{13}{22}$

50

Question 112:

Open cylinder A has a surface area of 8π cm² and a volume of 2π cm³. Open cylinder B is an enlargement of A and has a surface area of 32π cm². Calculate the volume of cylinder B.

A. 2π cm³
B. 8π cm³
C. 10π cm³
D. 14π cm³
E. 16π cm³
F. 32π cm³

Question 113:

Express $\frac{8}{x(3-x)} - \frac{6}{x}$ in its simplest form.

A. $\frac{3x-10}{x(3-x)}$
B. $\frac{3x+10}{x(3-x)}$
C. $\frac{6x-10}{x(3-2x)}$
D. $\frac{6x-10}{x(3+2x)}$
E. $\frac{6x-10}{x(3-x)}$
F. $\frac{6x+10}{x(3-x)}$

Question 114:

A bag contains 10 balls, where 9 are white and 1 is black. What is the probability that the black ball is drawn in the tenth and final draw if the drawn balls are not replaced?

A. 0
B. $\frac{1}{10}$
C. $\frac{1}{100}$
D. $\frac{1}{10^{10}}$
E. $\frac{1}{362,880}$

Question 115:

Gambit has an ordinary deck of 52 cards. What is the probability of Gambit drawing 2 Kings (without replacement)?

A. 0
B. $\frac{1}{169}$
C. $\frac{1}{221}$
D. $\frac{4}{663}$
E. None of the above.

Question 116:

There are two identical unfair dice, where the probability that the dice gets a 6 is twice as high as the probability of any other outcome, which are all equally likely. What is the probability that rolling both dice will give a total will be 12?

A. 0
B. $\frac{4}{49}$
C. $\frac{1}{9}$
D. $\frac{2}{7}$
E. None of the above.

Question 117:

A roulette wheel consists of 36 numbered spots and 1 zero spot (i.e. 37 spots in total).

What is the probability that the ball will stop in a spot either divisible by 3 or 2?

A. 0
B. $\frac{25}{37}$
C. $\frac{25}{36}$
D. $\frac{18}{37}$
E. $\frac{24}{37}$

Question 118:

An unbiased coin is flipped 4 times. What is the probability of obtaining 2 heads and 2 tails?

A. $\frac{1}{16}$

C. $\frac{3}{8}$

E.   None of the above.

B. $\frac{3}{16}$

D. $\frac{9}{16}$

Question 119:

Shivun rolls two fair dice. What is the probability that he obtains a total of 5, 6 or 7?

A. $\frac{9}{36}$

C. $\frac{1}{6}$

E.   None of the above.

B. $\frac{7}{12}$

D. $\frac{5}{12}$

Question 120:

Dr Savary has a bag that contains x red balls, y blue balls and z green balls. He pulls out a ball, replaces it, and then pulls out another. What is the probability that he picks one red ball and one green ball?

A. $\frac{2(x+y)}{x+y+z}$

C. $\frac{2xz}{(x+y+z)^2}$

E. $\frac{4xz}{(x+y+z)^4}$

B. $\frac{xz}{(x+y+z)^2}$

D. $\frac{(x+z)}{(x+y+z)^2}$

F.   More information needed.

Question 121:

Mr Kilbane has a bag that contains x red balls, y blue balls and z green balls. He pulls out a ball, does **NOT** replace it, and then pulls out another. What is the probability that he picks one red ball and one blue ball?

A. $\frac{2xy}{(x+y+z)^2}$

C. $\frac{2xy}{(x+y+z)^2}$

E. $\frac{4xy}{(x+y+z-1)^2}$

B. $\frac{2xy}{(x+y+z)(x+y+z-1)}$

D. $\frac{xy}{(x+y+z)(x+y+z-1)}$

F.   More information needed.

Question 122:

There are two tennis players. The first player wins the point with probability p, and the second player wins the point with probability 1- p. The first player to score four points wins the game, unless the score is 4 - 3. At this point, the first player to get two points ahead wins.

What is the probability that the first player wins in exactly 5 rounds?

A. $4p^4(1-p)$

C. $4p(1-p)$

E. $4p^5(1-p)$

B. $p^4(1-p)$

D. $4p(1-p)^4$

F.   More information needed.

Question 123:

Solve the equation $\frac{4x+7}{2} + 9x + 10 = 7$.

A. $\frac{22}{13}$

C. $\frac{10}{13}$

E. $\frac{13}{22}$

B. $-\frac{22}{13}$

D. $-\frac{10}{13}$

F. $-\frac{13}{22}$

50

Question 124:

The volume of a sphere is $V = \frac{4}{3}\pi r^3$, and the surface area of a sphere is $S = 4\pi r^2$. Express S in terms of V.

A. $S = (4\pi)^{2/3}(3V)^{2/3}$

B. $S = (8\pi)^{1/3}(3V)^{2/3}$

C. $S = (4\pi)^{1/3}(9V)^{2/3}$

D. $S = (4\pi)^{1/3}(3V)^{2/3}$

E. $S = (16\pi)^{1/3}(9V)^{2/}$

Question 125:

Express the volume of a cube, V, in terms of its surface area, S.

A. $V = (S/6)^{3/2}$

B. $V = S^{3/2}$

C. $V = (6/S)^{3/2}$

D. $V = (S/6)^{1/2}$

E. $V = (S/36)^{1/2}$

F. $V = (S/36)^{3/2}$

Question 126:

Solve the equations $4x + 3y = 7$ and $2x + 8y = 12$.

A. $(x,y) = \left(\frac{17}{13}, \frac{10}{13}\right)$

B. $(x,y) = \left(\frac{10}{13}, \frac{17}{13}\right)$

C. $(x,y) = (1,2)$

D. $(x,y) = (2,1)$

E. $(x,y) = (6,3)$

F. $(x,y) = (3,6)$

G. No solutions possible.

Question 127:

Rearrange $\frac{(7x+10)}{(9x+5)} = 3y^2 + 2$, to make x the subject.

A. $\frac{15\,y^2}{7 - 9(3y^2+2)}$

B. $\frac{15\,y^2}{7 + 9(3y^2+2)}$

C. $-\frac{15\,y^2}{7 - 9(3y^2+2)}$

D. $-\frac{15\,y^2}{7 + 9(3y^2+2)}$

E. $-\frac{5\,y^2}{7 + 9(3y^2+2)}$

F. $\frac{5\,y^2}{7 + 9(3y^2+2)}$

Question 128:

Simplify $3x\left(\frac{3x^7}{x^{\frac{1}{3}}}\right)^3$.

A. $9x^{20}$

B. $27x^{20}$

C. $87x^{20}$

D. $9x^{21}$

E. $27x^{21}$

F. $81x^{21}$

Question 129:

Simplify $2x[(2x)^7]^{\frac{1}{14}}$.

A. $2x\sqrt{2\,x^4}$

B. $2x\sqrt{2x^3}$

C. $2\sqrt{2\,x^4}$

D. $2\sqrt{2x^3}$

E. $8x^3$

F. $8x$

Question 130:

What is the circumference of a circle with an area of 10π?

A. $2\pi\sqrt{10}$

B. $\pi\sqrt{10}$

C. 10π

D. 20π

E. $\sqrt{10}$

F. More information needed.

Question 131:

Evaluate the value of $(3.4).5$, given that $a \cdot b = (ab) + (a + b)$.

A. 19

B. 54

C. 100

D. 119

E. 132

Question 132:

Calculate $(2.3).2$, given that $a.b = \dfrac{a^b}{a}$.

A. $\dfrac{16}{3}$

B. 1

C. 2

D. 4

E. 8

Question 133:

Solve $x^2 + 3x - 5 = 0$.

A. $x = -\dfrac{3}{2} \pm \dfrac{\sqrt{11}}{2}$

B. $x = \dfrac{3}{2} \pm \dfrac{\sqrt{11}}{2}$

C. $x = -\dfrac{3}{2} \pm \dfrac{\sqrt{11}}{4}$

D. $x = \dfrac{3}{2} \pm \dfrac{\sqrt{11}}{4}$

E. $x = \dfrac{3}{2} \pm \dfrac{\sqrt{29}}{2}$

F. $x = -\dfrac{3}{2} \pm \dfrac{\sqrt{29}}{2}$

Question 134:

How many times do the curves $y = x^3$ and $y = x^2 + 4x + 14$ intersect?

A. 0

B. 1

C. 2

D. 3

E. 4

Question 135:

Which of the following graphs **do not** intersect?

1. $y = x$
2. $y = x^2$
3. $y = 1 - x^2$
4. $y = 2$

A. 1 and 2

B. 2 and 3

C. 3 and 4

D. 1 and 3

E. 1 and 4

F. 2 and 4

Question 136:

Calculate the product of 897,653 and 0.009764.

A. 87646.8   C. 876.468   E. 8.76468

B. 8764.68   D. 87.6468   F. 0.876468

Question 137:

Solve for x: $\frac{7x+3}{10} + \frac{3x+1}{7} = 14$.

A. $\frac{929}{51}$ C. $\frac{949}{79}$

B. $\frac{949}{47}$ D. $\frac{980}{79}$

Question 138:

What is the area of an equilateral triangle with side length x.

A. $\frac{x^2\sqrt{3}}{4}$ C. $\frac{x^2}{2}$ E. x^2

B. $\frac{x\sqrt{3}}{4}$ D. $\frac{x}{2}$ F. x

Question 139:

Simplify $3 - \frac{7x(25x^2 - 1)}{49x^2(5x+1)}$.

A. $3 - \frac{5x-1}{7x}$ C. $3 + \frac{5x-1}{7x}$ E. $3 - \frac{5x^2}{49}$

B. $3 - \frac{5x+1}{7x}$ D. $3 + \frac{5x+1}{7x}$ F. $3 + \frac{5x^2}{49}$

Question 140:

Solve the equation $x^2 - 10x - 100 = 0$.

A. $-5 \pm 5\sqrt{5}$ C. $5 \pm 5\sqrt{5}$ E. $5 \pm 5\sqrt{125}$

B. $-5 \pm \sqrt{5}$ D. $5 \pm \sqrt{5}$ F. $-5 \pm \sqrt{125}$

Question 141:

Rearrange $x^2 - 4x + 7 = y^3 + 2$ to make x the subject.

A. $x = 2 \pm \sqrt{y^3 + 1}$ C. $x = -2 \pm \sqrt{y^3 - 1}$

B. $x = 2 \pm \sqrt{y^3 - 1}$ D. $x = -2 \pm \sqrt{y^3 + 1}$

Question 142:

Rearrange $3x + 2 = \sqrt{7x^2 + 2x + y}$ to make y the subject.

A. $y = 4x^2 + 8x + 2$ C. $y = 2x^2 + 10x + 2$ E. $y = x^2 + 10x + 2$

B. $y = 4x^2 + 8x + 4$ D. $y = 2x^2 + 10x + 4$ F. $y = x^2 + 10x + 4$

Question 143:

Rearrange $y^4 - 4y^3 + 6y^2 - 4y + 2 = x^5 + 7$ to make y the subject.

A. $y = 1 + (x^5 + 7)^{1/4}$ C. $y = 1 + (x^5 + 6)^{1/4}$

B. $y = -1 + (x^5 + 7)^{1/4}$ D. $y = -1 + (x^5 + 6)^{1/4}$

Question 144:

The aspect ratio of my television screen is 4:3 and the diagonal is 50 inches. What is the area of my television screen?

A. 1,200 inches²   C. 120 inches²   E. More information needed.

B. 1,000 inches²   D. 100 inches²

Question 145:

Rearrange the equation $\sqrt{1 + 3x^{-2}} = y^5 + 1$ to make x the subject.

A. $x = \dfrac{(y^{10} + 2y^5)}{3}$ C. $x = \sqrt{\dfrac{3}{y^{10} + 2y^5}}$ E. $x = \sqrt{\dfrac{y^{10} + 2y^5 + 2}{3}}$

B. $x = \dfrac{3}{(y^{10} + 2y^5)}$ D. $x = \sqrt{\dfrac{y^{10} + 2y^5}{3}}$ F. $x = \sqrt{\dfrac{3}{y^{10} + 2y^5 + 2}}$

Question 146:

Solve $3x - 5y = 10$ and $2x + 2y = 13$.

A. $(x, y) = \left(\dfrac{19}{16}, \dfrac{85}{16}\right)$ C. $(x, y) = \left(\dfrac{85}{16}, \dfrac{19}{16}\right)$ E. No solutions possible.

B. $(x, y) = \left(\dfrac{85}{16}, -\dfrac{19}{16}\right)$ D. $(x, y) = \left(-\dfrac{85}{16}, -\dfrac{19}{16}\right)$

Question 147:

The two inequalities $x + y \leq 3$ and $x^3 - y^2 < 3$ define a region on a plane. Which of the following points is inside the region?

A. $(2, 1)$ C. $(1, 2)$ E. $(1, 2.5)$

B. $(2.5, 1)$ D. $(3, 5)$ F. None of the above.

Question 148:

How many points of intersection do $y = x + 4$ and $y = 4x^2 + 5x + 5$ have?

A. 0   C. 2   E. 4

B. 1   D. 3

Question 149:

How many points of intersection do $y = x^3$ and $y = x$ have?

A. 0   C. 2   E. 4

B. 1   D. 3

Question 150:

A cube has unit length sides. What is the length of a line joining a vertex to the midpoint of the opposite side?

A. $\sqrt{2}$

B. $\sqrt{\frac{3}{2}}$

C. $\sqrt{3}$

D. $\sqrt{5}$

E. $\frac{\sqrt{5}}{2}$

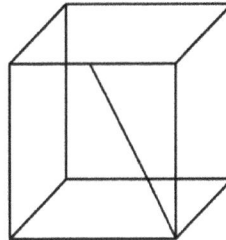

Question 151:

Solve for x, y, and z in the following system of equations.

1. $x + y - z = -1$
2. $2x - 2y + 3z = 8$
3. $2x - y + 2z = 9$

	x	y	z
A	2	-15	-14
B	15	2	14
C	14	15	-2
D	-2	15	14
E	2	-15	14
F	No solutions possible		

Question 152:

Fully factorise the following expression: $3a^3 - 30a^2 + 75a$.

A. $3a(a - 3)^3$

C. $3a(a^2 - 10a + 25)$

E. $3a(a + 5)^2$

B. $a(3a - 5)^2$

D. $3a(a - 5)^2$

Question 153:

Solve for x and y in the following simultaneous equations.

$4x + 3y = 48$

$3x + 2y = 34$

	x	y
A	8	6
B	6	8
C	3	4
D	4	3
E	30	12
F	12	30
G	No solutions possible	

Question 154:

Evaluate: $\dfrac{-\left(5^2 - 4 \times 7\right)^2}{-6^2 + 2 \times 7}$.

A. $-\dfrac{3}{50}$ C. $-\dfrac{3}{22}$ E. $\dfrac{9}{22}$

B. $\dfrac{11}{22}$ D. $\dfrac{9}{50}$ F. 0

Question 155:

All license plates are 6 characters long. The first 3 characters consist of letters and the next 3 characters of numbers. How many unique license plates are possible?

A. 676,000   C. 67,600,000   E. 17,576,000

B. 6,760,000   D. 1,757,600   F. 175,760,000

Question 156:

How many solutions are there for: $2(2(x^2 - 3x)) = -9$.

A. 0   C. 2   E. Infinite solutions.

B. 1   D. 3

Question 157:

Evaluate: $\left(x^{\frac{1}{2}} y^{-3}\right)^{\frac{1}{2}}$.

A. $\dfrac{x^{\frac{1}{2}}}{y}$ C. $\dfrac{x^{\frac{1}{4}}}{y^{\frac{3}{2}}}$

B. $\dfrac{x}{y^{\frac{3}{2}}}$ D. $\dfrac{y^{\frac{1}{4}}}{x^{\frac{3}{2}}}$

Question 158:

Bryan earned a total of £ 1,240 last week from renting out three flats. From this, he had to pay 10% of the rent from the 1-bedroom flat for repairs, 20% of the rent from the 2-bedroom flat for repairs, and 30% from the 3-bedroom flat for repairs. The 3-bedroom flat costs twice as much as the 1-bedroom flat. Given that the total repair bill was £276 calculate the rent for each apartment.

	1 Bedroom	2 Bedrooms	3 Bedrooms
A	280	400	560
B	140	200	280
C	420	600	840
D	250	300	500
E	500	600	1,000

Question 159:

Evaluate: $5\left[5(6^2 - 5 \times 3) + 400^{\frac{1}{2}}\right]^{1/3} + 7$.

A. 0   C. 32   E. 56

B. 25   D. 49   F. 200

Question 160:

What is the area of a regular hexagon with side length 1?

A. $3\sqrt{3}$ C. $\sqrt{3}$ E. 6

B. $\frac{3\sqrt{3}}{2}$ D. $\frac{\sqrt{3}}{2}$ F. More information needed

Question 161:

Dexter moves into a new rectangular room that is 19 metres longer than it is wide, and its total area is 780 square metres. What are the room's dimensions?

A. Width = 20 m; Length = -39 m   C. Width = 39 m; Length = 20 m   E. Width = -20 m; Length = 39 m

B. Width = 20 m; Length = 39 m   D. Width = -39 m; Length = 20 m

Question 162:

Tom uses 34 meters of fencing to enclose his rectangular lot. He measured the diagonals to 13 metres long. What is the length and width of the lot?

A. 3 m by 4 m   C. 6 m by 12 m   E. 9 m by 15 m

B. 5 m by 12 m   D. 8 m by 15 m   F. 10 m by 10 m

Question 163:

Solve $\frac{3x-5}{2} + \frac{x+5}{4} = x + 1$.

A. 1

B. 1.5

C. 3

D. 3.5

E. 4.5

F. None of the above

Question 164:

Calculate: $\frac{5.226 \times 10^6 + 5.226 \times 10^5}{1.742 \times 10^{10}}$.

A. 0.033

B. 0.0033

C. 0.00033

D. 0.00003

E. 0.0000033

Question 165:

Calculate the area of the triangle shown to the right:

A. $3 + \sqrt{2}$

B. $\frac{2 + 2\sqrt{2}}{2}$

C. $2 + 5\sqrt{2}$

D. $3 - \sqrt{2}$

E. 3

F. 6

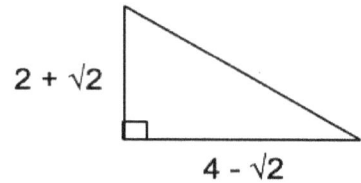

Question 166:

Rearrange $\sqrt{\frac{4}{x} + 9} = y - 2$ to make x the subject.

A. $x = \frac{11}{(y-2)^2}$

B. $x = \frac{9}{(y-2)^2}$

C. $x = \frac{4}{(y+1)(y-5)}$

D. $x = \frac{4}{(y-1)(y+5)}$

E. $x = \frac{4}{(y+1)(y+5)}$

F. $x = \frac{4}{(y-1)(y-5)}$

Question 167:

When 5 is subtracted from 5x the result is half the sum of 2 and 6x. What is the value of x?

A. 0

B. 1

C. 2

D. 3

E. 4

F. 6

Question 168:

Estimate $\frac{54.98 + 2.25^2}{\sqrt{905}}$.

A. 0

B. 1

C. 2

D. 3

E. 4

F. 5

Question 169:

At a Pizza Parlour, you can order single, double or triple cheese in the crust. You also have the option to include ham, olives, pepperoni, bell pepper, meat balls, tomato slices, and pineapples. How many different types of pizza are available at the Pizza Parlour?

A. 10

B. 96

C. 192

D. 384

E. 768

F. None of the above.

Question 170:

Solve the simultaneous equations $x^2 + y^2 = 1$ and $x + y = \sqrt{2}$, for x, y > 0.

A. $(x, y) = \left(\frac{\sqrt{2}}{2}, \frac{\sqrt{2}}{2}\right)$

B. $(x, y) = \left(\frac{1}{2}, \frac{\sqrt{3}}{2}\right)$

C. $(x, y) = \left(\sqrt{2} - 1, 1\right)$

D. $(x, y) = \left(\sqrt{2}, \frac{1}{2}\right)$

Question 171:

Which of the following statements is **FALSE**?

A. Congruent objects always have the same dimensions and shape.

B. Congruent objects can be mirror images of each other.

C. Congruent objects do not always have the same angles.

D. Congruent objects can be rotations of each other.

E. Two triangles are congruent if they have two sides and one angle of the same magnitude.

Question 172:

Solve the inequality: $x^2 \geq 6 - x$.

A. x ≤ -3 and x ≤ 2

B. x ≤ -3 and x ≥ 2

C. x ≥ -3 and x ≤ 2

D. x ≥ -3 and x ≥ 2

E. x ≥ 2 only

F. x ≥ -3 only

Question 173:

The hypotenuse of an isosceles, right-angled triangle is x cm. The other two sides are equal in length. What is the area of the triangle in terms of x?

A. $\frac{\sqrt{x}}{2}$

B. $\frac{x^2}{4}$

C. $\frac{x}{4}$

D. $\frac{3x^2}{4}$

E. $\frac{x^2}{10}$

Question 174:

Mr Heard derives a formula: $Q = \frac{(X+Y)^2 A}{3B}$. He doubles the values of X and Y, halves the value of A and triples the value of B. What happens to value of Q?

A. Decreases by $\frac{1}{3}$ C. Decreases by $\frac{2}{3}$ E. Increases by $\frac{4}{3}$

B. Increases by $\frac{1}{3}$ D. Increases by $\frac{2}{3}$ F. Decreases by $\frac{4}{3}$

Question 175:

Consider the graphs $y = x^2 - 2x + 3$, and $y = x^2 - 6x - 10$. Which of the following is true?

A. Both equations intersect the x-axis.

B. Neither equation intersects the x-axis.

C. The first equation does not intersect the x-axis; the second equation intersects the x-axis.

D. The first equation intersects the x-axis; the second equation does not intersect the x-axis.

ADVANCED MATHS QUESTIONS

Question 176:

The vertex of an equilateral triangle is covered by a circle whose radius is half the height of the triangle. What percentage of the triangle is covered by the circle?

A. 12%
B. 16%
C. 23%
D. 33%
E. 41%
F. 50%

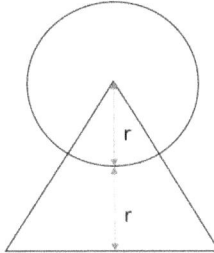

Question 177:

Three identical circles, which each have a radius *r*, fit into a quadrilateral as shown. Determine the height of the quadrilateral.

A. $2\sqrt{3}r$
B. $(2 + \sqrt{3})r$
C. $(4 - \sqrt{3})r$
D. $3r$
E. $4r$
F. More information needed.

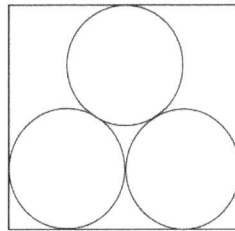

Question 178:

Two pyramids have equal volume and height, one with a square base of side length **a** and one with a hexagonal base of side length **b**. What is the ratio of the side length of the bases?

A. $\sqrt{\dfrac{3\sqrt{3}}{2}}$

B. $\sqrt{\dfrac{2\sqrt{3}}{3}}$

C. $\sqrt{\dfrac{3}{2}}$

D. $\dfrac{2\sqrt{3}}{3}$

E. $\dfrac{3\sqrt{3}}{2}$

Question 179:

One 9 cm cube is cut into 3 cm cubes. The total surface area increases by a factor of:

A. $\dfrac{1}{3}$
B. $\sqrt{3}$
C. 3
D. 9
E. 27

Question 180:

A cone has height twice its base width (four times the circle radius). What is the cone angle (half the angle at the vertex)?

A. $30°$

B. $\sin^{-1}\left(\frac{r}{2}\right)$

C. $\tan^{-1}\left(\frac{1}{4}\right)$

D. $\cos^{-1}\left(\sqrt{17}\right)$

Question 181:

A hemispherical speedometer has a maximum speed of 200 mph. What is the angle travelled by the needle at a speed of 70 mph?

A. $28°$

B. $49°$

C. $63°$

D. $88°$

E. $92°$

Question 182:

Two rhombuses, A and B, are similar. The area of A is 10 times that of B. What is the ratio of the smallest angles over the ratio of the shortest sides?

A. 0

B. $\frac{1}{10}$

C. $\frac{1}{\sqrt{10}}$

D. $\sqrt{10}$

E. ∞

Question 183:

If $f^{-1}(-x) = \ln(2x^2)$, what is $f(x)$?

A. $\sqrt{\frac{e^y}{2}}$

B. $\sqrt{\frac{e^{-y}}{2}}$

C. $\frac{e^y}{2}$

D. $\frac{-e^y}{2}$

E. $\sqrt{\frac{e^y}{2}}$

Question 184:

Which of the following has the largest value for $0 < x < 1$?

A. $\log_8(x)$

B. $\log_{10}(x)$

C. e^x

D. x^2

E. $\sin(x)$

Question 185:

The variable x is proportional to y cubed; y is proportional to the square root of z.

If z doubles, x changes by a factor of:

A. $\sqrt{2}$

B. 2

C. $2\sqrt{2}$

D. $\sqrt[3]{4}$

E. 4

Question 186:

The area between two concentric circles (shaded) is three times that of the inner circle.

What is the size of the gap?

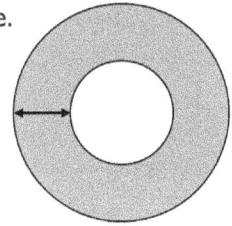

A. r

B. $\sqrt{2}r$

C. $\sqrt{3}r$

D. 2r

E. 3r

F. 4r

Question 187:

Solve $-x^2 \leq 3x - 4$.

A. $x \geq \frac{4}{3}$

B. $1 \leq x \leq 4$

C. $x \leq 2$

D. $x \geq 1$ or $x \geq -4$

E. $-1 \leq x \leq \frac{3}{4}$

Question 188:

The volume of a sphere is equal to its projected area. What is its radius?

A. $\frac{1}{2}$

B. $\frac{2}{3}$

C. $\frac{3}{4}$

D. $\frac{4}{3}$

E. $\frac{3}{2}$

Question 189:

What is the range where $x^2 < \frac{1}{x}$?

A. $x < 0$

B. $0 < x < 1$

C. $x > 0$

D. $x \geq 1$

E. None.

Question 190:

Simplify and solve the following expression:

$(e - a) (e + b) (e - c) (e + d)...(e - z)$.

A. 0

B. e^{26}

C. $e^{26} (a-b+c-d...+z)$

D. $e^{26} (a+b-c+d...-z)$

E. $e^{26} (abcd...z)$

F. None of the above.

Question 191:

Find the value of k such that the vectors $\mathbf{a} = -\mathbf{i} + 6\mathbf{j}$ and $\mathbf{b} = 2\mathbf{i} + k\mathbf{j}$ are perpendicular.

A. -2

B. $-\frac{1}{3}$

C. $\frac{1}{3}$

D. 2

Question 192:

What is the perpendicular distance between point **p** with position vector $4\mathbf{i} + 5\mathbf{j}$ and the line L given by vector equation $\mathbf{r} = -3\mathbf{i} + \mathbf{j} + \lambda(\mathbf{i} + 2\mathbf{j})$.

A. $2\sqrt{7}$ C. $2\sqrt{5}$

B. $5\sqrt{2}$ D. $7\sqrt{2}$

Question 193:

Find k such that point $\begin{pmatrix} 2 \\ k \\ -7 \end{pmatrix}$ lies within the plane $\mathbf{r} = \begin{pmatrix} 2 \\ 3 \\ -1 \end{pmatrix} + \lambda\begin{pmatrix} 4 \\ 1 \\ 0 \end{pmatrix} + \mu\begin{pmatrix} 2 \\ 1 \\ 3 \end{pmatrix}$.

A. -2 C. 0 E. 2

B. -1 D. 1

Question 194:

What is the largest solution to $\sin\left(\frac{\pi}{2} - 2\theta\right) = 0.5$ for $\frac{\pi}{2} \le x \le 2\pi$?

A. $\frac{5\pi}{3}$ C. $\frac{5\pi}{6}$ E. $\frac{11\pi}{6}$

B. $\frac{4\pi}{3}$ D. $\frac{7\pi}{6}$

Question 195:

$\cos^4(x) - \sin^4(x) \equiv$

A. $\cos(2x)$ C. $\sin(2x)$ E. $\tan(x)$

B. $2\cos(x)$ D. $\sin(x)\cos(x)$

Question 196:

How many real roots does $y = 2x^5 - 3x^4 + x^3 - 4x^2 - 6x + 4$ have?

A. 1   C. 3   E. 5

B. 2   D. 4

Question 197:

What is the sum of 8 terms, $\sum_1^8 u_n$, of an arithmetic progression with $u_1 = 2$ and $d = 3$?

A. 15   C. 100   E. 282

B. 82   D. 184

Question 198:

What is the coefficient of the x^2 term in the binomial expansion of $(2 - x)^5$?

A. -80   C. 40   E. 80

B. -48   D. 48

64

Question 199:

Given you have already thrown a 6, what is the probability of throwing three consecutive 6s using a fair die?

A. $\frac{1}{216}$

B. $\frac{1}{36}$

C. $\frac{1}{6}$

D. $\frac{1}{2}$

E. 1

Question 200:

Three people, A, B and C, play darts. The probability that they hit a bull's eye are respectively $\frac{1}{5}$, $\frac{1}{4}$, $\frac{1}{3}$. What is the probability that at least two shots hit the bullseye?

A. $\frac{1}{60}$

B. $\frac{1}{30}$

C. $\frac{1}{12}$

D. $\frac{1}{6}$

E. $\frac{3}{20}$

Question 201:

If probability of having blonde hair is 1 in 4, the probability of having brown eyes is 1 in 2, and the probability of having both is 1 in 8, what is the probability of having neither blonde hair nor brown eyes?

A. $\frac{1}{2}$

B. $\frac{3}{4}$

C. $\frac{3}{8}$

D. $\frac{5}{8}$

E. $\frac{7}{8}$

Question 202:

Differentiate and simplify the following expression: $y = x(x + 3)^4$.

A. $(x + 3)^3$

B. $(x + 3)^4$

C. $x(x + 3)^3$

D. $(5x + 3)(x + 3)^3$

E. $5x^3(x + 3)$

Question 203:

Evaluate $\int_1^2 \frac{2}{x^2} dx$.

A. -1

B. $\frac{1}{3}$

C. 1

D. $\frac{21}{4}$

E. 2

Question 204:

Express $\frac{5i}{1+2i}$ in the form $a + bi$.

A. $1 + 2i$

B. $4i$

C. $1 - 2i$

D. $2 + i$

E. $5 - i$

Question 205:

Simplify $7\log_a(2) - 3\log_a(12) + 5\log_a(3)$.

A. $\log_{2a}(18)$

B. $\log_a(18)$

C. $\log_a(7)$

D. $9\log_a(17)$

E. $-\log_a(7)$

Question 206:

What is the equation of the asymptote of the function $y = \frac{2x^2 - x + 3}{x^2 + x - 2}$?

A. $x = 0$ D. $y = 0$

B. $x = 2$ E. $y = 2$

C. $y = 0.5$

Question 207:

Find the intersection(s) of the functions $y = e^x - 3$ and $y = 1 - 3e^{-x}$.

A. 0 and $\ln(3)$ C. $\ln(4)$ and 1

B. 1 D. $\ln(3)$

Question 208:

Find the radius of the circle $x^2 + y^2 - 6x + 8y - 12 = 0$.

A. 3   C. 5   E. 12

B. $\sqrt{13}$ D. $\sqrt{37}$

Question 209:

Which value of a minimises the magnitude of $\int_0^a 2\sin(-x)\,dx$?

A. 0.5π C. 2π E. 5π

B. π D. 3π

Question 210:

When $\frac{2x+3}{(x-2)(x-3)^2}$ is expressed as a sum of partial fractions, what is the numerator in the $\frac{A}{(x-2)}$ term?

A. -7   C. 3   E. 7

B. -1   D. 6

SECTION 2

Section 2 of the ENGAA consists of 20 questions. You have 60 minutes to answer all the questions. Section 2 requires the same core knowledge as section 1.

Watch the Clock

The key to maximising your score is to have a strategy of how best to use your time. Each question is split into four parts, but not all parts are worth the same number of marks. Firstly, you should be aware of how long you should ideally be spending on each question on average. With 32 marks available, this equates to around one mark per minute, but some will take longer than others. An easier goal is to think that four questions, of 6-8 marks each, in 40 minutes means you should spend under 10 minutes per question, to (ideally) leave time for reading and checking.

Start with the 'Low Hanging Fruit'

If you spot a question you know immediately how to tackle, that is where you should start. This will quickly add to your marks, provide extra time, and also build your confidence before tackling the more challenging or unfamiliar questions. As the grouped questions follow on from each other, that might mean answering the beginning of all four questions or one question entirely, whichever you are most comfortable with. However, be warned that the more you jump around between different types of question, the more likely you are to make mistakes or waste time reacquainting yourself with the problem at hand. Once you have gone through the paper once and answered the 'obvious' questions, you can then reassess how long you have left and how many questions remain to best allocate your time before moving on.

Practice

The best way to prepare is to practice not only long answer questions of this level, but also different multiple-choice papers, to familiarise yourself with eliminating options and obtaining an answer as fast as possible. The questions in this book, although written in the style prior to the 2018 revision of the ENGAA specification, are a good starting point as the content of the questions and their difficulty is unchanged. As the exam is relatively new, there are not many past papers to use, however there are several exam boards as well as competitions like the Canadian Maths Competition that are multiple choice, which you can use for practice.

Thrive on Adversity

The ENGAA is specifically designed to be challenging and to take you out of your comfort zone. This is done in order to separate different tiers of students depending on their academic ability. The reason for this type of exam is that Cambridge attracts excellent students that will almost invariably score well in exams. If, for this reason, during your preparation, you come across questions that you find very difficult, use this as motivation to try and further your knowledge beyond the simple school syllabus. This is where the option for specialisation ties in.

Practice Calculus

Practicing calculus might seem unnecessary, but confidence in the core calculus techniques will allow you to answer questions significantly faster and with improved accuracy. As the math part is core knowledge for all Section 2 questions, you can be certain that to some degree or other you will be required to apply your calculus knowledge.

Think in Applied Formulas

Attempting to answer Section 2 Physics questions without first learning all of the standard formulae is like trying to run before you can walk – ensure you're completely confident with all the core formulas before starting the practice questions.

Variety

Physics is a very varied subject and the questions you may be asked in an exam will reflect this variety. Many students have a preferred area within physics, such as electronics, astrophysics, or mechanics. However, it is important to remember not to neglect any subject area in its entirety. It is entirely possible that of the questions in the ENGAA, several of them could be outside of your comfort zone – leaving you in a difficult position.

Graphs

Graph-sketching is usually a tricky area for many students. When tackling a graph-sketching problem, there are many approaches; however, it is useful to start with the basic features of the plot:

- What is the value of y when x is zero?
- What is the value of x when y is zero?
- Are there any special values of x and y?
- If there is a fraction involved, at what values of x is the numerator or denominator equal to zero?

If you asked to draw a function that is the sum, product or division of two functions, start by drawing out these sub-functions. Which function is the dominant function when x > 0 and x < 0?

Answering these basic questions will tell you where the asymptotes and intercepts are, which will help with drawing the function.

Remember the Basics

A surprisingly large number of students do not know what the properties of basic shapes are. For example, the area of a circle is πr^2, the surface area of a sphere is $4\pi r^2$ and the volume of a sphere is $\frac{4}{3}\pi r^3$. To 'go up' a dimension (i.e. to go from an area to a volume) you need to integrate and to 'go down' a dimension you need to differentiate. Learning this will make it easy to remember formulas for the areas and volumes of basic shapes – the constant at the front, however, will need to be memorised.

It is also important to remember important formulas, even though they may be included in formula booklets. The reason for this is that although you may have access to formula booklets during exams, this will not be the case in interviews which will follow the ENGAA. In addition, flicking through formula booklets takes up time during an exam and can be avoided if you are able to memorise important formulas.

QUESTIONS

Question 1.1
A golfer swings a club so that the head completes a (virtually) complete circle in T = 0.1 s. The length of the club is R = 1 m.

What angle from the horizontal should the club hit the ball at to maximise the distance travelled?

A. 0° C. 45° E. Any angle

B. 30° D. 90

Question 1.2
What is the (approximate) velocity of the club when it strikes the ball?

A. m/s C. 10 m/s E. 100 m/s

B. 3 m/ D. 60 m/

Question 1.3
Assuming that the ground is flat, what is the total time the golf ball is in the air?

A. $3\sqrt{2}$ s C. $6\sqrt{2}$ s E. 15 s

B. $5\sqrt{2}$ D. 12

Question 1.4
What therefore is the maximum horizontal distance the ball travels?

A. 180 m C. 360 m E. 720 m

B. $180\sqrt{2}$ D. $360\sqrt{2}$

Question 2.1
What is the mass of the moon in terms of the radius of the earth, r, if the density of the Moon is 75% of the density of earth, ρ, and that the radius of the Moon is 4 times smaller than the radius of Earth.

A. $\frac{1}{64}\rho\pi r^3$ C. $\frac{3}{64}\rho\pi r^2$ E. $\frac{1}{16}\rho\pi r^3$

B. $\frac{9}{16}\rho\pi r^3$ D. $\frac{3}{64}\rho\pi r^3$

Question 2.2
Given that gravitational force is determined by $F = G\frac{Mm}{R^2}$, which of the following is an expression for the acceleration due to gravity on earth, where G is the gravitational constant?

A. $G\rho\pi r^2$ C. $\frac{4}{3}G\rho\pi r^2$ E. $\frac{4}{3}G\rho\pi r$

B. $\frac{3}{4}G\rho\pi r$ D. $\frac{1}{16}G\rho\pi r$

Question 2.3
Estimate the gravitational acceleration at the surface of the Moon relative to g on earth.

A. 8g

B. $\frac{g}{8}$

C. $\frac{g}{16}$

D. $\frac{3g}{16}$

E. $\frac{7g}{16}$

Question 2.4
How would the orbital speed of a satellite, of equal mass and equal orbital radius, change between orbiting earth and the moon?

A. Decrease by a factor of $\sqrt{3}/4$.

B. Decrease by a factor of $\frac{7}{16}$.

C. It will remain unchanged.

D. Increase by a factor of $\frac{1}{16}$.

E. Increase by a factor of 8.

Question 3.1
Identical resistors are connected using wire of negligible resistance to a 1.4 V power supply. What would the resistance be of two resistors in parallel?

A. 2R

B. $\frac{R}{2}$

C. $\frac{3R}{2}$

D. $\frac{2R}{3}$

E. R

Question 3.2
What would the total resistance in this circuit be, considering all the resistors to be identical?

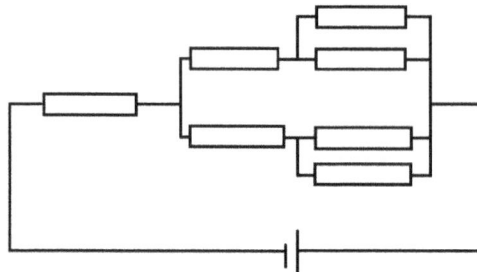

A. 7R

B. 3R

C. $\frac{7R}{3}$

D. $\frac{3R}{4}$

E. $\frac{7R}{4}$

Question 3.3
What would R be to produce a total current of 2 A?

A. Ω

B. 0.4 Ω

C. Ω

D. 0.5 Ω

E. Ω

Question 3.4
What would be the power dissipated in the circuit?

A. 0.6 W

B. 1.2 W

C. 2.3 W

D. 2.8 W

E. W

70

Question 4.1: The position of a particle the moves along the x axis is given by the equation $x = 10 + 1.5t^3$. What expression would represent the velocity of the particle?

A. $v = 1 + 1.5t^2$ C. $v = 10 + 4.5t^2$ E. $v = 3.0t^3$
B. $v = 4.5t^2$ D. $v = 45 + 4.5t^3$

Question 4.2: At which time, t, the acceleration of the particle is equal to zero?

A. $t = 20$ s   C. $t = 4.5$ s   E. $t = 1.5$ s
B. $t = 10$ s   D. $t = 0$

Question 4.3: What is the average velocity of the particle for the time interval t = 2 to 10 seconds?

A. $v = 150$ m/s   C. $v = 190$ m/s   E. $v = 3.0t^3$
B. $v = 400$ m/s   D. $v = 250$ m/s

Question 4.4: If the equation represents the motion of the ball, of mass m, that is thrown against a wall. What will be the energy transferred to the wall in an elastic collision, considering that the ball travels for 5 seconds until it hits the wall?

A. $E = 780m$ J   B. $E = 500m$ J   C. $E = 1300m$ J

Question 5.1
A lift with a mass of 800 kg can carry up to 700 kg of passengers. Calculate the total energy needed for an electric motor is used to raise the elevator with a full load from the ground floor to the third floor which is 7 m higher. You may assume g =10 m/s².

A. 105 J   C. 105 kJ   E. 105 MJ
B. 210 J   D. 210 kJ

Question 5.2
Calculate the power of motor required to do this in 30 s.

A. 700 W   C. 7 kW   E. 70 kW
B. 3500 W   D. 35 kW

Question 5.3
Find the average kinetic energy of the lift and passengers during the ascent.

A. 41 J   C. 67 J   E. 105 J
B. 52 J   D. 82 J

Question 5.4
The elevator takes 10 seconds to accelerate and 10 seconds to decelerate using the same magnitude of constant force in each case. Between acceleration and deceleration, no force is used. Calculate the average speed which the lift attains.

A. 0.04 ms⁻¹   C. 0.35 ms⁻¹   E. 2.3 ms⁻¹
B. 0.14 ms⁻¹   D. 1.3 ms⁻¹

Question 6.1

If the friction coefficient between m_1 and m_2 is μ_1, and between m_2 and the inclined plane is μ_2, (where α = the angle of inclination) determine the acceleration of m_2.

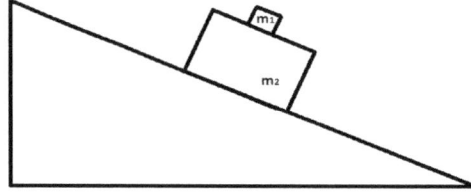

A. $g(1 - \mu_2\cos\alpha)$ C. $mg(\sin\alpha - \mu_2\cos\alpha)$ E. $g(\sin\alpha - \mu_2\cos\alpha)$

B. $g(\mu_2\sin\alpha - \mu_1\cos\alpha)$ D. $g(\sin\alpha - \mu_1\cos\alpha)$

Question 6.2

Determine the acceleration of m_1 with respect to the plane.

A. $g(\sin\alpha - \mu_1\cos\alpha)$ C. $g\sin\alpha(\mu_1 - \mu_2)$ E. $g(\mu_2\sin\alpha - \mu_1\cos\alpha)$

B. $mg(\sin\alpha - \mu_1\cos\alpha)$ D. $g(\sin\alpha - \cos\alpha)$

Question 6.3

Determine the acceleration of m_1 with respect to m_2.

A. $g\cos\alpha(\mu_1 + \mu_2)$ C. $g\sin\alpha(\mu_1 - \mu_2)$ E. $mg\cos\alpha(\mu_1 + \mu_2)$

B. $g\cos\alpha(\mu_2 - \mu_1)$ D. $mg\cos\alpha(\mu_1 - \mu_2)$

Question 6.4

Which of the following would give the coefficient of friction, μ_2?

A. $\tan(\alpha)$ C. $\sin(\alpha)$ E. $mg\tan(\alpha)$

B. $\cos(\alpha)$ D. mg

Question 7.1:

The graph below represents the velocity as a function of time of a car (with mass 1000 kg) moving in a straight line. How far has the car travelled after 30 min?

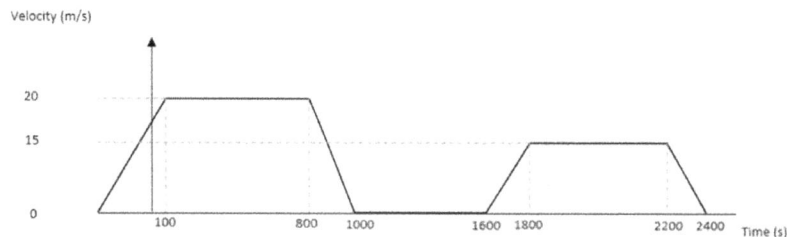

A. 4900 m   C. 20000 m   E. 5900 m

B. 9000 m   D. 18500 m

72

Question 7.2:

Compute the following calculation $\int_{1600}^{1800}$ v dt. Select the option that represents the correct corresponding value and its physical meaning.

A. Acceleration: 1500 ms^{-2}
B. Acceleration: 2300 ms^{-2}
C. Velocity: 1500 ms^{-1}
D. Distance: 1000 m.
E. Distance: 1500 m.

Question 7.3:

Calculate the force exerted on the car at t = 900s.

A. 110 N
B. 100 N
C. 75 N
D. 500 N
E. 765 N

Question 7.4:

What is the power delivered by the car's engine in the interval of 0 – 100 s?

A. 2 kW
B. 900 W
C. 1.5 kW
D. 2.5 kW
E. 1.05 kW

Question 8.1:

The following diagram represents the horizontal component of a force acting on a particle as a function of its position. If the particle is at rest at x = 0, what is its position (along the x-axis) when its velocity is maximum?

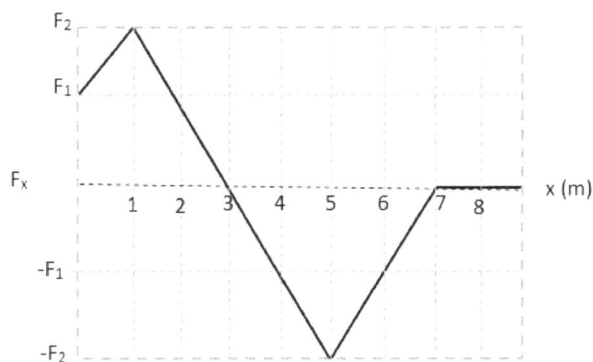

A. 1 m
B. 5 m
C. 3 m
D. 4 m
E. 8 m

Question 8.2: What is the particle's position (along the x-axis) when its kinetic energy is maximum?

A. 1 m
B. 5 m
C. 7 m
D. 3 m
E. 8 m

Question 8.3: What is the particle' position (along the x-axis) when its velocity is zero?

A. 6 m   C. 7 m   E. 8 m
B. 1 m   D. 3 m

Question 8.4: After x = 6 metres, which of the following statements is true?

A. The velocity vector is pointing in the same direction as it is in x = 3m.
B. The velocity vector is pointing in the same direction as it is in x = 2m.
C. The acceleration vector is pointing in the same direction as it is in x = 2m.
D. The acceleration vector is point in the opposite direction as it is in x = 2m.

Question 9.1: A little girl throws vertically a ball, of mass 0.5 kg, into the air at t =0. Which of the following graphs best describe the velocity of the ball as a function of time? Disregard the influence of air in the movement of the ball.

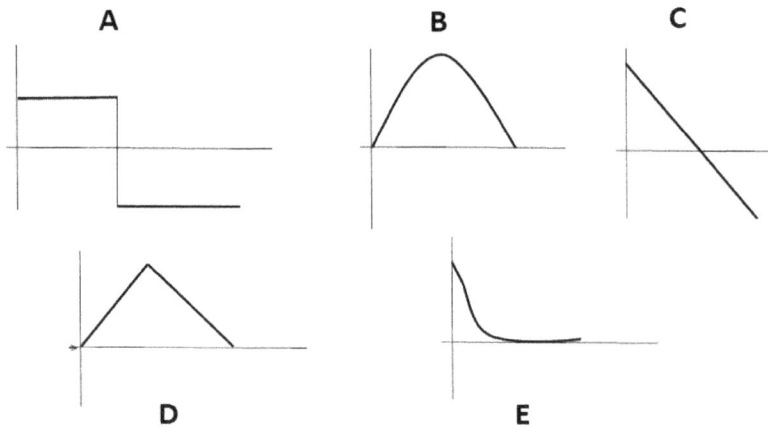

Question 9.2: Given that the ball leaves the girl's hand at 15 m/s, what is the approximate highest point from the ground the ball can reach? Disregard the influence of air.

A. 6 m   C. 25 m   E. 8 m
B. 12 m   D. 3.5m

Question 9.3: Approximately, how long does the ball take to reach the highest point of its trajectory?

A. 1.5 s   C. 5 s   E. 9.5 s
B. 10 s   D. 4 s

Question 9.4: What would the potential and kinetic energy of the ball be, respectively, 5.0 m from the ground?

A. 31.75 J and 24.5 J   C. 24.5 J and 31.75 J   E. 11 J and 27 J
B. 25.5 J and 37.75 J   D. 30 J and 26 J

74

Question 10.1: A horizontal force is applied to the ramp (Object 1), of mass 10 kg, as shown in the figure. There is no friction between the ramp and the ground. The coefficient of friction between the ramp and Object 2, also of mass 10 kg, is $\mu = 0.2$. What option correctly demonstrates the forces applied on Object 2?

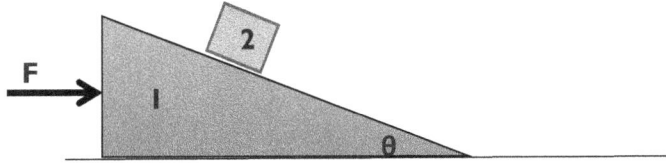

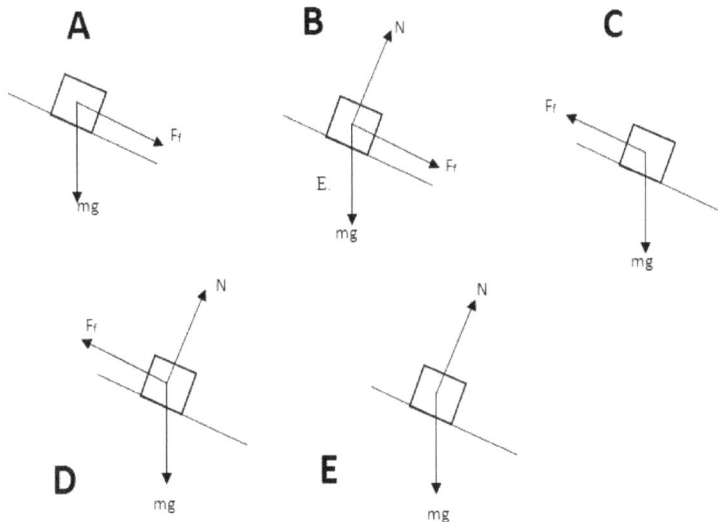

Question 10.2: Express the weight of Object 2 as a function of the angle θ and the coefficient of friction μ.

A. $mg = N\mu(\sin\theta - \cos\theta)$ C. $mg = N(\mu\sin\theta - \cos\theta)$ E. $mg = N(\cos\theta - \mu\sin\theta)$

B. $mg = N\mu(\cos\theta + \sin\theta)$ D. $mg = \mu(\sin\theta + N\cos\theta)$

Question 10.3: Express the acceleration of Object 2, in relation to the Object 1, as a function of the angle θ and the coefficient of friction μ.

A. $a = N(\sin\theta - \mu\cos\theta)$ C. $a = \dfrac{N}{m_2}(\cos\theta + \mu\cos\theta)$ E. $a = \dfrac{N}{m_2}(\cos\theta + \mu\sin\theta)$

B. $a = \dfrac{N}{m_2}(\sin\theta - \mu\cos\theta)$ D. $a = \dfrac{N}{m_2}(\sin\theta + \mu\cos\theta)$

Question 10.4: What is the maximum value of the force F, in Newtons, that can be applied without provoking the movement of Object 2 upwards along the ramp? Consider $\sin\theta = 0.6$ and $\cos\theta = 0.8$.

A. 219 N   C. 200N   E. 149 N

B. 350 N   D. 155 N

75

ANSWERS

ANSWER KEY

Q	A	Q	A	Q	A	Q	A	Q	A	Q	A	Q	A
1	F	31	C	61	D	91	C	121	B	151	D	181	C
2	A	32	G	62	F	92	A	122	A	152	D	182	C
3	D	33	D	63	B	93	C	123	F	153	B	183	E
4	E	34	D	64	A	94	B	124	D	154	E	184	C
5	G	35	E	65	C	95	A	125	A	155	E	185	C
6	C	36	B	66	C	96		126	B	156	B	186	A
7	D	37	A	67	B	97	D	127	A	157	C	187	D
8	E	38	E	68	C	98	C	128	F	158	A	188	C
9	D	39	F	69	B	99	C	129	D	159	C	189	B
10	D	40	F	70	B	100	C	130	A	160	B	190	A
11	F	41	E	71	E	101	B	131	D	161	B	191	C
12	B	42	A	72	C	102	C	132	D	162	B	192	C
13	C	43	B	73	E	103	C	133	F	163	C	193	E
14	F	44	C	74	E	104	C	134	B	164	C	194	E
15	E	45	D	75	D	105	E	135	C	165	A	195	A
16	E	46	F	76	E	106	A	136	B	166	C	196	C
17	A	47	E	77	E	107	C	137	C	167	D	197	C
18	E	48	C	78	C	108	E	138	A	168	C	198	E
19	G	49	C	79	A	109	E	139	A	169	D	199	A
20	E	50	B	80	D	110	C	140	C	170	A	200	D
21	H	51	D	81	E	111	E	141	B	171	C	201	C
22	E	52	F	82	C	112	E	142	D	172	B	202	D
23	F	53	G	83	C	113	E	143	C	173	B	203	C`
24	D	54	D	84	C	114	B	144	A	174	A	204	D
25	E	55	D	85	B	115	C	145	C	175	C	205	B
26	D	56	A	86	D	116	B	146	C	176	C	206	E
27	G	57	E	87	B	117	B	147	C	177	B	207	A
28	F	58	C	88	E	118	C	148	B	178	A	208	D
29	B	59	D	89	C	119	D	149	D	179	C	209	C
30	A	60	C	90	A	120	C	150	E	180	E	210	E

SECTION 1: WORKED ANSWERS

Question 1: F
Statement F, that the amplitude of a wave determines its mass, is false. Waves are continuous, not particulate and do not have mass.

Question 2: A
We are given that displacement s = 30 m, initial speed u = 0 ms^{-1} and acceleration a = 5.4 ms^{-2}. We are asked to find the final speed v, and the time taken t.

We require the SUVAT equation without time to find final velocity, which is $v^2 = u^2 + 2as$.

$$v^2 = u^2 + 2as \Rightarrow v^2 = 0^2 + 2 \times 5.4 \times 30$$

$$\Rightarrow v^2 = 324 \therefore v = 18 \, ms^{-1}$$

Next, to find the time, we can use:

$$s = ut + \frac{1}{2}at^2 \Rightarrow 30 = 0 \times t + \frac{1}{2} \times 5.4 \times t^2$$

$$\Rightarrow t^2 = \frac{30}{2.7} = 3.3 \, s$$

Question 3: D
The time period is calculated using the fact that the canoe rises and falls 7 times within 49 seconds. Therefore, T = 49/7 = 7 seconds.

To find the wavelength, we need to use the $v = f\lambda$ equation. Frequency is the reciprocal of T, so it is given by 1/T = 1/7 s^{-1}. Therefore:

$$\lambda = \frac{v}{f} = \frac{5}{1/7} = 35 \text{ m.}$$

Question 4: E
This is a straightforward question, as there is an equation which directly links all the provided quantities. The only challenging aspect is the numbers being more difficult in the calculations.

$$\text{Power} = \frac{\text{Force x Distance}}{\text{Time}} = \frac{375 \text{ N x } 1.3 \text{ m}}{5 \text{ s}} = 75 \times 1.3 = 97.5 \text{ W}$$

Question 5: G

As there is constant acceleration, we need to use a SUVAT equation. We have been given the initial velocity, time taken and the acceleration.

Therefore, we need to use $v = u + at$ to find the final velocity.

$$v = u + at = 0 + 5.6 \times 8 = 44.8\,ms^{-1}$$

And, to find distance, we need to use:

$$s = ut + \frac{at^2}{2} = 0 + 5.6 \times \frac{8^2}{2} = 179.2\ m$$

Question 6: C

The sky diver leaves the plane and will accelerate until the air resistance equals their weight – this is their terminal velocity. The sky diver will accelerate under the force of gravity. If the air resistance force exceeded the force of gravity the sky diver would accelerate away from the ground, and if it were less than the force of gravity they would continue to accelerate towards the ground.

Question 7: D

We have been given that s = 20 m, u = 0 ms⁻¹ and a = 10 ms⁻². To find the momentum before impact, we need to know the apple's velocity before impact. Therefore, we wish to find the final velocity and the equation required is:

$$v^2 = u^2 + 2as \Rightarrow v^2 = 0 + 2 \times 10 \times 20$$

$$\Rightarrow v^2 = 400 \therefore v = 20\,ms^{-1}.$$

Therefore, the momentum is given by:

$$p = mv = 20 \times 0.1 = 2\,kgms^{-1}.$$

Question 8: E

Electromagnetic waves have varying wavelengths and frequencies, and, as their energy is proportional to their frequency, have varying energies. Therefore, only statements 1, 2 and 5 are correct.

Question 9: D

The total resistance = R + r = 0.8 + 1 = 1.8 Ω. Therefore, the current can be calculated as follows:

$$I = \frac{EMF}{Total\ Resistance} = \frac{36}{1.8} = 20\ A.$$

Question 10: D

Use Newton's second law and remember to work in SI units:

$$F = \frac{\Delta p}{\Delta t} = \frac{\Delta(mv)}{\Delta t} = m\frac{\Delta v}{\Delta t}.$$

The change in velocity is given in the question – it is 100 ms⁻¹ as the dart is brought to rest. Therefore, the average force is given by:

$$\Rightarrow F = 20 \times 10^{-3} \times \frac{100}{10\ x\ 10^{-3}} = 200\ N.$$

Question 11: F

As the Professor is lifting the bag against gravity, work is being done. This work done is given by:

Work Done $= F \times d =$ Bag's Weight x Distance $= 50 \times 10 \times 0.7 = 350$ N.

The average power, therefore, is given by:

Power $= \dfrac{\text{Work}}{\text{Time}} = \dfrac{350}{3} = 116.7$ W.

This gives a value of 117 W to 3 significant figures, so the correct answer is F.

Question 12: B

The engine is supplying the driving force – therefore, to calculate the current, we are only concerned with the driving force. To find the power supplied by the engine, we need to use:

$P = Fv = 300$ N $\times 30$ ms$^{-1} = 9000$ W.

To calculate the current, we need to use:

$P = IV \Rightarrow I = \dfrac{P}{V} = \dfrac{9000}{200} = 45$ A.

Question 13: C

Work is defined as the magnitude of the force multiplied by the displacement in the direction of the force. Therefore, statement 3 is correct. Statement 1, therefore, is not correct as it suggests you divide the force by the displacement. For statement 2, we need to work out the units of work. Using the equation, we can see that:

$W = F \times d \Rightarrow W = ma \times d$

By considering the units, we can clearly see that:

$[W] = kg \times ms^{-2} \times m = kgm^2s^{-2}.$

Therefore, statement 2 is incorrect. Thus, only statement 3 is correct.

Question 14: F

Joules are the unit of energy – using the equation W = F × d, we can obtain that 1 Joule = 1 N x 1 m.

Pa is the unit of Pressure - therefore, by using the following equation, we obtain that:

$P = \dfrac{Force}{Area} = \dfrac{N}{m^2}.$

$\therefore Pa \times m^3 = \dfrac{Nm^3}{m^2} = Nm.$

Therefore, statement 2 is indeed true. Statement 1 is also true as this is the defining equation for kinetic energy.

The kinetic energy that a body carries is the total energy it has due to its movement. Therefore, if we were to bring the object to rest, it would lose exactly its kinetic energy – as this is the energy 'locked-up' in its motion. Therefore, all 3 statements are true.

Question 15: E

Radiation can indeed be in the form of waves or particles and may be ionising depending on its energy. Gamma radiation does have very high energy, followed by beta radiation and then alpha radiation. Statement E, however, is not true as X-rays are waves and, therefore, are not an example of particle radiation.

Question 16: E

Statement 1 is false as the half-life depends on the physical properties of the atom – the atom type and isotope clearly affect the physical properties of the atom. Statement 2 is the correct definition of half-life.

Statement 3 is also correct: half-life in exponential decay will always have the same duration, independent of the quantity of the matter in question – this is a defining feature of exponential decay. In non-exponential decay, half-life is dependent on the quantity of matter in question.

Question 17: A

In contrast to nuclear fission, where neutrons are shot at unstable atoms, nuclear fusion is based on the high speed, high-temperature collision of molecules, most commonly hydrogen, to form a new, heavier atoms while releasing energy. Therefore, all statements are true except statement A, which is the case for nuclear fission.

Question 18: E

Nuclear fission releases a significant amount of energy, which is the basis of many nuclear weapons. Shooting neutrons at unstable atoms destabilises the nuclei which in turn leads to a chain reaction and fission. Nuclear fission can lead to the release of ionizing gamma radiation. Therefore, all of the statements are true.

Question 19: G

As the two resistors are connected in a series circuit, the current is the same at all points – therefore, statement 1 is true. Due to Ohm's Law, the potential difference across the two resistors must also be the same – as the current and resistance of each resistor are equal. Therefore, statement 2 is also true. Ohm's Law can also be applied to the two resistors together, so statement 3 is also correct.

Question 20: E

To find the circumference, the radius of the orbit needs to be evaluated first. To determine the radius, we have to convert the time into seconds and then use the following equation:

$$Speed = \frac{Distance}{Time} \Rightarrow Distance = Speed \times Time.$$

$$\therefore Distance = 3 \times 10^8 \times (60 \times 8) = 1440 \times 10^8 \approx 1.5 \times 10^{11} \, m$$

As the answer only requires the power of ten, we can take this as 10^{11} metres. To find the circumference, we need to use:

$$C = 2\pi r = 2\pi \times (10^{11}) \approx 10^{12} \text{ m.}$$

Question 21: H

Speed is a scalar quantity whilst velocity is a vector, which means it describes both magnitude and direction. Speed describes the distance a moving object covers over time (i.e. speed = distance/time), whereas velocity describes the rate of change of the displacement of an object (i.e. velocity = displacement/time). The SI unit for speed is meters per second (ms^{-1}), while ms^{-2} is the standard unit of acceleration.

Question 22: E

Ohm's Law only applies to conductors as insulators cannot carry a current. The law can be mathematically expressed as $V = IR$. This suggest that potential difference is directly proportional to current, which corresponds to statement E.

Question 23: F

Any object at rest is not accelerating and therefore has no resultant force acting upon it. Therefore, statement 1 is indeed true.

Strictly speaking, Newton's second law is given by:

$$F = \frac{dp}{dt} = \text{Rate of change of momentum.}$$

This can be rearranged as follows:

$$F = \frac{dp}{dt} = \frac{d(mv)}{dt} = m\frac{dv}{dt} + v\frac{dm}{dt}.$$

In the case of constant mass, this reduces to:

$$F = m\frac{dv}{dt} = ma.$$

Therefore, this is not the most general form of Newton's Second Law and is only applicable in the cases of constant mass. Statement 2, therefore, is false.

Question 24: D

Equation 1 is correct, as charge equals current multiplied by time. By substituting in Ohm's Law, which is $I = \frac{V}{R}$, equation 1 is obtained. Equally, the equation $I = \frac{P}{V}$ may be substituted to reproduce equation 2. Equation 3, however, is incorrect as charge is simply equal to the numerator of the equation – the division by resistance is incorrect.

Question 25: E

The total weight of the elevator and the people within it is given by:

$$W = mg = 10 \times (1600 + 200) = 18,000 \text{ N.}$$

The total downwards force on the elevator is the sum of the frictional force and the weight. The motor force has to act in the upwards direction. Applying Newton's second law of motion on the car gives:

$$F_M = \text{Motor Force} - [\text{Frictional Force} + \text{Weight}]$$

$$F_M = M - 4,000 - 18,000 = M - 22,000$$

To get an acceleration of 1 ms⁻² upwards, this must be equal to:

$$F_M = ma = 1800 \times 1 = 1800 \, N$$

$$\Rightarrow 1800 \, N = M - 22{,}000 \quad \therefore M = 23{,}800 \, N.$$

Question 26: D

The total distance travelled is the sum of the distance travelled during acceleration phase and the distance travelled during the braking phase. The distance during <u>acceleration phase</u> is given by:

$$s = ut + \frac{at^2}{2} = 0 + \frac{5 \times 10^2}{2} = 250 \text{ m.}$$

To find the distance during the second phase, we need to find the final velocity:

$$v = u + at = 0 + 5 \times 10 = 50 \text{ ms}^{-1}.$$

Now, we can use $a = \frac{v-u}{t}$ to calculate the deceleration:

$$a = \frac{0 - 50}{20} = -2.5 \text{ ms}^{-2}.$$

Therefore, the distance travelled during the <u>deceleration phase</u> is given by:

$$s = ut + \frac{at^2}{2} = 50 \times 20 + \frac{-2.5 \times 20^2}{2} = 1000 - \frac{2.5 \times 400}{2}$$

$$\Rightarrow s = 1000 - 500 = 500 \text{ m.}$$

Therefore, the total distance is the sum of these two distances, which gives:

$$\text{Total Distance} = 250 + 500 = 750 \text{ m.}$$

Question 27: G

It is not possible to calculate the power of the heater as we do not know the current that flows through it or its internal resistance. The 8 ohms refers to the external copper wire and not the heater. Whilst it is important that you know how to use equations like P = IV, it's more important that you know when you *can't* use them!

Question 28: F

This question has a lot of information but there is no mention of the duration of each pulse. The quantity of time for which the electrons are accelerated is necessary to calculate power. Similarly, you cannot calculate power by using P = IV as you do not know how many electrons are accelerated through the potential difference per unit time. Thus, more information is required to calculate the power.

Question 29: B

When an object is in equilibrium with its surroundings, it radiates and absorbs energy at the same rate and so its temperature remains constant. In other words, there is no *net* energy transfer. Therefore, statement A is false.

Radiation is slower than conduction and convection. Therefore, statements C and D are both false.

Question 30: A

The work done by the force is given by:

Work Done = Force $\times$ Distance = $12\,N \times 3\,m = 36\,J$.

Since the surface is frictionless, we can further state that:

Work Done = Kinetic Energy $\Rightarrow E_k = \dfrac{mv^2}{2} = \dfrac{6v^2}{2}$.

$\therefore 36 = 3v^2 \Rightarrow v = \sqrt{12} = \sqrt{4}\sqrt{3} = 2\sqrt{3}\,ms^{-1}$.

Question 31: C

The total energy supplied to the water is the product of the change in the temperature, the mass of the water and its specific heat capacity, which is 4000 J in this case. Therefore:

$Q = mc\,\Delta T = $ Change in temperature $\times$ Mass of water $\times$ 4,000 J.

$\Rightarrow Q = 40 \times 1.5 \times 4,000 = 240,000\,J$.

The power of the heater can also be evaluated with the quantities given, which allows us to determine the resistance using the $P = \dfrac{V^2}{R}$ relation:

$P = \dfrac{\text{Work Done}}{\text{time}} = \dfrac{240,000}{50 \times 60} = \dfrac{240,000}{3,000} = 80\,W$.

$P = IV = \dfrac{V^2}{R} \Rightarrow R = \dfrac{V^2}{P} = \dfrac{100^2}{80} = \dfrac{10,000}{80} = 125\,\Omega$.

Question 32: G

Nuclear power plants utilise the large amount of energy released during atomic fission, therefore statement I is indeed true. Splitting an atom into two or more parts will, by definition, produce molecules of different sizes than the original atom – as these are the constituents of the parent particle. The free neutrons and photons produced by the splitting of atoms form the basis of the energy release.

Question 33: D

Gravitational potential energy is just an extension of the equation work done equals force multiplied by distance (force is the weight of the object, *mg,* and distance is the height, *h*). Therefore, statements 1 and 2 are both correct. The reservoir in statement 3 would have a potential energy of 10^{10} Joules, which is equivalent to 10 Giga Joules, as:

$$E_p = 10^6 \text{ kg} \times 10 \text{ N} \times 10^3 \text{ m} = 10 \text{ GJ}.$$

Question 34: D

Statement 1 is the common formulation of Newton's third law. Statement 2 presents a consequence of the application of Newton's third law.

Statement 3 is false: rockets can still accelerate because the products of burning fuel are ejected in the opposite direction from which the rocket needs to accelerate. Therefore, the fuel experiences the equal and opposite force to the rocket.

Question 35: E

Positively charged objects have lost electrons, as electrons are negatively charged particles. Therefore, statement 1 is false. For the second statement, we can recall the following equation:

$$\text{Charge} = \text{Current x Time} = \frac{\text{Voltage}}{\text{Resistance}} \times \text{Time}.$$

Therefore, we can evaluate the amount of charge when provided with the period of time, voltage and resistance. Lastly, objects can become charged by friction as electrons are transferred from one object to the other.

Question 36: B

Each body of mass exerts a gravitational force on another body with mass. This is true for all planets, as well as all bodies with mass in the universe. Gravitational force is indeed dependent on the mass of both objects, which is evident from the gravitational force equation.

Satellites stay in orbit due to centripetal force that acts tangentially to gravity (not because of the thrust from their engines). Two objects will only land at the same time if they also have the same shape as otherwise air resistance would result in different terminal velocities. Note that this is not relevant in space as there is no air resistance.

Question 37: A

Metals conduct electrical charge easily and provide little resistance to the flow of electrons. Charge can also flow in several directions, although will typically have a general direction of motion in a circuit. Lastly, all conductors have an internal resistance and therefore provide *some* resistance to electrical charge.

Question 38: E

First, we can calculate the rate of petrol consumption:

$$\frac{Speed}{Consumption} = \frac{60 \, miles/hour}{30 \, miles/gallon} = 2 \, gallons/hour.$$

Therefore, the amount of energy required in one hour is: $2 \, gallons = 2 \times 9 \times 10^8 = 18 \times 10^8$ J. As $1 \, hour = 60 \times 60 = 3600$ seconds, we can use the following equation to determine the power:

$$Power = \frac{Energy}{Time} = \frac{18 \times 10^8}{3600} \Rightarrow P = \frac{18}{36} \times 10^6 = 5 \times 10^5 \, W.$$

As only 20% of the power is delivered to the wheels, we need to find 20% of the total power.

$$\therefore P_{wheels} = 5 \times 10^5 \times 0.2 = 10^5 W = 100 \, kW.$$

Question 39: F

Beta radiation is stopped by a few millimetres of aluminium, but not by paper. Therefore, statement 2 is correct. In β^- radiation, a neutron changes into a proton plus an emitted electron. This means the atomic mass number remains unchanged. Therefore, statement 1 is also correct. Finally, statement 3 is also correct – as beta particles are charged, they are deflected within electric fields. As they are also in motion, they experience a magnetic force and so also deflect within a magnetic field.

Question 40: F

Firstly, we can calculate the mass of the car using the following relation: $Mass = \frac{Weight}{g} = \frac{15,000}{10} = 1,500 \, kg.$

The average braking force is the product of the mass of the car and its deceleration. To calculate the deceleration of the car, we can use the following equation, where v = 0 ms⁻¹, u = 15 ms⁻¹ and t = 10 × 10⁻³ s:

$$v = u + at \Rightarrow a = \frac{v-u}{t} = \frac{0-15}{0.01} = 1500 \, ms^{-2}.$$

Therefore, the average braking force is given by:

$$F = ma = 1500 \times 1500 = 2\,250\,000 \, N.$$

Question 41: E

Electrical insulators offer high resistance to the flow of charge. Insulators are usually non-metals, as metals conduct charge very easily due to their sea of delocalised electrons. Insulators can indeed be charged by friction as charge does not flow easily within the material – therefore, it cannot cancel the charge out immediately like a conductor.

Question 42: A

The car accelerates for the first 10 seconds at a constant rate, which is given by the gradient of the first linear region, and then decelerates after t = 30 seconds, which is given by the gradient of the second linear region. It does not reverse, as the velocity is not negative. Therefore, only statement 1 is incorrect.

Question 43: B
The distance travelled by the car is represented by the area under the curve (integral of velocity) which is given by the area of two triangles and a rectangle:

$$\text{Area} = \left(\frac{1}{2} \times 10 \times 10\right) + (20 \times 10) + \left(\frac{1}{2} \times 10 \times 10\right)$$

$$\Rightarrow \text{Area} = 50 + 200 + 50 = 300 \text{ m}.$$

Question 44: C
We can use Newton's Second Law to determine the change in velocity. We know that F = 10,000 N, mass = 1,000 kg and change in time is 5 seconds, so these can be substituted.

$$F = \frac{\Delta p}{\Delta t} = m\frac{\Delta v}{\Delta t} \Rightarrow \Delta v = \frac{F \times \Delta t}{m} = \frac{10,000 \times 5}{1,000} = 50 \ ms^{-1}.$$

Question 45: D
This question tests both your ability to convert unusual units into SI units and to select the relevant values (for example, the crane's mass is not important here). Firstly, the conversions are: 0.01 tonnes = 10 kg, 100 cm = 1 metre and 5,000 milliseconds = 5 seconds.

To calculate power, we need to use the following equation:

$$\text{Power} = \frac{\text{Work Done}}{\text{Time}} = \frac{\text{Force x Distance}}{\text{Time}}$$

The force that the crane is working against is the weight of the car. This is easily evaluated:

Weight of the car = $10 \times g = 10 \times 10 = 100N \Rightarrow \text{Power} = \frac{100 \times 1}{5} = 20$ W.

Question 46: F
The total resistance, denoted R_T, in a parallel circuit is given by:

$$\frac{1}{R_T} = \frac{1}{R_1} + \frac{1}{R_2} + \cdots$$

Thus, we can substitute in the values provided for the two resistors:

$$\Rightarrow \frac{1}{R_T} = \frac{1}{1} + \frac{1}{2} = \frac{3}{2} \therefore R_T = \frac{2}{3}\Omega$$

Using Ohm's Law, we can obtain the current:

$$I = \frac{20\,V}{\frac{2}{3}\Omega} = 20 \times \frac{3}{2} = 30 \text{ A}.$$

Question 47: E
Water has a greater refractive index than air; therefore, the speed of light decreases when it enters water and increases when it leaves water. The direction of light also changes when light enters or leaves water. This phenomenon is known as refraction and is governed by Snell's Law.

Question 48: C

The voltage in a parallel circuit is equal across each branch. Therefore, the voltage of branch A equals that of branch B. The resistance of the two branches are obtained by multiplying the number of resistors in each branch by the individual resistance:

$R_A = 6 \times 5 = 30 \ \Omega, \quad R_B = 10 \times 2 = 20 \ \Omega.$

Using Ohm's Law, the current across each branch is easily evaluated:

$V = IR \Rightarrow I = \dfrac{V}{R} \quad \therefore \quad I_A = \dfrac{60}{30} = 2 \ A, \ I_B = \dfrac{60}{20} = 3 \ A.$

Question 49: C

This is a very straightforward question, but it is slightly challenging due to the awkward units provided. Ensure you are able to work comfortably with prefixes of 10^9 and 10^{-9} and convert between the relevant units correctly. The conversions required are:

50,000,000,000 nanowatts = 50 W, 0.000000004 giga-amps = 4 A.

Using the $P = IV$ relation, it is straightforward to determine the voltage of the circuit:

$V = \dfrac{P}{I} = \dfrac{50}{4} = 12.5 \ V = 0.0125 \ kV.$

Question 50: B

Radioactive decay of a single atom is highly random and unpredictable - the behaviour of a large group of atoms can be successfully modelled using a half-life, but it is not possible to predict when a single atom will decay. Therefore, statement A is false.

Only gamma decay releases gamma rays and few types of decay release X-rays. Therefore, both statements C and D are incorrect. The electrical charge of an atom's nucleus decreases after alpha decay as two protons are lost. Therefore, statement E is also incorrect.

Question 51: D

There is more than one way to approach this question. You can calculate current and use that to find the total resistance, or you can use algebra by denoting the unknown resistance and using the relevant power equation. The former method is as follows:

$P = IV \Rightarrow I = \dfrac{P}{V} = \dfrac{60}{15} = 4 \ A.$

This is the current through the circuit, so we can utilise Ohm's Law to find the total resistance of the circuit as follows:

$R = \dfrac{V}{I} = \dfrac{15}{4} = 3.75 \ \Omega.$

So, each resistor has a resistance of $\dfrac{3.75}{3} = 1.25 \ \Omega.$

If two more resistors are added, the overall resistance $R_T = 1.25 \times 5 = 6.25 \ \Omega.$

Question 52: F

To calculate the useful work done, and hence the efficiency, we require the useful output over the total energy inputted. In this case, we are not provided with the resistive forces on the tractor, whether it is stationary or moving at the end point, and if there is any change in vertical height.

Question 53: G

Electromagnetic induction is defined by statements 1 and 2. An electrical current is generated when a coil moves in a magnetic field. In essence, there needs to be relative motion between a conductor, such as the wire, and a magnetic field, which is satisfied in all three cases.

Question 54: D

An ammeter will always give the same reading in a series circuit, in which the current is the same at all points, but not in a parallel circuit where current splits at each branch in accordance with Ohm's Law. Therefore, the current does indeed depend on the resistance of that branch. It is also conserved – so the sum of the currents in the branches is equal to the total current within the circuit.

Question 55: D

Electrons move in the opposite direction to the conventional current (they move from negative to positive) as they are attracted to the positive terminal due to their negative charge. Therefore, statement 3 is incorrect.

Question 56: A

For a fixed resistor, the current is directly proportional to the potential difference. For a filament lamp, the metal filament becomes hotter as the current increases. This causes the metal atoms to vibrate and move more, resulting in more collisions with the flow of electrons. This makes it harder for the electrons to move through the lamp and results in increased resistance. Therefore, the graph's gradient decreases as current increases.

Question 57: E

Vector quantities consist of both direction and magnitude and can be added by taking account of the direction in the sum. They can also be subtracted by reversing their direction. Lastly, displacement is indeed a vector.

Question 58: C

The gravity on the moon is 6 times weaker than that on Earth. Therefore:

$$g_{Earth} = 10 \text{ ms}^{-2} \Rightarrow g_{moon} = \frac{10}{6} = \frac{5}{3} \text{ ms}^{-2}.$$

Since Weight = Mass × Gravity, the mass of the rock is given by:

$$m_{rock} = \frac{250}{\frac{5}{3}} = \frac{750}{5} = 150 \text{ kg}.$$

Therefore, the density of the rock is given by the following:

$$\text{Density} = \frac{\text{mass}}{\text{volume}} = \frac{150}{250} = 0.6 \text{ kg/cm}^3.$$

Question 59: D

An alpha particle consists of a helium nucleus. Thus, alpha decay causes the mass number to decrease by 4 and the atomic number to decrease by 2. Five iterations of this would decrease the mass number by 20 and the atomic number by 10. This would result in a mass number of 225 − (4 × 5) = 205 and an atomic number of 78 − (2 × 5) = 68.

Question 60: C

The circuit is connected to the primary coil in the transformer. Therefore, to find the potential difference in the primary coil, we can use Ohm's law:

$V_{primary}$ = IR = 20 × 10 = 200 V.

Now the equation provided can to be utilised to determine the potential difference exiting the transformer:

$\frac{N1}{N2} = \frac{V1}{V2} \Rightarrow \frac{5}{10} = \frac{200}{V2} \Rightarrow V_{secondary} = \frac{2,000}{5}$ = 400 V.

Question 61: D

For objects in free fall that have reached terminal velocity, the weight is exactly equal and opposite to the air resistance. To find the mass of the sphere, the weight needs to be determined. This can be done using the work done by the resistive forces as follows:

Work Done = Force × Distance $\Rightarrow$ Force = $\frac{10,000\,J}{100\,m}$ = 100 N.

Therefore, the sphere's weight = 100 N, as it balances this resistive force. Since g = 10 ms^{-2}, the sphere's mass is simply:

Mass = $\frac{Weight}{g} = \frac{100}{10} = 10$ kg.

Question 62: F

The wavelength of ultraviolet waves is longer than that of x-rays. Wavelength is inversely proportional to frequency. Most electromagnetic waves are not stopped with aluminium (and require thick lead to stop them), and they travel at the speed of light. Humans can only see a very small part of the spectrum. Therefore, all of the statements are incorrect.

Question 63: B

If an object moves towards the sensor, the wavelength will appear to decrease and the frequency increase. The faster the rate of approach, the faster the increase in frequency and decrease in wavelength. Alternatively, when the object moves away, we expect the frequency to decrease at that rate.

Question 64: A

The bullet has an initial velocity of 1000 m/s, and is brought to rest in a time on 0.1 s. Therefore, the deceleration of the bullet, due the wall, is given by:

Deceleration = $\frac{\text{Change in Velocity}}{\text{Time}} = \frac{v - u}{t} = \frac{1,000}{0.1} = 10,000$ ms^{-2}.

Using Newton's second law, the braking force provided by the brick wall can be evaluated:

Braking Force = Mass × Acceleration = 10,000 x 0.005 = 50 N.

Question 65: C

Polonium has undergone alpha decay, as its mass number and atomic number have both changed due to the decay. Only alpha decays have this property. Thus, Y is a helium nucleus and contains 2 protons and 2 neutrons.

Therefore, as there are 6×10^{23} particles in a single mole, each with two protons, 10 moles of Y contain:

Number of protons = $2 \times 10 \times (6 \times 10^{23})$ protons = $120 \times 10^{23} = 1.2 \times 10^{25}$ protons.

Question 66: C

The rod's activity is less than 1,000 Bq after 300 days. To calculate the longest possible half-life, we must assume that the activity is just below 1,000 Bq after 300 days. Thus, the half-life has decreased activity from 16,000 Bq to 1,000 Bq during this period of time.

After one half-life, we would expect the activity to decrease to 8,000 Bq. After two half-lives, it should be 4,000 Bq. Similarly:

After three half-lives: Activity = 2,000 Bq.

After four half-lives: Activity = 1,000 Bq.

Thus, the rod has halved its activity a minimum of 4 times in 300 days. Therefore, the largest possible estimate for the half-life is given by:

300/4 = 75 days.

Question 67: B

There is no change in the atomic mass or proton numbers in gamma radiation. In alpha decay, a particle with two protons and two neutrons is emitted. This results in a decrease in proton number by 2 and neutron number by 2.

Thus, after 3 rounds of alpha decay, the proton number will be 89 − (3 × 2) = 83. Initially, there are 200 − 89 = 111 neutrons. After the alpha decay rounds, the number of neutrons is 111 − (3 × 2) = 105.

Question 68: C

The speed of the wave can be calculated by using the distance and time given:

$$\text{Speed} = \frac{\text{distance}}{\text{time}} = \frac{500}{1.5} = 333 \text{ ms}^{-1}.$$

As the wavelength is linked to wave speed, we can utilise this equation:

$$\text{Wavelength} = \frac{\text{Speed}}{\text{Frequency}} = \frac{333}{440}.$$

If you approximate 333 to 330, it is easier to reduce the fraction:

$$\frac{330}{440} = \frac{3}{4} = 0.75 \text{ m}.$$

Question 69: B

Firstly, it should be noted that all the options are a magnitude of 10 apart. Therefore, you do not have to worry about getting the correct value as long as you get the correct power of 10. You can therefore make your life easier by approximating.

The area of the shell is given by the following as it has a circular face:

$A_{shell} = \pi r^2 = \pi \times (50 \times 10^{-3})^2 = \pi \times (5 \times 10^{-2})^2 = \pi \times 25 \times 10^{-4} = 7.5 \times 10^{-3} \, m^2$

The deceleration of the shell is easy to evaluate using the following equation:

$a = \dfrac{u-v}{t} = \dfrac{200}{500 \times 10^{-6}} = 0.4 \times 10^6 \, ms^{-2}$.

Then, using Newton's Second Law:

$Braking \, force = Mass \times Acceleration = 1 \times (0.4 \times 10^6) = 4 \times 10^5 \, N$.

Therefore, we can now calculate the pressure:

$Pressure = \dfrac{Force}{Area} = \dfrac{4 \times 10^5}{7.5 \times 10^{-3}} = \dfrac{8}{15} \times 10^8 \, Pa \approx 5 \times 10^7 \, Pa$.

Question 70: B

The fountain transfers 10% of 1,000 J of energy per second into lifting the 120 litres of water per minute. Thus, it transfers 100 J into 2 litres of water per second. Therefore, we can equate this to the gravitational energy equation:

Total Gravitational Potential Energy, $E_p = mg\Delta h$

$\Rightarrow 100 \, J = 2 \times 10 \times h \Rightarrow h = \dfrac{100}{20} = 5 \, m$.

Question 71: E

In step-down transformers, the number of turns of the primary coil is larger than that of the secondary coil to decrease the voltage. If a transformer is 100% efficient, the electrical power input is equal to the electrical power output ($P = IV$). Therefore, both statements 1 and 3 are true.

Question 72: C

The percentage of C^{14} in the bone halves every 5,730 years. Since it has decreased from 100% to 6.25%, it has undergone 4 half-lives – this can be seen by continuously halving 100. On the fourth iteration, 6.25 % is achieved, therefore there have been 4 half-lives during this period. Therefore, the bone is 4 × 5,730 years old = 22,920 years.

Question 73: E

This is a straightforward question in principle, as it just requires you to plug the values into the following equation - just ensure you work in SI units to get the correct answer.

$Velocity = Wavelength \times Frequency$.

$\Rightarrow Frequency = \dfrac{2 \, m/s}{2.5 \, m} = 0.8 \, Hz = 0.8 \times 10^{-6} \, MHz = 8 \times 10^{-7} \, MHz$.

Question 74: E

If an element has a half-life of 25 days, its population will be halved every 25 days.

A total of 350/25 = 14 half-lives have elapsed. Thus, the count rate has halved 14 times. Therefore, to calculate the original rate, the final count rate must be doubled 14 times = 50×2^{14}.

$2^{14} = 2^5 \times 2^5 \times 2^4 = 32 \times 32 \times 16 = 16,384$.

Therefore, the original count rate is given by:

Original count rate = $50 \times 2^{14} = 16,384 \times 50 = 819,200$.

Question 75: D

Recall the following relations for voltage and power:

$$\text{Voltage} = \text{Current} \times \text{Resistance} = \frac{\text{Power}}{\text{Current}},$$

$$\text{Power} = \frac{\text{Work Done}}{\text{Time}} = \frac{\text{Force} \times \text{Distance}}{\text{Time}} = \text{Force} \times \text{Velocity}.$$

The first relation, $V = IR$, can directly be used to derive A. The second relation, $V = \frac{P}{I}$, and the final equation for power, can be used to derive B. C is derived from: $\text{Voltage} = \frac{\text{Power}}{\text{Current}} = \frac{\text{Force} \times \text{Velocity}}{\text{Current}}$.

As charge is the product of the current and time, E and F are derived from:

$$\text{Voltage} = \frac{\text{Power}}{\text{Current}} = \frac{\text{Force} \times \text{Distance}}{\text{Time} \times \text{Current}} = \frac{J}{As} = \frac{J}{C}.$$

D is incorrect as NmC = JC which, by the expression above is not equal to the Volt. By comparing this to the expression, we can see that the correct variant would instead be NmC^{-1}.

Question 76: E

The forces acting on the ball are weight, which is constant and equal to the mass times the gravitational acceleration, and tension T which varies with position. By substituting the expression for acceleration given into Newton's Second Law, we obtain:

$$F = ma, \qquad a = \frac{v^2}{r}$$

$$\Rightarrow T + mg = m\frac{v^2}{r}$$

The minimum speed at the top of the arc is when it just manages to reach the top, which is when the tension is completely provided by the weight. Therefore, the force keeping the object moving in the circle is instantaneously its weight. Therefore:

$$T = 0 \Rightarrow mg = m\frac{v^2}{r} \Rightarrow v = \sqrt{gr}.$$

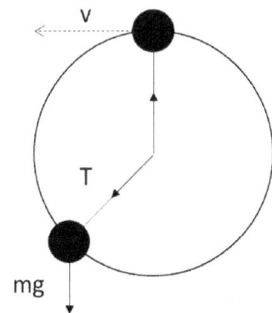

Question 77: E

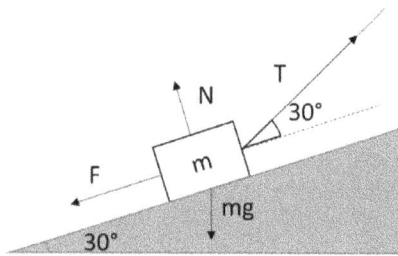

To move at a steady velocity, there must be zero acceleration, so the forces on the object must be balanced. By resolving the forces along the slope, we obtain the following equation:

$$F + mg\sin(30) = T\cos(30) \Rightarrow \frac{mg}{2} + F = \frac{T\sqrt{3}}{2}$$

$$\therefore T = \frac{2}{\sqrt{3}}\left(\frac{mg}{2} + F\right)$$

Work done in pulling the box is given by the relation W = F × d, where the distance can be expressed as a function of velocity and time, $d = v\Delta t$, so:

$$W = vF\Delta t$$

Since power can expressed as $P = \frac{W}{\Delta t}$, we have that $P = \frac{vF\Delta t}{\Delta t}$ and we can rearrange as follows:

$$P = Fv \Rightarrow P = vT\cos(30) \Rightarrow P = \left(\frac{mg}{2} + F\right)v.$$

Question 78: C

This is a conservation of energy problem. In the absence of friction, there is no dissipation of energy – therefore, the sum of the potential and kinetic energy must be constant: $\frac{1}{2}mv^2 + mgh = E$.

At its highest point, the velocity and kinetic energy are both zero so $E = mgh_1$.

At the bottom of the swing, the potential energy is converted to kinetic energy. Therefore:

$$\frac{1}{2}mv^2 = E = mgh_1 \therefore v = \sqrt{2gh_1}$$

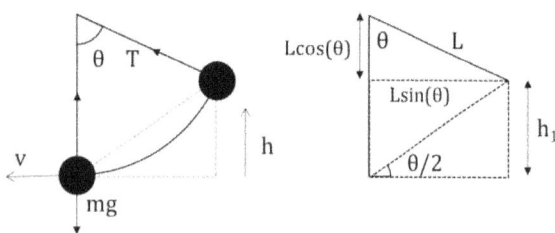

From the diagrams on the left, this initial height can easily be found:

$$h_1 = l(1 - \cos(\theta))$$

$$\Rightarrow v = \sqrt{2gh_1} = \sqrt{2gl(1 - \cos(\theta))}$$

Question 79: A

As the light intensity increases, the resistance of the LDR will decrease. As the potential difference is proportional to the resistance, according to Ohm's Law, this will increase the share of the potential across the normal resistor. Therefore, V_{out} will ultimately increase.

Question 80: D

This is essentially a modified Newton's pendulum. An elastic collision is one in which both kinetic energy and momentum are conserved. The options provided indicate that the balls which move will have the velocity. Therefore, we can use this information to answer the question.

Using the conservation of momentum, we can see that if three balls move after the collision, then:

$$p_i = (3m)v, \ p_f = mu + mu + mu = 3mu$$

$$\therefore 3mv = 3mu \Rightarrow v = u.$$

Therefore, D is a possible solution. However, if there had been 1 ball moving after the collision, then A is also clearly possible from the conservation of momentum. Therefore, we have to use the kinetic energy conservation to determine the answer. This has been done, using the u = v result above for three bodies:

$$\frac{1}{2}(3m)v^2 = \frac{1}{2}mv^2 + \frac{1}{2}mv^2 + \frac{1}{2}mv^2 = \frac{3}{2}mv^2$$

Therefore, the kinetic conservation is indeed conserved in this case. This can be shown to not be true for A using an identical calculation. Therefore, three balls will swing at a velocity equal to the velocity of the first to conserve both momentum and kinetic energy.

N.B – although calculating from first principles would not be expected, it is possible to get the answer by checking which of the answer choices agree with the conservation of momentum and energy.

Question 81: E

The displacement is the difference between the final position and the initial position. The distance, however, is the total length of the path travelled by the ball. Therefore, the displacement will be zero, as the initial and final position are the same. For the distance, SUVAT or energy conservation may be used. Using energy conservation, the kinetic energy equals the gravitational potential energy.

$$\frac{1}{2}mv^2 = mgh \Rightarrow h = \frac{1}{2g}(5)^2 = \frac{1}{2} \times 2.5 = 1.25 \text{ m}$$

The total distance will be twice this as the object traverses the path again during its descent. Therefore, the final answer is 1.25 × 2 = 2.5 m for the distance.

Question 82: C

The centre of the mass of the two blocks will be at their geometric centres. So, in this case, that is halfway across each rectangle. The moments can be taken about the black point indicated on the diagram, where the weights of the blocks are used:

1 × F = (1 × (20 × 10)) + (2.5 × (20 × 10)) = 700 N.

This corresponds to a mass of 70 kg and so this will be shown on the scale.

Question 83: C

At the top of the bounce, the kinetic energy is zero as velocity is zero. Highest velocity will be downwards before impact, where the potential energy lost is equal to the kinetic energy gained (assuming no air resistance, the conservation of energy can be applied).

Therefore, the kinetic energy before hitting the ground is given by:

$$\frac{1}{2}mv^2 = mgh \Rightarrow E_k = m \times 10 \times 3 = 30m$$

The highest velocity is given by:

$$v^2 = 2gh = 60 \Rightarrow v = 2\sqrt{15}.$$

Question 84: C

A body can only be in equilibrium if all the forces are parallel or they all pass through one point, so 1 and 3 fulfil this. Both 2 and 4 cannot be in equilibrium as the balances are not balanced regardless of the magnitude of the forces involved.

Question 85: B

The initial kinetic energy must equal the work done, by the braking force, to stop the car. Therefore, we can equate the following, where the first step makes use of the fact that the force is half the weight of the car:

$$\frac{1}{2}mv^2 = Fd \Rightarrow \frac{1}{2}mv^2 = \frac{mg}{2}d$$
$$\therefore v^2 = gd \Rightarrow d = \frac{v^2}{g}.$$

Question 86: D

We can use the proportion of amplitude left to work out how many half-lives, the time taken for the amplitude to half, have passed:

$$\frac{25}{200} = \frac{1}{8} = \frac{1}{2^3}.$$

This suggests that 3 half-lives have passed in this duration. Therefore, 12 seconds is three half-lives and $t_{1/2} = 12/3 = 4s$.

Question 87: B

The question asks for the power dissipated – this occurs due to the frictional force. The equation $P = Fv$ will be utilised, where F is the frictional force. The normal force is *mg*, so the frictional force is given by *μmg*. The power, therefore, is *μmgv*. The acceleration is a red herring here.

Question 88: E

The two waves would interfere destructively as they are half a wavelength phase difference. A wave would reflect back onto itself in this way if reflected from a plane, perpendicular surface. These two waves travelling in opposite directions (incident and reflected) would produce a standing wave, with this exact point in time corresponding to zero amplitude. There are 5 nodes with two fixed ends making it the 4th harmonic of a standing wave. Thus, all the statements are true.

Question 89: C

Beta decay changes a neutron to a proton – therefore, a doesn't change but b increases by one. Then, in alpha decay, the product emits an alpha particle which is two protons and two neutrons - therefore, a decreases by 4 and b decreases by two. This results in a decrease of 4 for a, and 1 for b.

Question 90: A

Use a free body diagram of half the sphere, essentially cutting it down the middle, as shown on the right. The p indicates the internal fluid pressure.

The pressure force acting against the plane is uniform and is given by:

$F = P \times A = P \times \pi r^2$

The stress is uniform around the circumference – where the area of the strip is $2\pi rt$ – it is given by:

Stress force $= \sigma\, 2\pi rt$

By setting the two terms equal for equilibrium, you immediately obtain the answer A.

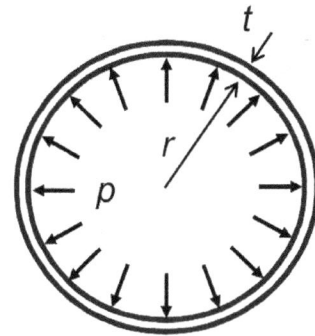

Question 91: C

Springs behave like capacitors in series or parallel (the opposite of resistors), so we find the effective spring constant of the springs in series to be: $\frac{1}{k} + \frac{1}{k} = \frac{k^2}{2k} = \frac{k}{2}$. Then, we can add the constant of the parallel spring to obtain $\frac{3k}{2}$.

Question 92: A

A free-body diagram on the trailer gives a force exerted by the spring, causing the trailer to accelerate at 2 ms⁻². The force is given by:

$F = ma = 10,000 \times 2 = 20,000$ N.

As this is the force exerted by the spring, it is also equal to:

$F = kx \Rightarrow x = F/k = 20,000/100,000 = 0.2$ m.

Therefore, the energy stored in the spring is given by:

$E = \frac{1}{2} kx^2 = \frac{1}{2} \times 100,000 \times (0.2)^2 = 2000$ J.

Question 93: C

The current in the circuit can be determined using the power and the EMF. This gives a value of:

$$P = IV \Rightarrow I = \frac{P}{V} = \frac{10}{5} = 2 \text{ A}.$$

This can be used to determine the number of electrons passing through the LED in 10 seconds:

$$Q = It = 2 \times 10 = 20 \text{ C} \Rightarrow N = 20/(1.6 \times 10^{-19}) \approx \frac{200}{16} \times 10^{19} \approx 1.25 \times 10^{20}$$

The energy of each electron is easily determined using:

$$W = Vq = 5 \times (1.6 \times 10^{-19}) = 8 \times 10^{-19} \text{ J}.$$

Question 94: B

Moments taken with the pivot at the wall must balance. Therefore, if we label the distance from the wall to the flowerpot as L, we can write:

$$\frac{2}{3} LT \sin \theta = Lmg \qquad \Rightarrow T = \frac{3mg}{2 \sin \theta}$$

Question 95: A

The superposed signal will have a frequency of 150 kHz, the lowest common factor of 30 kHz and 50 kHz. Therefore, the time period is given by:

$$T = 1/f = 6.67 \text{ μs}.$$

Question 96: B

The Young's Modulus, E, is from the gradient before the elastic limit, which is readily evaluated from the graph:

$$E = \frac{\Delta \ Stress}{\Delta \ Strain} = \frac{500 \text{ MPa}}{0.05} = 10 \text{ GPa}.$$

The strain energy is evaluated from the area beneath the graph up to x, the elastic limit:

$$\text{Strain Energy} = \frac{1}{2} \times 0.05 \times 500 \text{ MPa} = 12.5 \text{ MJ}.$$

Question 97: D

The impulse is the change in momentum. Therefore, we can use the impulse and the mass to determine the change in velocity of the ball:

$$I = \Delta mv \Rightarrow \Delta v = \frac{I}{m} = \frac{0.27}{0.1} = 2.7 \text{ m/s}.$$

Question 98: C

This problem relies on knowledge of projectiles and can be derived by differentiating an expression for the distance in terms of θ. The velocity component in the horizontal direction is vcosθ. Therefore:

Speed = Distance / Time ⇒ Distance = Speed × Time = vcosθ × t

The time, however, is obtained using SUVAT. This can be substituted to give:

Distance = Speed × Time = vcosθ × $\frac{2v\sin\theta}{g}$ = $\frac{v^2\sin2\theta}{g}$.

This is maximised at 45 degrees.

Question 99: C

Only 90% of the motor's power is used to provide the driving force on the car. Therefore, we can use this and the equation which relates power to speed to give:

Power = 0.9P ⇒ P = Fv ⇒ F=0.9P/v

Given the force, we can easily calculate the work done by the resistive force using d = 1000m:

W = Fd ⇒ W = 900P/v.

Question 100: C

There will be an inertial force on the block. A free body diagram reveals a vertical force mg and horizontal force ma acting on the block. The components of these must balance in the direction parallel to the wedge.

By equating the component of weight parallel to the wedge and the component of the driving force parallel to the wedge, we obtain:

ma cos θ = mg sin θ ⇒ a = g tan θ.

Question 101: B·

Each three-block combination is mutually exclusive to any other combination, so the probabilities are added. Each block pick is independent of all other picks, so the probabilities can be multiplied. For this scenario, there are three possible combinations:

P(2 red blocks and 1 yellow block) = P(Red, red, yellow) + P(Red, yellow, red) + P(Yellow, red, red)

∴ P (2 red blocks and 1 yellow block) = $(\frac{12}{20} \times \frac{11}{19} \times \frac{8}{18})$ + $(\frac{12}{20} \times \frac{8}{19} \times \frac{11}{18})$ + $(\frac{8}{20} \times \frac{12}{19} \times \frac{11}{18})$

⇒ P (2 red blocks and 1 yellow block) = $\frac{3 \times 12 \times 11 \times 8}{20 \times 19 \times 18}$ = $\frac{44}{95}$.

Question 102: C

This is simple algebraic manipulation. Multiply through by 15, then rearrange for x:

$$3(3x + 5) + 5(2x - 2) = 18 \times 15$$

$$\Rightarrow 9x + 15 + 10x - 10 = 270$$

$$\Rightarrow 9x + 10x = 270 - 15 + 10 \Rightarrow 19x = 265$$

$$\therefore x \approx 13.95.$$

Question 103: C

This is a rare case where you need to factorise a complex polynomial. The first term is clearly the product of $3x$ and x, therefore, we can insert these into the brackets immediately. Now, we can consider possible pairings with sum to give 11 and produce a product of -20.

$(3x + a)(x + b) = 0$, possible pairs: 2×10, 10×2, 4×5, 5 4

$$\therefore (3x - 4)(x + 5) = 0$$

$$\Rightarrow 3x - 4 = 0, \text{ so } x = \frac{4}{3}.$$

$$\Rightarrow x + 5 = 0, \text{ so } x = -5.$$

Therefore, these are the two possible solutions for x.

Question 104: C

This question just requires some basic algebraic manipulation – make sure you don't make silly mistakes in these questions! The steps are as follows:

$$\frac{5(x-4)}{(x+2)(x-4)} + \frac{3(x+2)}{(x+2)(x-4)}$$

$$= \frac{5x-20+3x+6}{(x+2)(x-4)} = \frac{8x-14}{(x+2)(x-4)}.$$

Question 105: E

As p is directly proportional to the cube root of q, we can write the following relation, where k is the constant of proportionality:

$$p \propto \sqrt[3]{q} \Rightarrow p = k\sqrt[3]{q}$$

Therefore, we can use the fact that when $p = 12$, $q = 27$ to determine the value of k.

$$12 = k(\sqrt[3]{27}) = 3k \Rightarrow k = 4.$$

Now, at $p = 24$, we can simply substitute in the values for k and p to determine q:

$$p = 4\sqrt[3]{q} \Rightarrow 24 = 4\sqrt[3]{q},$$

$$\therefore 6 = \sqrt[3]{q} \Rightarrow q = 6^3 = 216.$$

Question 106: A

As we are looking for the prime factors of 72^2, we just need to find the prime factors of 72 and then square them. This is done as follows:

$8 \times 9 = 72$

$\Rightarrow 8 = (4 \times 2) = 2 \times 2 \times 2 = 2^3$

$\Rightarrow 9 = 3 \times 3 = 3^2$

$\therefore (2^3 \times 3^2)^2 = 2^6 \times 3^4$

Question 107: C

Note that $1.151 \times 2 = 2.302$. Therefore, we can divide the numerator and denominator of the expression by 1.151, which produces:

$$\frac{2 \times 10^5 + 2 \times 10^2}{10^{10}} = 2 \times 10^{-5} + 2 \times 10^{-8}$$

$\Rightarrow 0.00002 + 0.00000002 = 0.00002002.$

Question 108: E

We can simply expand the expression given, and compare the coefficients to determine the values of **a** and **b**. This is done as follows:

$(y+2)^2 - 5 = y^2 + 4y + 4 - 5 = y^2 + 4y + 4 - 5 = y^2 + 4y - 1.$

$\therefore y^2 + ay + b = y^2 + 4y - 1.$

Therefore, by directly equating the coefficients, we can see that: **a** = 4 and **b** = -1.

Question 109: E

To simplify the expression, take $5(m + 4n)$ as a common factor to give:

$$\frac{4(m+4n)}{5(m+4n)} + \frac{5(m-2n)}{5(m+4n)}.$$

This can be further simplified to give:

$$\frac{4m+16n+5m-10n}{5(m+4n)} = \frac{9m+6n}{5(m+4n)} = \frac{3(3m+2n)}{5(m+4n)}.$$

Question 110: C

As A is inversely proportional to the square root of B, we can write the following relation, where k is the constant of proportionality:

$$A \propto \frac{1}{\sqrt{B}} \Rightarrow A = \frac{k}{\sqrt{B}}.$$

Therefore, we can use the fact that when A = 4, B = 25 to determine the value of k. Substitute the values in to give:

$$4 = \frac{k}{\sqrt{25}} \Rightarrow k = 4 \times 5 = 20.$$

$$\therefore A = \frac{20}{\sqrt{B}}.$$

So, when B = 16, A is given by:

$$A = \frac{20}{\sqrt{16}} = \frac{20}{4} = 5.$$

Question 111: E

This question tests your knowledge of circle theorems. Angles SVU and STU are opposites and add up to 180° - therefore, STU = 91°.

The angle of the centre of a circle is twice the angle at the circumference - therefore, SOU = 2 x 91° = 182°.

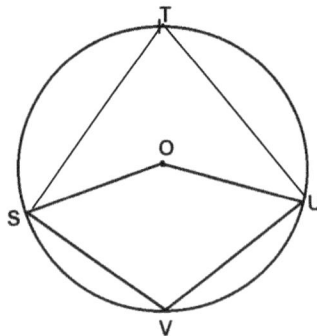

Question 112: E

Cylinder B is an enlargement of A, so the increases in radius (r) and height (h) will be proportional: $\frac{r_A}{r_B} = \frac{h_A}{h_B}$. Let us call the constant of proportionality n, where $n = \frac{r_A}{r_B} = \frac{h_A}{h_B}$. Therefore, we can write the following, where the surface area of the open cylinder is given by A = 2πrh:

$$\Rightarrow \frac{\text{Area A}}{\text{Area B}} = \frac{2\pi r_A h_A}{2\pi r_B h_B} = n \times n = n^2 \Rightarrow \frac{\text{Area A}}{\text{Area B}} = \frac{32\pi}{8\pi} = 4 \therefore n = 2.$$

The constant of proportionality, n = 2, also applies to the volume, where it must be accounted for three times due to the three lengths. The cylinder's volumes are therefore related by $n^3 = 8$.

If the smaller cylinder has volume 2π cm³, then the larger will have volume 2π × n³ = 2π × 8 = 16π cm³.

Question 113: E

This is another simple algebraic manipulation question. The steps are as follows, where we have cross-multiplied to obtain a common denominator:

$$\Rightarrow \frac{8}{x(3-x)} - \frac{6(3-x)}{x(3-x)} \Rightarrow \frac{8-18+6x}{x(3-x)} = \frac{6x-10}{x(3-x)}$$

Question 114: B

For the black ball to be drawn in the last round, white balls must be drawn every round. Therefore, the probability of obtaining 9 white balls in a row is given by:

$$P = \frac{9}{10} \times \frac{8}{9} \times \frac{7}{8} \times \frac{6}{7} \times \frac{5}{6} \times \frac{4}{5} \times \frac{3}{4} \times \frac{2}{3} \times \frac{1}{2}.$$

$$\Rightarrow P = \frac{9 \times 8 \times 7 \times 6 \times 5 \times 4 \times 3 \times 2 \times 1}{10 \times 9 \times 8 \times 7 \times 6 \times 5 \times 4 \times 3 \times 2 \times 1} = \frac{1}{10}.$$

Question 115: C

There are 4 Kings within a standard deck. Therefore, the probability of obtaining a King is $\frac{4}{52} = \frac{1}{13}$, and the probability of obtaining a King again, after the first card, is given by $\frac{3}{51}$. These are independent events – therefore, the probability of drawing two Kings is their product:

$$P = \frac{1}{13} \times \frac{3}{51} = \frac{3}{663} = \frac{1}{221}.$$

Question 116: B

The probabilities of all outcomes must sum to one, so if the probability of rolling a 1 is x, then:

$$\Rightarrow x + x + x + x + 2x = 1 \therefore x = \frac{1}{7}.$$

The probability of obtaining two sixes, therefore, is given by:

$$P_{12} = \frac{2}{7} \times \frac{2}{7} = \frac{4}{49}.$$

Question 117: B

There are plenty of ways of obtaining the answer, however the easiest is as follows: 0 is divisible by both 2 and 3. Half of the numbers from 1 to 36 are even (i.e. 18 of them). 3, 9, 15, 21, 27, 33 are the only numbers divisible by 3 that we've missed in our even numbers. Therefore, there are 25 outcomes divisible by 2 or 3, out of 37.

Question 118: C

To approach this question, list the six ways of achieving this outcome: HHTT, HTHT, HTTH, TTHH, THTH and THHT. The probability of each of these six outcomes is given by:

$$\frac{1}{2} \times \frac{1}{2} \times \frac{1}{2} \times \frac{1}{2} = \left(\frac{1}{2}\right)^4 = \frac{1}{2^4}$$

Therefore, the probability of two heads and two tails is:

$$P = 6 \times \frac{1}{2^4} = \frac{6}{16} = \frac{3}{8}.$$

Question 119: D

Firstly, we need to count the number of ways to get a 5, 6 or 7 (draw the square below if helpful). The ways to get a 5 are: 1, 4; 2, 3; 3, 2; 4, 1. The ways to get a 6 are: 1, 5; 2, 4; 3, 3; 4, 2; 5, 1. The ways to get a 7 are: 1, 6; 2, 5; 3, 4; 4, 3; 5, 2; 6, 1. That is 15 out of 36 possible outcomes, which reduces to 5 out of 12.

	1	**2**	**3**	**4**	**5**	**6**
1	2	3	4	5	6	7
2	3	4	5	6	7	8
3	4	5	6	7	8	9
4	5	6	7	8	9	10
5	6	7	8	9	10	11
6	7	8	9	10	11	12

Question 120: C

In total, there are $x + y + z$ balls in the bag, and the probability of picking a red ball is $\frac{x}{(x+y+z)}$ and the probability of picking a green ball is $\frac{z}{(x+y+z)}$. These are independent events, so the probability of picking red then green is the product of their individual probabilities: $\frac{xz}{(x+y+z)^2}$.

The probability of picking green then red is the same. These outcomes are mutually exclusive, so are added. This gives an answer of $\frac{2xz}{(x+y+z)^2}$.

Question 121: B

There are two ways of obtaining a red ball and a blue ball: pulling out a red ball then a blue ball or pulling out a blue ball and then a red ball. Let us work out the probability of the first:

$$P = \frac{x}{(x+y+z)} \times \frac{y}{(x+y+z-1)}$$

The probability of the second option will be the same. These are mutually exclusive options, so the probabilities may be summed. This gives the answer $\frac{2xy}{(x+y+z)(x+y+z-1)}$.

Question 122: A

Let x correspond to when Player 1 wins the point, and y correspond to when Player 2 wins the point.

Player 1 wins the game in five rounds in the following scenarios: *yxxxx, xyxxx, xxyxx, xxxyx*. (Note the case of *xxxxy* would lead to player 1 winning in 4 rounds, which the question forbids.)

Each of these have a probability of $p^4 \times (1-p)$. Thus, the solution is this multiplied by four as there are four ways for Player 1 to win in 5 rounds; therefore, the final answer is $4p^4(1-p)$.

Question 123: F

This question requires some simple algebraic manipulation, as follows:

$$4x + 7 + 18x + 20 = 14$$

$$\Rightarrow 22x + 27 = 14 \Rightarrow 22x = -13$$

$$\therefore x = -\frac{13}{22}.$$

Question 124: D

Rearrange the expression for volume to obtain an expression for r:

$$r^3 = \frac{3V}{4\pi} \Rightarrow r = \left(\frac{3V}{4\pi}\right)^{1/3}$$

We can substitute this expression into the equation for surface area to find a relationship between S and V.

$$S = 4\pi \left[\left(\frac{3V}{4\pi}\right)^{\frac{1}{3}}\right]^2 = 4\pi \left(\frac{3V}{4\pi}\right)^{\frac{2}{3}}$$

$$= \frac{4\pi(3V)^{\frac{2}{3}}}{(4\pi)^{\frac{2}{3}}} = (3V)^{\frac{2}{3}} \times \frac{(4\pi)^1}{(4\pi)^{\frac{2}{3}}}$$

$$= (3V)^{\frac{2}{3}}(4\pi)^{1-\frac{2}{3}} = (4\pi)^{\frac{1}{3}}(3V)^{\frac{2}{3}}.$$

Question 125: A

Let one side of the cube be of length *x*. Therefore:

$$S = 6x^2 \Rightarrow x = \left(\frac{S}{6}\right)^{\frac{1}{2}}$$

We ca substitute this expression into the equation for V to find the relation between S and V:

$$\Rightarrow V = x^3 \Rightarrow V = \left[\left(\frac{S}{6}\right)^{\frac{1}{2}}\right]^3$$

$$\therefore V = \left(\frac{S}{6}\right)^{\frac{3}{2}}$$

Question 126: B

By multiplying the second equation by 2, we obtain 4x + 16y = 24. Subtracting the first equation from this, we get:

$13y = 17 \Rightarrow y = \frac{17}{13}$.

Then, by substituting the expression for y into the first equation, we obtain $x = \frac{10}{13}$. You could also try substituting possible solutions one by one, although given that the equations are both linear and contain easy numbers, it is quicker to solve them algebraically.

Question 127: A

Firstly, we multiply by the denominator and partially expand the brackets on the right side:

$(7x + 10) = (3y^2 + 2)(9x + 5) \Rightarrow (7x + 10) = 9x(3y^2 + 2) + 5(3y^2 + 2)$

Next, gather the x terms:

$7x - 9x(3y^2 + 2) = 5(3y^2 + 2) - 10$

Take x outside the brackets and rearrange to arrive at the final answer:

$x[7 - 9(3y^2 + 2)] = 5(3y^2 + 2) - 10$

$\therefore x = \frac{5(3y^2 + 2) - 10}{7 - 9(3y^2 + 2)} = \frac{(15y^2)}{(7 - 9(3y^2 + 2))}$.

Question 128: F

This tests your ability to manipulate indices. The steps are as follows:

$$3x\left(\frac{3x^7}{x^{\frac{1}{3}}}\right)^3 = 3x\left(\frac{3^3 x^{21}}{x^{\frac{3}{3}}}\right) = 3x\,\frac{27x^{21}}{x} = 81x^{21}$$

Question 129: D

The expression given can be rearranged to give:

$2x \times [2^{\frac{7}{14}} x^{\frac{7}{14}}] = 2x \times [2^{\frac{1}{2}} x^{\frac{1}{2}}]$

$= 2x (\sqrt{2}\,\sqrt{x}) = 2\left[\sqrt{x}\sqrt{x}\right]\left[\sqrt{2}\sqrt{x}\right] = 2\sqrt{2x^3}$.

Question 130: A

The equation for the area of a circle is given by: $A = \pi r^2$. Therefore, we can equate the following:

$10\pi = \pi r^2$

$\therefore r = \sqrt{10}$

The circumference, therefore, is given by:

$C = 2\pi r = 2\pi\sqrt{10}$.

Question 131: D

This can be evaluated simply using the rule given:

$3.4 = 12 + (3 + 4) = 19$

$19.5 = 95 + (19 + 5) = 119$

$\therefore (3.4).5 = 119$

Question 132: D

This can be evaluated simply using the rule given:

$2.3 = \dfrac{2^3}{2} = 4$

$4.2 = \dfrac{4^2}{4} = 4$

$\therefore (2.3).2 = 4.$

Question 133: F

This is a tricky question that requires you to know how to 'complete the square'. Alternatively, you could use the quadratic formula. The steps are as follows:

$(x + 1.5)(x + 1.5) = x^2 + 3x + 2.25$

$\Rightarrow (x + 1.5)^2 - 7.25 = x^2 + 3x - 5 = 0$

$\therefore (x + 1.5)^2 = 7.25 = \dfrac{29}{4}$

We can now rearrange this equation for *x*:

$x + 1.5 = \sqrt{\dfrac{29}{4}} \Rightarrow x = -\dfrac{3}{2} \pm \sqrt{\dfrac{29}{4}} = -\dfrac{3}{2} \pm \dfrac{\sqrt{29}}{2}.$

Question 134: B

Whilst you definitely need to solve this graphically, it is necessary to complete the square for the first equation to allow you to draw it more easily:

$\Rightarrow (x + 2)^2 = x^2 + 4x + 4$

$\therefore y = (x + 2)^2 + 10 = x^2 + 4x + 14$

This is now an easy curve to draw (it is the quadratic $y = x^2$ shifted 2 units left and 10 units up). The turning point of this quadratic is to the left and well above anything in x^3, so the only solution is the first intersection of the two curves in the upper right quadrant around (3.4, 39). Therefore, there is only one intersection.

Question 135: C

By far the easiest way to solve this is to sketch the graphs (don't waste time solving them algebraically). From the graphs, it is evident that $y = 2$ and $y = 1 - x^2$ do not intersect, since the latter has its turning point at $(0, 1)$ and zero points at $x = -1$ and 1. The first two graphs, $y = x$ and $y = x^2$, intersect at the origin and $(1, 1)$, and $y = 2$ runs through both. Therefore, only 3 and 4 do not intersect.

Question 136: B

Notice that you're not required to get the actual values – just the number's magnitude. Thus, 897653 can be approximated to 900,000 and 0.009764 to 0.01. Therefore, the estimate is simply:

$900,000 \times 0.01 = 9,000.$

This has an order of magnitude of 10^4 which corresponds to B.

Question 137: C

To begin this, multiply the expression through by 70:

$7(7x + 3) + 10(3x + 1) = 14 \times 70$

We can then expand the brackets and simplify:

$49x + 21 + 30x + 10 = 980$

$\Rightarrow 79x + 31 = 980$

$\Rightarrow x = \frac{949}{79}.$

Question 138: A

Firstly, split the equilateral triangle into 2 right-angled triangles and apply Pythagoras' theorem:

$\Rightarrow x^2 = \left(\frac{x}{2}\right)^2 + h^2 \Rightarrow h^2 = \frac{3}{4}x^2$

$\therefore h = \sqrt{\frac{3x^2}{4}} = \frac{\sqrt{3x^2}}{2}$

The area of a triangle is given by:

$A = \frac{1}{2} \times \text{Base} \times \text{Height} = \frac{1}{2} \times \frac{\sqrt{3x^2}}{2}$

$\Rightarrow A = x\frac{\sqrt{3x^2}}{4} = x\frac{\sqrt{3}\sqrt{x^2}}{4} = \frac{x^2\sqrt{3}}{4}.$

Question 139: A

This is a question testing your ability to spot 'the difference between two squares.' We can factorise the expression to give the following, where the common term has been cancelled out:

$3 - \frac{7x(5x - 1)(5x+1)}{(7x)^2(5x+1)} \Rightarrow 3 - \frac{(5x - 1)}{7x}.$

Question 140: C

The easiest way to do this is to 'complete the square':

$(x - 5)^2 = x^2 - 10x + 25$

$\Rightarrow (x - 5)^2 - 125 = x^2 - 10x - 100 = 0$

$\Rightarrow (x - 5)^2 = 125$

$\Rightarrow x - 5 = \pm\sqrt{125} = \pm\sqrt{25}\,\sqrt{5} = \pm 5\sqrt{5}$

$\therefore x = 5 \pm 5\sqrt{5}$.

Question 141: B

Firstly, factorise by completing the square and then simply rearrange for x:

$x^2 - 4x + 7 = (x - 2)^2 + 3$

$\Rightarrow (x - 2)^2 = y^3 + 2 - 3$

$\Rightarrow x - 2 = \pm\sqrt{y^3 - 1}$

$\Rightarrow x = 2 \pm \sqrt{y^3 - 1}$.

Question 142: D

Begin by squaring both sides of the expression, and then rearrange for y:

$(3x + 2)^2 = 7x^2 + 2x + y$

$\Rightarrow y = (3x + 2)^2 - 7x^2 - 2x = (9x^2 + 12x + 4) - 7x^2 - 2x$

$\therefore y = 2x^2 + 10x + 4$.

Question 143: C

This is a fourth order polynomial, which you aren't expected to be able to factorise at GCSE. This is where looking at the options makes your life a lot easier. In all of them, opening the bracket on the right side involves making $(y \pm 1)^4$ on the left side, i.e. the answers are hinting that $(y \pm 1)^4$ is the solution to the fourth order polynomial.

Since there are negative terms in the equations (e.g. $-4y^3$), the solution has to be:

$(y - 1)^4 = y^4 - 4y^3 + 6y^2 - 4y + 1$

$\Rightarrow (y - 1)^4 + 1 = x^5 + 7$

$\Rightarrow y - 1 = (x^5 + 6)^{\frac{1}{4}}$

$\therefore y = 1 + (x^5 + 6)^{1/4}$.

Question 144: A

Let the width of the television be $4x$ and the height of the television be $3x$. By Pythagoras's theorem:

$$(4x)^2 + (3x)^2 = 50^2 \Rightarrow 25x^2 = 2500$$

$$\therefore x = 10.$$

Therefore, the screen is 30 inches by 40 inches. So, the area is 1,200 inches^2.

Question 145: C

Firstly, square both sides and then multiply out the brackets:

$$1 + \frac{3}{x^2} = (y^5 + 1)^2$$

$$\Rightarrow \frac{3}{x^2} = (y^{10} + 2y^5 + 1) - 1$$

$$\Rightarrow x^2 = \frac{3}{y^{10} + 2y^5} \Rightarrow x = \sqrt{\frac{3}{y^{10} + 2y^5}}.$$

Question 146: C

The easiest method is to double the first equation and triple the second to obtain:

$$\Rightarrow 6x - 10y = 20 \text{ and } 6x + 6y = 39.$$

By subtracting the first equation from the second, we obtain:

$$16y = 19 \Rightarrow y = \frac{19}{16}.$$

By substituting this value of y into the first equation, the value of x can be obtained: $x = \frac{85}{16}$.

Question 147: C

This is fairly straightforward; the first inequality is the easier one to work with. By inserting test values, it is clear to see that B, D and E violate it, so we just need to check A and C in the second inequality.

C: $1^3 - 2^2 < 3$, but A: $2^3 - 1^2 > 3$.

Therefore, the answer is given by C.

Question 148: B

Whilst this can be done graphically, it's quicker to do algebraically (because the second equation is not as easy to sketch). Intersections occur where the curves have the same coordinates.

$$\Rightarrow x + 4 = 4x^2 + 5x + 5$$

$$\Rightarrow 4x^2 + 4x + 1 = 0$$

$$\Rightarrow (2x + 1)(2x + 1) = 0$$

Thus, the two graphs only intersect once at $x = -\frac{1}{2}$.

Question 149: D

It is better to approach this algebraically, as the equations are easy to work with and you would need to sketch very accurately to get the answer. Intersections occur where the curves have the same coordinates.

$$x^3 = x \Rightarrow x^3 - x = 0 \Rightarrow x(x^2 - 1) = 0$$

You need to spot the 'difference between two squares' to obtain:

$$x(x + 1)(x - 1) = 0.$$

Thus, there are 3 intersections - at $x = 0, 1$ and -1.

Question 150: E

Note that the line is the hypotenuse of a right-angled triangle with one side unit length and one side of length ½. By Pythagoras's theorem on the triangle shown in the diagram, we can equate:

$$\left(\tfrac{1}{2}\right)^2 + 1^2 = x^2 \Rightarrow x^2 = \tfrac{1}{4} + 1 = \tfrac{5}{4}$$

$$\Rightarrow x = \sqrt{\tfrac{5}{4}} = \tfrac{\sqrt{5}}{\sqrt{4}} = \tfrac{\sqrt{5}}{2}.$$

Question 151: D

We can eliminate z from equation (1) and (2) by multiplying equation (1) by 3 and adding it to equation (2):

3x + 3y − 3z = -3	Equation (1) multiplied by 3.
2x − 2y +3z = 8	Equation (2) then add both equations.
5x + y = 5	We label this as equation (4).

Now we must eliminate the same variable z from another pair of equations by using equation (1) and (3):

2x + 2y − 2z = -2	Equation (1) multiplied by 2.
2x − y + 2z = 9	Equation (3) then add both equations.
4x + y = 7	We label this as equation (5).

We now use both equations (4) and (5) to obtain the value of x:

5x + y = 5	Equation (4).
- 4x - y = -7	Equation (5) multiplied by -1.
x = -2	

Substitute x back in to calculate y:

$$4x + y = 7 \Rightarrow 4(-2) + y = 7 \Rightarrow -8 + y = 7 \Rightarrow y = 15$$

Substitute x and y back in to calculate z:

$$x + y - z = -1 \Rightarrow -2 + 15 - z = -1 \Rightarrow 13 - z = -1$$

$$-z = -14 \Rightarrow z = 14 \therefore x = -2, y = 15, z = 14.$$

Question 152: D

It is evident that 3a is common to all terms and so, can be factored out to give:

$3a(a^2 - 10a + 25)$ $= 3a(a - 5)(a - 5) = 3a(a - 5)^2$.

Question 153: B

Note that 12 is the LCM (Lowest Common Multiple) of 3 and 4. Thus:

-3 (4x + 3y) = -3 (48)   Multiply each side by -3.

 4 (3x + 2y) = 4 (34)   Multiply each side by 4.

-12x – 9y = -144

<u>12x + 8y = 136</u>   Add together.

$\Rightarrow$ -y = -8 $\Rightarrow$ y = 8.

Substitute y back in:

4x + 3y = 48

4x + 3(8) = 48

4x + 24 = 48

4x = 24

$\Rightarrow$ x = 6.

Question 154: E

Don't be fooled, this is an easy question - just obey BODMAS and do not skip steps.

$\frac{-(25-28)^2}{-36+14} = \frac{-(-3)^2}{-22} \Rightarrow \frac{-(9)}{-22} = \frac{9}{22}$.

Question 155: E

As there are 26 possible letters for each of the 3 letters in the license plate, and there are 10 possible numbers (0-9) for each of the 3 numbers in the same plate, the number of license plates would be:

$(26) \times (26) \times (26) \times (10) \times (10) \times (10) = 17,576,000$.

Question 156: B

Expand the brackets of the expression given and then factorise:

$\Rightarrow 4x^2 - 12x + 9 = 0 \Rightarrow (2x - 3)(2x - 3) = 0$.

Thus, only one solution exists, x = 1.5. Note that you could also use the fact that the discriminant, $b^2 - 4ac = 0$ to get the answer.

Question 157: C

The expression can be rewritten as:

$$\left(x^{\frac{1}{2}}\right)^{\frac{1}{2}} (y^{-3})^{\frac{1}{2}} \Rightarrow y^{-\frac{3}{2}} = \frac{x^{\frac{1}{4}}}{y^{\frac{3}{2}}}.$$

Question 158: A

Let x, y, and z represent the rent for the 1-bedroom, 2-bedroom, and 3-bedroom flats, respectively. We can form 3 different equations using the information given: 1 for the rent, 1 for the repairs, and the last one for the statement that the 3-bedroom unit costs twice as much as the 1-bedroom unit.

(1) $x + y + z = 1240$

(2) $0.1x + 0.2y + 0.3z = 276$

(3) $z = 2x$

Substitute z = 2x in both of the two other equations to eliminate z:

(4) $x + y + 2x = 3x + y = 1240$

(5) $0.1x + 0.2y + 0.3(2x) = 0.7x + 0.2y = 276$

$-2(3x + y) = -2(1240)$ Multiply each side of (4) by -2.

$10(0.7x + 0.2y) = 10(276)$ Multiply each side of (5) by 10.

(6) $-6x -2y = -2480$ Add these 2 equations.

(7) $\underline{7x + 2y = 2760}$

$x = 280$

$z = 2(280) = 560$ Because z = 2x.

$280 + y + 560 = 1240$ Because x + y + z = 1240.

$\Rightarrow y = 400.$

Thus, the units rent for £280, £400, £560 per week respectively.

Question 159: C

Following the rules of BODMAS, the expression can be reduced as shown:

$$5\left[5(6^2 - 5 \times 3) + 400^{\frac{1}{2}}\right]^{1/3} + 7$$

$$\Rightarrow 5\left[5(36 - 15) + 20\right]^{\frac{1}{3}} + 7 \Rightarrow 5\left[5(21) + 20\right]^{\frac{1}{3}} + 7$$

$$\Rightarrow 5\left(105 + 20\right)^{\frac{1}{3}} + 7 \Rightarrow (125)^{\frac{1}{3}} + 7 \Rightarrow 5(5) + 7$$

$$\Rightarrow 25 + 7 = 32.$$

Question 160: B
Consider a triangle formed by joining the centre to two adjacent vertices. Six similar triangles can be made around the centre – thus, the central angle is 60 degrees. Since the two lines forming the triangle are of equal length, we have 6 identical equilateral triangles in the hexagon.

Now split the triangle in half and apply Pythagoras' theorem: $1^2 = 0.5^2 + h^2$

Thus, $h = \sqrt{\dfrac{3}{4}} = \dfrac{\sqrt{3}}{2}$.

Thus, the area of the triangle is: $\dfrac{1}{2}bh = \dfrac{1}{2} \times 1 \times \dfrac{\sqrt{3}}{2} = \dfrac{\sqrt{3}}{4}$.

Therefore, the area of the hexagon is: $\dfrac{\sqrt{3}}{4} \times 6 = \dfrac{3\sqrt{3}}{2}$.

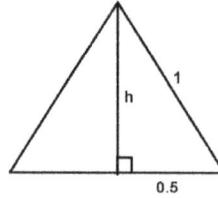

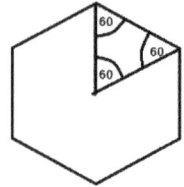

Question 161: B
Let x be the width and $x + 19$ be the length.

Thus, the area of a rectangle is $x(x + 19) = 780$.

Therefore:

$x^2 + 19x - 780 = 0 \Rightarrow (x - 20)(x + 39) = 0$

$\Rightarrow x - 20 = 0$ or $x + 39 = 0$

$\therefore x = 20$ or $x = -39$.

As length cannot be a negative number, we disregard $x = -39$ and use $x = 20$ instead.

Thus, the width is 20 metres and the length is 39 metres.

Question 162: B
The quickest way to solve is by trial and error, substituting the provided options. However, if you're keen to do this algebraically, you can do the following:

Start by setting up the equations:

Perimeter $= 2L + 2W = 34 \Rightarrow L + W = 17 \Rightarrow W = 17 - L$.

Using Pythagoras: $L^2 + W^2 = 13^2$

$\Rightarrow L^2 + (17 - L)^2 = 169$

$\Rightarrow L^2 + 289 - 34L + L^2 = 169$

$\Rightarrow 2L^2 - 34L + 120 = 0$

$\Rightarrow L^2 - 17L + 60 = 0$

$\Rightarrow (L - 5)(L - 12) = 0$

$\therefore L = 5$ and $L = 12$, $W = 12$ and $W = 5$.

Question 163: C

Multiply both sides by 8, expand and then rearrange for x:

$4(3x - 5) + 2(x + 5) = 8(x + 1)$

$\Rightarrow 12x - 20 + 2x + 10 = 8x + 8$

$\Rightarrow 14x - 10 = 8x + 8$

$\Rightarrow 14x = 8x + 18$

$6x = 18 \therefore x = 3.$

Question 164: C

Note that 1.742×3 is 5.226. Therefore, the original equation can be simplified to:

$\Rightarrow \dfrac{3 \times 10^6 + 3 \times 10^5}{10^{10}}$

$= 3 \times 10^{-4} + 3 \times 10^{-5} = 3.3 \times 10^{-4}.$

Question 165: A

The area of a triangle is given by half its base times the height. Therefore, in this case, it is given by:

$\text{Area} = \dfrac{(2 + \sqrt{2})(4 - \sqrt{2})}{2} = \dfrac{8 - 2\sqrt{2} + 4\sqrt{2} - 2}{2}$

$\Rightarrow \text{Area} = \dfrac{6 + 2\sqrt{2}}{2} = 3 + \sqrt{2}$.

Question 166: C

The aim here is to isolate x, so the first step will be to square both sides, which gives:

$\dfrac{4}{x} + 9 = (y - 2)^2 \Rightarrow \dfrac{4}{x} = (y - 2)^2 - 9$

Then you cross multiply, followed by factorisation:

$\dfrac{x}{4} = \dfrac{1}{(y-2)^2 - 9} \Rightarrow x = \dfrac{4}{y^2 - 4y + 4 - 9} \Rightarrow x = \dfrac{4}{y^2 - 4y - 5}$

$\therefore x = \dfrac{4}{(y+1)(y-5)}.$

Question 167: D

Set up the equation using the statement provided. This leads to:

$5x - 5 = 0.5(6x + 2)$

$\Rightarrow 10x - 10 = 6x + 2$

$\Rightarrow 4x = 12$

$\therefore x = 3$

Question 168: C

Firstly, you should round the numbers appropriately:

$$\Rightarrow \frac{55 + \left(\frac{9}{4}\right)^2}{\sqrt{900}} = \frac{55 + \frac{81}{16}}{30}$$

As 81 rounds to 80, this can be approximated to give:

$$\frac{55 + 5}{30} = \frac{60}{30} = 2.$$

Question 169: D

There are three outcomes from choosing the type of cheese in the crust. For each of the additional toppings to possibly add, there are 2 outcomes: to include or not to include a certain topping, for each of the 7 toppings.

Thus, the number of different kinds of pizza is:

$3 \times 2 \times 2 \times 2 \times 2 \times 2 \times 2 \times 2 = 3 \times 2^7 = 3 \times 128 = 384$.

Question 170: A

Although it is possible to do this algebraically, by far the easiest way is via trial and error. This is indicated by the fact that rearranging the first equation to make x or y the subject leaves you with a difficult equation to work with (e.g. $x = \sqrt{1 - y^2}$), when you try to substitute it into the second.

An exceptionally good student might notice that the equations are symmetric in x and y, i.e. the solution is when x = y. Thus $2x^2 = 1$ and $2x = \sqrt{2}$ which gives $\frac{\sqrt{2}}{2}$ as the answer.

Question 171: C

If two shapes are congruent, then they are the same size and shape. Thus, congruent objects can be rotations and mirror images of each other. The two triangles in E are indeed congruent (SAS). Congruent objects must, by definition, have the same angles.

Question 172: B

Firstly, rearrange the equation and then factorise the expression:

$x^2 + x - 6 \geq 0$

$\Rightarrow (x + 3)(x - 2) \geq 0$

Remember that this is a quadratic inequality, so it requires a quick sketch to ensure you do not make a silly mistake with which way the sign is.

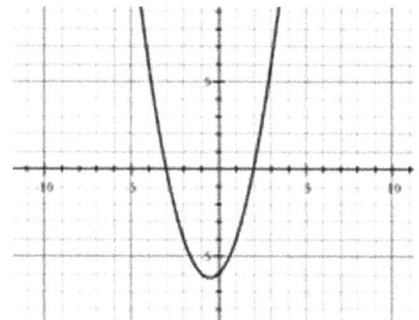

Using the roots of the quadratic, we can see that $y = 0$ when $x = 2$ and $x = -3$. Therefore:

$y > 0$ when $x > 2$ or $x < -3 \Rightarrow x \leq -3$ and $x \geq 2$.

Question 173: B

Using Pythagoras, and the fact that the other two sides are equal in length, we can obtain the following relation between x and the length of the two sides a:

$$a^2 + b^2 = x^2 \Rightarrow a = b \Rightarrow 2a^2 = x^2.$$

$$\text{Area} = \frac{1}{2}\text{base} \times \text{height} = \frac{1}{2}a^2.$$

As $a^2 = \frac{x^2}{2}$, we can rewrite the area in terms of x:

$$\text{Area} = \frac{1}{2} \times \frac{x^2}{2} = \frac{x^2}{4}.$$

Question 174: A

If X and Y are doubled, the value of Q increases by 4. Halving the value of A reduces this to 2. Finally, tripling the value of B reduces this to ⅔, i.e. the value decreases by ⅓. Increases by $\frac{2}{3}$ is incorrect as the expression is scaled by $\frac{2}{3}$, it does not increase by $\frac{2}{3}$ – this would be equivalent to scaling by a factor of by $\frac{5}{3}$.

Question 175: C

The quickest way to do this is to sketch the curves. This requires you to factorise both equations by completing the square:

$$x^2 - 2x + 3 = (x - 1)^2 + 2 \Rightarrow x^2 - 6x - 10 = (x - 3)^2 - 19$$

Thus, the first equation has a turning point at (1, 2) and doesn't cross the x-axis. The second equation has a turning point at (3, -19) and crosses the x-axis twice. The first equation is shown in green on the right, and the second equation is in purple.

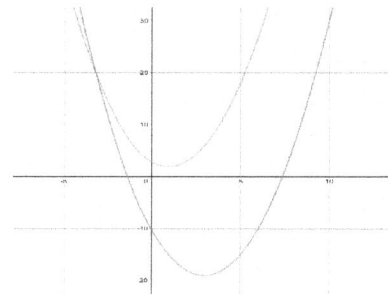

Question 176: C

As it is an equilateral triangle, the angles must be equal to 60°. The sector area is given by the following:

$$\theta = 60° \Rightarrow A = \frac{60}{360}\pi r^2 = \frac{1}{6}\pi r^2$$

$$\Rightarrow \frac{x}{\sin 30°} = \frac{2r}{\sin 60°} \therefore x = \frac{2r}{\sqrt{3}}$$

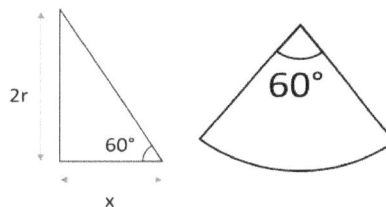

Therefore, the total triangle area is given by:

$$\Rightarrow 2 \times \frac{1}{2} \times \frac{2r}{\sqrt{3}} \times 2r = \frac{4r^2}{\sqrt{3}}.$$

Thus, the proportion covered is given by:

$$\frac{\frac{1}{6}\pi r^2}{\frac{4r^2}{\sqrt{3}}} = \frac{\sqrt{3}\pi}{24} \approx 23\%.$$

Question 177: B

To approach this question, construct the triangle shown in the diagram on the right. This allows you to determine the length of the vertical side.

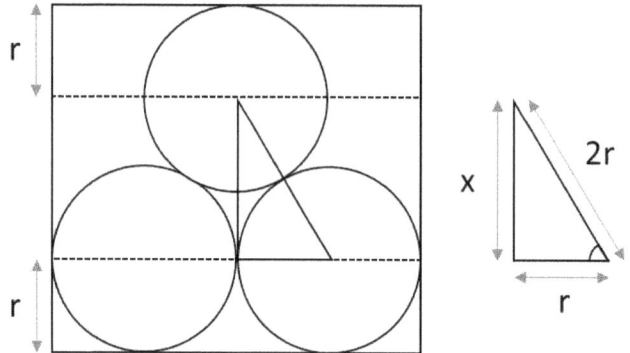

$\Rightarrow (2r)^2 = r^2 + x^2 \Rightarrow 3r^2 = x^2$

$\therefore x = \sqrt{3}r$

Total height $= 2r + x$

$\therefore H = (2 + \sqrt{3})r$

Question 178: A

The volume of a pyramid is given by: $\text{Volume} = \frac{1}{3}\text{Height} \times \text{Base area.}$

Therefore, the base area of both pyramids must be equal if h and V are the same. To find the area of the hexagon, we need the internal angle. This is given by:

External angle $= 360°/6 = 60°$

$\Rightarrow$ Internal angle $= 180° -$ External angle $= 120°$

A hexagon can be broken up into two trapezia of height h, as shown in the diagram, where:

$\frac{b}{\sin 90°} = \frac{h}{\sin 60°} \Rightarrow h = \frac{\sqrt{3}}{2}b$

$\therefore$ Trapezium area $= \frac{(2b+b)}{2}\frac{\sqrt{3}}{2}b = \frac{3\sqrt{3}}{4}b^2$

$\therefore$ Total hexagon area $= \frac{3\sqrt{3}}{2}b^2$

Now we can equate the areas:

$a^2 = \frac{3\sqrt{3}}{2}b^2 \Rightarrow \text{Ratio} = \sqrt{\frac{3\sqrt{3}}{2}}.$

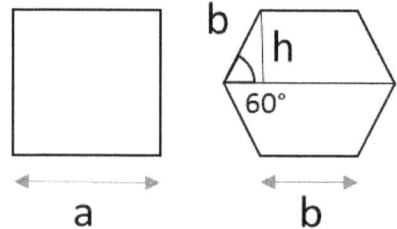

Question 179: C

A cube has 6 sides so the total surface area of a 9 cm cube is given by: 6×9^2.

The 9 cm cube splits into 3 cm cubes. The area of one 3 cm cube is $= 6 \times 3^2$. The total area is 3^3 multiplied by this area. Therefore, the ratio is evaluated as:

$\Rightarrow \frac{6 \times 3^2 \times 3^3}{6 \times 3^2 \times 3^2} = 3.$

Question 180: E

The cone is shown in the diagram. To determine θ, trigonometry can be used:

$$\tan \theta = \frac{\text{Opposite}}{\text{Adjacent}} = \frac{r}{4r} = \frac{1}{4}$$

$$\Rightarrow \theta = \tan^{-1}\left(\frac{1}{4}\right).$$

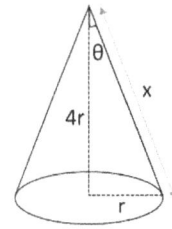

Question 181: C

A hemisphere has an angle of 180 degrees. In this case, this max angle would correspond to the max speed: 200 mph. Therefore, for 70 mph:

$$\frac{\theta}{180} = \frac{70}{200} \Rightarrow \theta = \frac{7 \times 180}{20} = 63°.$$

Question 182: C

Since the two rhombuses are similar, they have identical angles. Therefore, the ratio of their angles is 1.

The length scales with square root of area, therefore:

$$\text{Area}_A = 10 \times \text{Area}_B \Rightarrow \text{Length}_A = \sqrt{10} \times \text{Length}_B$$

$$\Rightarrow \frac{\text{angle A}/\text{angle B}}{\text{length A}/\text{length B}} = \frac{1}{\sqrt{10}/1} = \frac{1}{\sqrt{10}}.$$

Question 183: E

Finding an inverse function is equal to reflecting about the y = x line. Therefore, we can simply interchange the two variables to determine the inverse:

$$y = \ln(2x^2) \Rightarrow e^y = 2x^2 \Rightarrow x = \sqrt{\frac{e^y}{2}}$$

The -x input does not affect the original function, as the value of x is squared. Therefore, f(-x) = f(x). So, the answer remains unchanged.

$$\Rightarrow f(x) = \sqrt{\frac{e^y}{2}}.$$

Question 184: C

Firstly, we can approximate the values provided and then compare then with one another:

$$\Rightarrow \log_8(x) \text{ and } \log_{10}(x) < 0$$

$$\Rightarrow x^2 < 1$$

$$\Rightarrow \sin(x) \leq 1$$

$$\Rightarrow 1 < e^x < 2.72.$$

Therefore, e^x is largest over this range.

Question 185: C

The two relationships given are: $x \propto y^3, y \propto \sqrt{z}$. Therefore, the relationship between x and z is given by:

$$\Rightarrow x \propto \sqrt{z}^3$$

Therefore, if z is doubled, then: $\sqrt{2}^3 = 2\sqrt{2}$.

Question 186: A

The area of the shaded part, that is the difference between the area of the larger and smaller circles, is three times the area of the smaller so: $\pi r^2 - \pi x^2 = 3\pi x^2$. From this, we can see that the area of the larger circle, radius x, must be 4x the smaller one. Therefore:

$$4\pi r^2 = \pi x^2 \Rightarrow 4r^2 = x^2$$

$$\therefore x = 2r$$

The gap, therefore, is given by: $x - r = 2r - r = r$.

Question 187: D

Firstly, rearrange the equation and solve the quadratic:

$$x^2 + 3x - 4 \geq 0 \Rightarrow (x - 1)(x + 4) \geq 0$$

$$\Rightarrow x - 1 \geq 0 \text{ or } x + 4 \geq 0$$

$$\therefore x \geq 1 \text{ or } x \geq -4.$$

Question 188: C

The expressions for the volume of the sphere and its projected area, a circle, are equal:

$$\frac{4}{3}\pi r^3 = \pi r^2 \Rightarrow \frac{4}{3}r = 1$$

$$\therefore r = \frac{3}{4}.$$

Question 189: B

The two graphs have been sketched on the right. Clearly, the two graphs intersect when $x^2 = \frac{1}{x}$, which is at x = 1. Now we can consider the inequality.

When $x > 1$, $x^2 > 1$, $\frac{1}{x} < 1$.

When $x < 1$, $x^2 < 1$, $\frac{1}{x} > 1$.

In the $x < 0$ range, the $\frac{1}{x}$ graph is negative. Therefore, this region does not satisfy the inequality. Thus, the range is $0 < x < 1$.

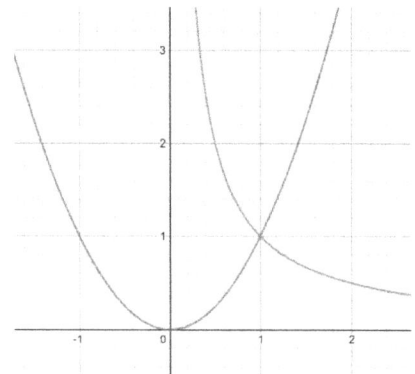

Question 190: A

Don't be afraid of how difficult this initially looks. If you follow the pattern, you get (e - e) which equals 0. Anything multiplied by 0 gives zero.

Question 191: C

For two vectors to be perpendicular their scalar product must be equal to 0. Therefore:

$$\begin{pmatrix} -1 \\ 6 \end{pmatrix} \cdot \begin{pmatrix} 2 \\ k \end{pmatrix} = 0 \Rightarrow -2 + 6k = 0$$

$$\therefore k = \frac{1}{3}.$$

Question 192: C

The point q connects the perpendicular line from the plane to the point p.

$$q = -3i + j + \lambda_1(i + 2j)$$

$$\overrightarrow{PQ} = -3i + j + \lambda_1(i + 2j) - 4i - 5j = \begin{pmatrix} -7 + \lambda_1 \\ -4 + 2\lambda_1 \end{pmatrix}$$

PQ is perpendicular to the plane r – therefore, the dot product of $\overrightarrow{PQ}$ and the vector along the line, the direction vector, must be 0. Therefore:

$$\begin{pmatrix} -7 + \lambda_1 \\ -4 + 2\lambda_1 \end{pmatrix} \cdot \begin{pmatrix} 1 \\ 2 \end{pmatrix} = 0 \qquad\qquad \Rightarrow -7 + \lambda_1 - 8 + 4 + \lambda_1 = 0$$

$$\Rightarrow \lambda_1 = 3 \quad \therefore \overrightarrow{PQ} = \begin{pmatrix} -4 \\ 2 \end{pmatrix}$$

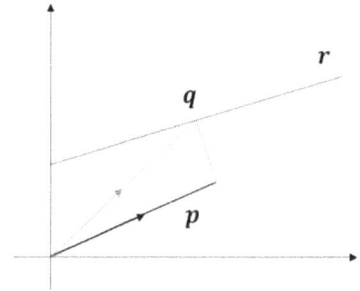

The perpendicular distance from the plane to point p is therefore the modulus of the vector joining the two $\overrightarrow{PQ}$:

$$|\overrightarrow{PQ}| = \sqrt{(-4)^2 + 2^2} = \sqrt{20} = 2\sqrt{5}.$$

Question 193: E

This is essentially a system of equations, and you approach it just like a set of simultaneous equations. You obtain these equations by equating each component (x, y and z). There are three equations, and three unknowns, so it is straightforward to solve them.

$$-1 + 3\mu = -7 \quad \therefore \mu = -2$$

$$2 + 4\lambda + 2\mu = 2 \quad \therefore \lambda = 1$$

$$3 + \lambda + \mu = k \quad \therefore k = 2.$$

Question 194: E

Recall the trigonometric identity: $\sin\left(\frac{\pi}{2} - 2\theta\right) = \cos(2\theta)$.

Root solution to $\cos(\theta) = 0.5 \Rightarrow \theta = \frac{\pi}{3}$

Solution to $\cos(2\theta) = 0.5 \Rightarrow \theta = \frac{\pi}{6}$

Largest solution within range is: $2\pi - \frac{\pi}{6} = \frac{(12-1)\pi}{6} = \frac{11\pi}{6}$.

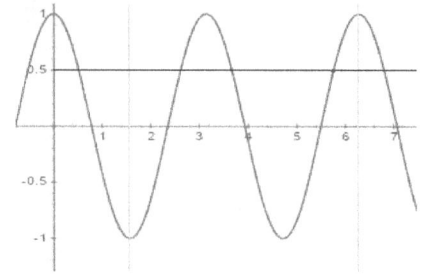

Question 195: A

This requires some knowledge of identities and algebraic manipulation:

$$\cos^4(x) - \sin^4(x) \equiv \{\cos^2(x) - \sin^2(x)\}\{\cos^2(x) + \sin^2(x)\}$$

From difference of two squares and using Pythagorean identity: $\cos^2(x) + \sin^2(x) = 1$.

$$\Rightarrow \cos^4(x) - \sin^4(x) \equiv \cos^2(x) - \sin^2(x)$$

The double angle formula states that: $\cos(A + B) = \cos(A)\cos(B) - \sin(A)\sin(B)$.

$\therefore$ If $A = B$, $\cos(2A) = \cos(A)\cos(A) - \sin(A)\sin(A) = \cos^2(A) - \sin^2(A)$.

Therefore, $\cos^4(x) - \sin^4(x) \equiv \cos(2x)$.

Question 196: C

The easiest method here is to insert values and find some factors of the polynomial. This leads to the following factorisation:

$$(x + 1)(x + 2)(2x - 1)(x^2 + 2) = 0$$

Therefore, there are three real roots at $x = -1, x = -2, x = 0.5$ and two imaginary roots at 2i and -2i.

Question 197: C

An arithmetic sequence has constant difference d - so the sum increases by a constant amount, d, each time:

$$u_n = u_1 + (n - 1)d \Rightarrow \sum_{1}^{n} u_n = \frac{n}{2}\{2u_1 + (n - 1)d\}$$

$$\Rightarrow \sum_{1}^{8} u_n = \frac{8}{2}\{4 + (8 - 1)3\} = 100.$$

Question 198: E

The binomial equation needs to be utilised here:

$$\binom{n}{k} 2^{n-k}(-x)^k$$

In this case, n = 5 and k = 2. Therefore, this gives:

$$\Rightarrow \binom{5}{2} 2^{5-2}(-x)^2 = 10 \times 2^3 x^2 = 80x^2.$$

Question 199: A

Having already thrown a 6 is irrelevant. A fair die has equal probability $P = \frac{1}{6}$ for every throw.

For three throws:

$$P(6 \cap 6 \cap 6) = \left(\frac{1}{6}\right)^3 = \frac{1}{216}.$$

Question 200: D

The situation is depicted in the probability tree diagram below:

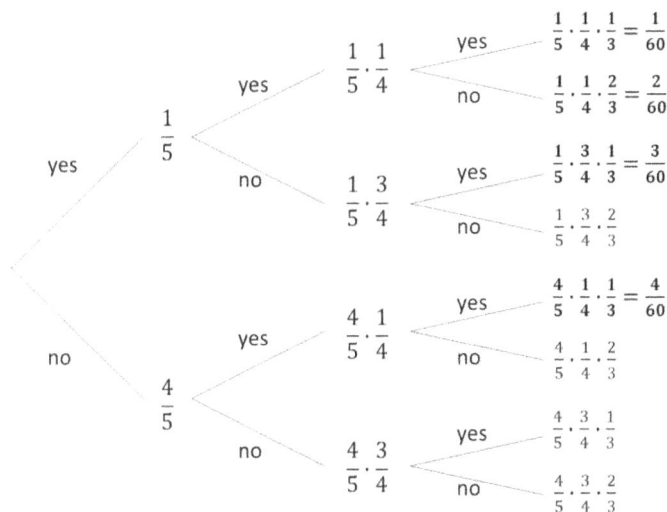

Total probability is sum of all probabilities:

$$P_{Total} = P(Y \cap Y \cap Y) + P(Y \cap Y \cap N) + P(Y \cap N \cap Y) + P(N \cap Y \cap Y)$$

$$P_{Total} = \frac{1}{60} + \frac{2}{60} + \frac{3}{60} + \frac{4}{60} = \frac{10}{60} = \frac{1}{6}.$$

Question 201: C

If A represents the probability of having blonde hair, and B the probability of having brown eyes, then we can state the probability of having neither is given by:

$$P[(A \cup B)'] = 1 - P[(A \cup B)] = 1 - \{P(A) + P(B) - P(A \cap B)\} = 1 - \frac{2+6-1}{8} = \frac{3}{8}.$$

Question 202: D

Using the product rule, the following is obtained:

$$\frac{dy}{dx} = x \cdot 4(x+3)^3 + 1 \cdot (x+3)^4 = 4x(x+3)^3 + (x+3)(x+3)^3$$

$$\Rightarrow \frac{dy}{dx} = (5x+3)(x+3)^3.$$

Question 203: C

This is a straightforward definite integral to compute. The steps are as follows:

$$\Rightarrow \int_1^2 \frac{2}{x^2}\,dx = \int_1^2 2x^{-2}\,dx$$

$$\Rightarrow \left[\frac{2x^{-1}}{-1}\right]_1^2 = \left[\frac{-2}{x}\right]_1^2 = \frac{-2}{2} - \frac{-2}{1} = 1.$$

Question 204: D

To express the expression in the desired form, you have to multiply by the complex conjugate, as follows:

$$\frac{5i}{1+2i} \cdot \frac{1-2i}{1-2i}$$

$$\Rightarrow \frac{5i+10}{1+4} = \frac{5i+10}{5} = i + 2.$$

Question 205: B

Firstly, we can use the rules of logarithms to simplify each term:

$$7\log_a(2) - 3\log_a(12) + 5\log_a(3)$$

$$\Rightarrow 7\log_a(2) = \log_a(2^7) = \log_a(128)$$

$$\Rightarrow 3\log_a(12) = \log_a(1728)$$

$$\Rightarrow 5\log_a(3) = \log_a(243)$$

Therefore, the original equation becomes:

$$\log_a(128) - \log_a(1728) + \log_a(243)$$

$$= \log_a\left(\frac{128 \times 243}{1728}\right) = \log_a(18).$$

Question 206: E

Rational functions which are a ratio of two quadratic functions have a horizontal asymptote. This can be determined by dividing each term by the highest order in the polynomial i.e. x^2, as shown:

$$\frac{2x^2 - x + 3}{x^2 + x - 2} = \frac{2 - \frac{1}{x} + \frac{3}{x^2}}{1 + \frac{1}{x} - \frac{2}{x^2}}$$

$$\lim_{x \to \infty}\left(\frac{2 - \frac{1}{x} + \frac{3}{x^2}}{1 + \frac{1}{x} - \frac{2}{x^2}}\right) = \frac{2}{1} \text{ i.e. } y \to 2.$$

So, the equation of the asymptote is $y = 2$.

Question 207: A

The intersection points are found by equating the two expressions:

$1 - 3e^{-x} = e^x - 3$

$\Rightarrow 4 = e^x + 3e^{-x} = \dfrac{(e^x)^2}{e^x} + \dfrac{3}{e^x} = \dfrac{(e^x)^2 + 3}{e^x}$

This is a quadratic equation in e^x:

$(e^x)^2 - 4(e^x) + 3 = 0 \Rightarrow (e^x - 3)(e^x - 1) = 0$

Therefore, $e^x = 3, x = \ln(3)$ or $e^x = 1 \Rightarrow x = 0$.

Question 208: D

To determine the radius, the equation must be rearranged into the format: $(x + a)^2 + (y + b)^2 = r^2$.

$(x - 3)^2 + (y + 4)^2 - 25 = 12$

$\Rightarrow (x - 3)^2 + (y + 4)^2 = 37$

$\therefore r = \sqrt{37}$.

Question 209: C

This question can be attempted graphically or algebraically. For the latter, note that $\sin(-x) = -\sin(x)$. Then the integral becomes:

$$\Rightarrow \int_0^a 2\sin(-x)\,dx = -2\int_0^a \sin(x)\,dx = -2[\cos(x)]_0^a = \cos(a) - 1$$

This is equal to zero at the following values:

$\cos(a) - 1 = 0 \therefore a = 2k\pi$.

Therefore, the integral of any whole period of $\sin(x) = 0$, i.e. $a = 2k\pi$, will be 0.

Question 210: E

Firstly, break the expression into the partial fraction terms:

$$\frac{2x + 3}{(x - 2)(x - 3)^2} = \frac{A}{(x - 2)} + \frac{B}{(x - 3)} + \frac{C}{(x - 3)^2}$$

$$\Rightarrow 2x + 3 = A(x - 3)^2 + B(x - 2)(x - 3) + C(x - 2)$$

When $x = 3, (x - 3) = 0 \quad \Rightarrow C = 9$. When $x = 2, (x - 2) = 0 \Rightarrow A = 7$.

$$\Rightarrow 2x + 3 = 7(x - 3)^2 + B(x - 2)(x - 3) + 9(x - 2)$$

Equating coefficients of x^2 on either side gives: $0 = 7 + B$ which gives: $B = -7$.

SECTION 2: WORKED SOLUTIONS

Q	A	Q	A	Q	A	Q	A	Q	A
1.1	C	3.1	B	5.1	C	7.1	E	9.1	C
1.2	D	3.2	E	5.2	B	7.2	D	9.2	B
1.3	C	3.3	B	5.3	A	7.3	B	9.3	A
1.4	C	3.4	D	5.4	C	7.4	D	9.4	C
2.1	A	4.1	B	6.1	E	8.1	C	10.1	C
2.2	E	4.2	D	6.2	A	8.2	D	10.2	E
2.3	D	4.3	C	6.3	B	8.3	A	10.3	D
2.4	A	4.4	A	6.4	A	8.4	D	10.4	A

Question 1.1: C

The distance the ball travels is maximised when the angle to the horizontal at which the golfer hits the ball is $45°$. This is shown as follows: the velocity component in the horizontal direction is $v\cos\theta$, where v is the speed of the ball. Therefore:

Speed = Distance / Time $\Rightarrow$ Distance = Speed × Time = $v\cos\theta$ × t = $v\cos\theta$ × $\frac{2v\sin\theta}{g}$ = $\frac{v^2\sin2\theta}{g}$

The time, however, is obtained using SUVAT. This expression is maximised at 45 degrees.

Question 1.2: D

The velocity at which the golfer hits the ball is the distance traversed by the club head divided by the time:

$v = \frac{2\pi L}{T} = 62.5 \text{m/s}$.

Question 1.3: C

The vertical velocity is $v\sin(45)$, and the acceleration downwards is $g = -10 \text{ m/s}^2$. The initial velocity is known from the answer of 1.2. Therefore, a SUVAT equation can be used, where the time is twice the time taken for the ball to reach $v = 0$ in the air.

Assuming that the initial velocity of the ball is approximately $v = 60$ m/s, we obtain:

$v = u + at \Rightarrow t = \frac{v-u}{a} = \frac{0 - 60\sin(45)}{-10} = 3\sqrt{2} \text{ s}$

As the time t the ball spends in the air is twice the time to reach peak height when $v = 0$, we simply double this value to give a total time of $6\sqrt{2}$ s.

Question 1.4: C

Time spent for vertical motion is the same as that for horizontal. Assuming air resistance is negligible, the distance D the ball travels in that time is then:

$D = v_h t = v\cos(45°)t = 360 \text{ m}$.

Question 2.1: A

The mass of the Moon is given by:

$$M_{Moon} = \rho_{Moon} \times V_{Moon} = \rho_{Moon} \times 4r_{Moon}^3 \times \pi/3.$$

As density and radius are different for the Moon that that of Earth, substitute in for $\left(\frac{3}{4}\right)\rho$ and $\left(\frac{1}{4}\right)r$ to make it in terms of the earth's data:

$$M_{Moon} = \frac{3}{4}\rho_{(earth)} \times \frac{4}{3} \times \left(\frac{r_{earth}}{4}\right)^3 \times \pi = \frac{\rho_{earth} \times r_{earth}^3 \times \pi}{64}.$$

Question 2.2: E

The gravitational force is $F = G\frac{Mm}{R^2}$. According to Newton's second law $F = ma$. Therefore, the gravitational acceleration is $= \frac{GM}{R^2}$. Using concepts applied in question 2.1, one can write:

$$M = \rho V = \frac{4}{3}\rho\pi r^3.$$

By substituting this into the gravitational acceleration equation, it is clear that:

$$g_{Earth} = G\rho\frac{4}{3}\pi r.$$

Question 2.3: D

Repeating the same calculations done in the Question 2.2, but this time for the moon, we obtain:

$$g_{moon} = \frac{GM_{moon}}{R^2}$$

Expressing the acceleration of the moon in terms of the density and radius of the earth (as in Question 2.1), one could write:

$$g_{moon} = \frac{G*\rho_{earth}*R_{earth}^3*\pi}{\frac{R_{earth}^2}{16}} = 16G\rho_{earth}R_{earth}\pi \implies g_{moon} = (3/16)g_{earth}$$

Question 2.4: A

For a satellite, we can equate the following:

$$F = mg = \frac{mv^2}{r}.$$

So, v is proportional to the square root of g – therefore, the speed will decrease with decreased g by a factor of $\sqrt{\left(\frac{3}{16}\right)}$.

Question 3.1: B

The resistance for resistors in parallel is the sum of the reciprocals:

$$\frac{1}{R_{total}} = \frac{1}{R} + \frac{1}{R} \implies R_{total} = \frac{R}{2}.$$

Question 3.2: E

Evaluate each group in parallel as one component and then add in series to first – remember to use 3.1 to make the calculation easier:

$$R_{total} = R + 1/\left(1/\left(R + 1/\left(\frac{1}{R} + \frac{1}{R}\right)\right) + 1/\left(R + 1/\left(\frac{1}{R} + \frac{1}{R}\right)\right)\right)$$

$$\Rightarrow R_{total} = R + 1/\left(2/\left(R + \frac{R}{2}\right)\right) = R + 1/\left(2/\left(\frac{3R}{2}\right)\right) = R + \frac{3R}{4} = \frac{7R}{4}.$$

Question 3.3: B

Using the equation for the voltage in the circuit, $V = IR_{total}$, one can use total resistance from previous question.

$$V = IR_{total} \Rightarrow R_{total} = \frac{V}{I} = \frac{1.4}{2} = 0.7 \ \Omega.$$

$$\Rightarrow R_{total} = \frac{7}{4}R \Rightarrow R = 0.4 \ \Omega.$$

Question 3.4: D

Using the equation for the power in terms of voltage and current $P = VI$, one can use the total current and voltage from question 3.1.

$$\Rightarrow P = 1.4 \times 2 = 2.8 \ W.$$

Question 4.1: B

Since the position is given by the equation $x = 10 + 1.5t^3$, and knowing that the velocity is a rate of change of the position, that is $v = \frac{dx}{dt}$, we have:

$$\Rightarrow v = \frac{dx}{dt} = \frac{d(10 + 1.5t^3)}{dt} = 4.5t^2$$

Question 4.2 D

The acceleration represents the rate of change of the velocity, so if one takes the derivative of the velocity in relation to time, one will have expression for acceleration.

$$a = \frac{dv}{dt} = \frac{d(4.5t^2)}{dt} = 9.0t.$$

Equating the acceleration to zero gives:

$$a = 9.0t^2 = 0 \Rightarrow t = 0.$$

Question 4.3 C

One can apply the equation for the position of the particle to figure out its total displacement for the given interval.

$$t = 2s \Rightarrow x_1 = 22 \ m, \quad t = 10s \Rightarrow x_2 = 1510 \ m$$

$$\Rightarrow \text{Total displacement: } \Delta x = x_2 - x_1 = 1488 \ m$$

$$\therefore \text{Average velocity: } \Delta v = \frac{\Delta x}{\Delta t} = \frac{1488}{8} = 186 \ m/s.$$

Question 4.4 A

First step is to calculate the average velocity of the ball for the first 5 seconds.

$t = 5s \Rightarrow x = 197.5 \text{ m} \Rightarrow v = \frac{\Delta x}{\Delta t} = \frac{197.5}{5} = 39.5 \text{ m/s}.$

Considering that an elastic collision is in place, the total energy transferred to the wall is equal to the kinetic of the ball in the moment of collision. Thus:

$E = \frac{mv^2}{2} = \frac{(m*39.5^2)}{2} \approx 780 \text{ mJ}.$

Question 5.1: C

The gravitational potential energy is given by the following:

$\Rightarrow E = mgh = (700 + 800) \times 10 \times 7 = 105{,}000 \text{ J}.$

Question 5.2: B

This is a straightforward calculation:

Power = Energy / Time = 3500 W.

Question 5.3: A

The average velocity needs to be determined to find the kinetic energy:

Average velocity = Distance / Time = 0.23 ms^{-1}

$\Rightarrow$ Kinetic energy $= \frac{mv^2}{2} = 40.8 \text{ J}.$

Question 5.4: C

The lift accelerates with the acceleration a for $t = 10$ s, then it uniformly moves for $t = 10$ s with the speed $v = at$ and, finally, it decelerates with the same acceleration for $t = 10$ s. Therefore:

$h = \frac{at^2}{2} + vt + \frac{at^2}{2} = 2at^2 = 2vt$

$\Rightarrow v = \frac{h}{2t} = 0.35 \text{ ms}^{-1}.$

Question 6.1: E

One can apply Newton's second law to block 2 and solve the equation: $F_2 = m_2a_2$. The component of the weight pulling the block down can be expressed as: $F_{x2} = m_2g\sin\alpha$.

The component of the weight in the y direct (equals to the normal reaction against the ramp) can be expressed as: $F_{y2} = m_2g\cos\alpha$. Therefore, the friction force can be expressed as: $F_{f1-2} = F_{y2}\mu_2 = m_2g\cos\alpha\mu_2$.

Finally, applying Newton's second law, the acceleration of block 2 can be expressed as:

$m_2a_2 = m_2g\sin\alpha - \mu_2m_2g\cos\alpha \Rightarrow a_2 = g(\sin\alpha - \mu_2\cos\alpha).$

Question 6.2: A

Same reasoning applied in question 6.1 can be applied to block 1 to express the acceleration of block. The component of the weight pulling the block down can be expressed as: $F_{x1} = m_1 g \sin\alpha$.

The component of the weight in the y direct (equals to the normal reaction against the ramp) can be expressed as: $F_{y1} = m_1 g \cos\alpha$. Therefore, the friction force can be expressed as: $F_{f1-2} = F_{y1}\mu_1 = m_1 g \cos\alpha \mu_1$

Applying Newton's second law:

$m_1 a_1 = m_1 g \sin\alpha - \mu_1 m_1 g \cos\alpha \Rightarrow a_1 = g(\sin\alpha - \mu_1 \cos\alpha)$.

Question 6.3: B

To determine the acceleration of m_1 in respect to m_2, one can simply subtract the acceleration of block 1 with respect to the plane from the acceleration of block 2 with respect to the plane:

$a_1' = a_1 - a_2 = g(\sin\alpha - \mu_1 \cos\alpha) - g(\sin\alpha - \mu_2 \cos\alpha) = g\cos\alpha(\mu_2 - \mu_1)$.

Question 6.4: A

When the forces on block 2 are balanced, we can equate the following:

$m_2 g \sin\alpha = \mu_2 m_2 g \cos\alpha$

$\Rightarrow \mu_2 = \sin\alpha / \cos\alpha = \tan\alpha$.

Question 7.1 D

This question can be easily solved if the total time is divided in small intervals of time and analyzed individually. 30 min corresponds to 1800s. According to the graph:

Interval: 0 – 100 s: Accelerated movement: $a = \frac{\Delta v}{\Delta t} = \frac{20}{100} = 0.2$ m/s^2. Distance can be then calculated:

$v^2 = v_o^2 + 2ad \Rightarrow d_{0-100} = \frac{v^2 - v_o^2}{2a} = \frac{20^2 - 0}{2*0.2} = 1000$ m.

Interval: 100 – 800 s: Constant velocity: $v = 20\frac{m}{s} \Rightarrow d_{100-800} = v * \Delta t = 20 * 700 = 14000$ m.

Interval 800 – 1000 s: Decelerated movement: $a = \frac{\Delta v}{\Delta t} = -\frac{20}{200} = -0.1\frac{m}{s^2}$. Distance can be calculated: $v^2 = v_o^2 + 2ad \Rightarrow d_{800-1000} = \frac{v^2 - v_o^2}{2a} = \frac{0 - 20^2}{2*-0.1} = 2000$ m.

Interval 1000 – 1600 s: The velocity is zero which means that the car is not moving.

Interval 1600 – 1800 s: Accelerated movement: $a = \frac{\Delta v}{at} = \frac{15}{200} = 0.075$ m/s^2. Distance can be calculated: $v^2 = v_o^2 + 2ad \Rightarrow d_{1600-1800} = \frac{15^2 - 0}{2*0.075} = 1500$ m.

Total Distance: $D = 1000 + 14{,}000 + 2000 + 1500 = 18{,}500$ m.

Question 7.2: D

Using information from previous question one can write the expression for the velocity in the interval 1600 − 1800 s. $v = v_0 + at = 0 + 0.075t$.

No calculation is needed for this problem if the student realizes that the integration of the velocity for that in the interval will give you the distanced travelled by the car in the interval, which has already been computed in the previous question and it 1500 m.

The integration then will be:

$$\int_{1600}^{1800} v \, dt = \int_{0}^{200} 0.075 \, t \, dt = 0.0375t^2 \big|_{0}^{200} = 0.0375(200^2) = 1500 \text{ m}.$$

Question 7.3: B

At t = 900s, the car is subjected to an acceleration of $|a| = 0.1$ m/s^2. This is easily determined by the gradient of the velocity-time graph.

According to Newton's second law:

$F = ma \Rightarrow 1000 * 0.1 = 100$N.

Question 7.4: A

At t = 50s, the car is subjected to an acceleration of $|a| = 0.2$ m/s^2, leading to an applied force of:

$F = ma = 1000 * 0.2 = 200$ N.

The force applied to the car by the engine is constant throughout the whole interval. You can calculate the average work done by the engine as follows:

$W = Fd = 200 * 1000 = 200000$ J.

Finally, one can calculate the power:

$P = \frac{w}{\Delta t} = \frac{200000}{100} = 2000$ W.

Question 8.1: C

One can observe that the force applied to the particle is positive up to x = 3 m, which means that the particle is being accelerated up to that point, and its velocity is increasing.

After x = 3 m, the force applied to the particle starts to be negative, which means that the particle is being decelerated and its velocity decreasing after x = 3m. Therefore, the velocity is maximum at x = 3 m.

Question 8.2: D

Following similar reasoning as in the previous question, and knowing the maximum velocity happens at x = 3m, consequently, one can affirm the kinetic energy is maximum at x = 3 m as it is equal to ½ mv^2.

Question 8.3: A

One can get this answer by analysing the graph. The area under the curve of a graph (force x distance) will give the work that has been done to the particle. Since the particle is at rest at x = 0, its kinetic energy and velocity are zero at x = 0.

The area from x = 0 to 3 represents the work done to the particle by the positive force. The area from x = 3 to 6, represents the work done to the particle by the negative force; that is, the force in the opposite direction of the positive force.

Since the two areas cancel each other exactly at x = 6, all the work that has previous been done the particle by the positive force will have been cancelled by the work done by the negative force (acting in the opposite direction). Therefore, the kinetic energy of the particle at x = 0 is zero and, consequently, its velocity is also zero.

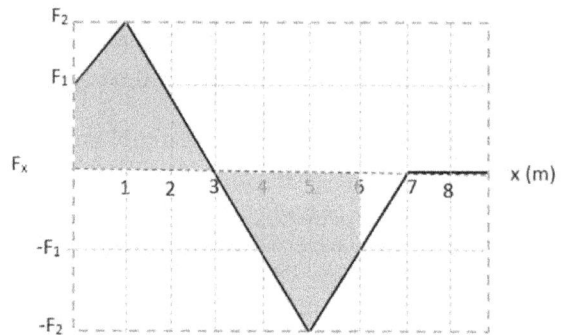

Question 8.4: D

After x = 6 metres, the force is acting in the opposite direction compared to at x = 2 m. Therefore, the acceleration vector is also pointing the opposite direction as it was in x = 2 m.

Question 9.1: C

This set of questions tests your knowledge of accelerated and decelerated motion and conservation of energy. When the ball is thrown up in the air, it is subjected to the gravitational acceleration, pointing downwards towards the centre of the Earth. Therefore, the acceleration is constant and negative, with the velocity reaching zero at the highest point and the same velocity, in modulus, when it returns to the starting point. Option C is the only one that represents such behaviour.

Question 9.2: B

To solve this question, one should consider the conservation of energy, since the resistance of air can be neglected.

Considering the position from which the ball leaves as $h = 0$, the potential energy of the ball will be zero. Since the ball leaves the girl's hand at $v = 15m/s$, its initial kinetic energy can be calculated:

$$E_k = \frac{1}{2} m. v^2 = \frac{1}{2} * 0.5 * 15^2 = 56.25 \, J.$$

Considering conservation of energy and knowing that, at the highest position, the kinetic energy of the ball is zero, one can assume that the kinetic energy the ball had at the beginning of its trajectory will be entirely converted in potential energy. From this assumption, one can calculate the high reached by the ball:

$$E_k = E_p \Rightarrow 56.25 = mgh \Rightarrow h \approx 12 \, m.$$

Question 9.3: A

One can assume this represents a case of decelerated movement with $a = -9.8 \, m/s^2$, and easily calculate the time that takes for the ball's velocity be reduced from 15 m/s to 0:

$$v = v_0 + a. t \Rightarrow t = \frac{(v-v_0)}{a} = \frac{(0-15)}{-9.8} = 1.53 \, s.$$

Question 9.4: C

First, on can calculate the potential energy of the ball 5.0m from the ground:

$$E_{P(5.0m)} = mgh = 0.5 * 9.8 * 5 = 24.5 \, J.$$

Since, there is conservation of energy, one can subtract the potential energy at 5.0 m from the kinetic energy of the ball at the starting point (calculated in question 9.1):

$$E_{k(5.0m)} = 56.25 - 24.5 = 31.75 \, J.$$

Question 10.1: C

Option B displays the correct representation of the forces with the weight of Object 2 pointing downwards towards the centre of Earth, the normal reaction against the surface of the ramp and friction force pointing downwards along the surface of the ramp as a reaction against the eminent movement of Object 2 upwards along surface of the ramp.

Question 10.2: E

Considering that Object 2 stays at rest in relation to the ramp, one can equate the vertical components of the forces acting on the object:

$$F_f sin\theta + mg = Ncos\theta$$

$$mg = Ncos\theta - F_f sin\theta = Ncos\theta - N\mu sin\theta = N(cos\theta - \mu sin\theta).$$

Question 10.3: D

Following similar reasoning as in the previous question, one can equate the horizontal components of the forces acting on the object:

$$Nsin\theta + F_f cos\theta = m_2 a$$

$$\Rightarrow m_2 a = Nsin\theta + N\mu cos\theta = N(sin\theta + \mu cos\theta)$$

$$\Rightarrow a = \frac{N}{m_2}(sin\theta + \mu cos\theta).$$

Question 10.4: A

According to Newton's second law:

$$F = ma = (m_1 + m_2)a \Rightarrow F = 20a \Rightarrow a = \frac{F}{20}.$$

Dividing the answers for question 10.3 and 10.2 leads to:

$$\frac{a}{g} = \frac{(sin\theta + \mu cos\theta)}{cos\theta - \mu sin\theta}$$

$$\frac{F}{20g} = \frac{(sin\theta + \mu cos\theta)}{cos\theta - \mu sin\theta} \Rightarrow F = 20g * \frac{(sin\theta + \mu cos\theta)}{cos\theta - \mu sin\theta} = 219N.$$

ENGAA PAST PAPER WORKED SOLUTIONS

Hundreds of students take the ENGAA exam each year. These exam papers are then released online to help future students prepare for the exam. Since the ENGAA is a relatively new exam, past papers have become an invaluable resource in students' preparation.

Where can I get ENGAA Past Papers?

This book does not include ENGAA past paper questions because it would be over 500 pages long if it did! However, all ENGAA past papers since 2016 (including the specimen paper) are available for free from the official ENGAA website. To save you the hassle of downloading lots of files, we've put them all into one easy-to-access folder for you at https://www.uniadmissions.co.uk/application-guides/every-past-papers-answer-sheets/.

How should I use ENGAA Past Papers?

ENGAA Past papers are one the best ways to prepare for the ENGAA. Their use can dramatically boost your scores in a short period of time. The way you use them will depend on your learning style and how much time you have until the exam date but here are some general pointers:

- Four to eight weeks of preparation is usually sufficient for most students.
- Make sure you are completely comfortable with the ENGAA syllabus before attempting past papers – they are a scare resource and you shouldn't 'waste them' if you're not fully prepared to take them.
- It's best to start working through practice questions before tackling full papers under time conditions.
- You can find two additional mock papers in the *ENGAA Practice Papers* Book (flick to the back to get a free copy).

How should I use this book?

This book is designed to accelerate your learning beyond ENGAA past papers. Avoid the urge to have this book open alongside a past paper you're seeing for the first time. The ENGAA is difficult because of the intense time pressure it puts you under – the best way of replicating this is by doing past papers under strict exam conditions (no half measures!). Don't start out by doing past papers (see previous section) as this 'wastes' papers.

Once you have finished, take a break and then mark your answers. Review the questions that you got wrong followed by ones which you found tough/spent too much time on. This is the best way to learn and, with practice, you should find yourself steadily improving. You should keep a track of your scores on the next page so you can track your progress.

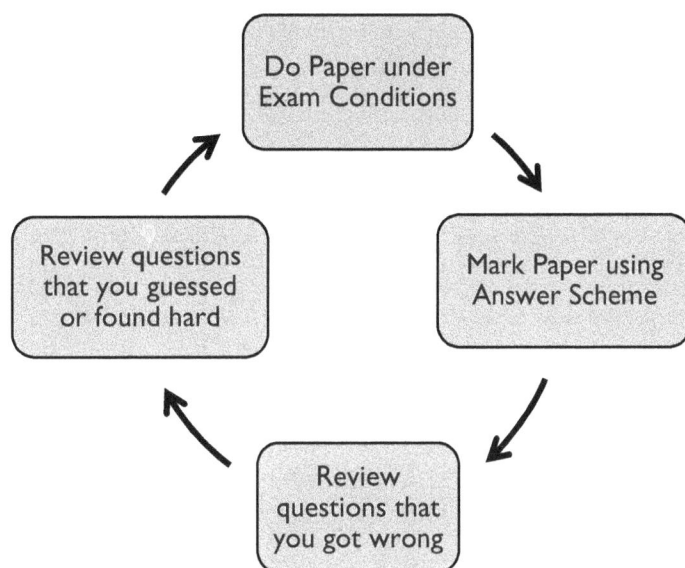

Do Paper under Exam Conditions → Mark Paper using Answer Scheme → Review questions that you got wrong → Review questions that you guessed or found hard → (back to Do Paper under Exam Conditions)

Scoring Tables

Use these to keep a record of your scores – you can then easily see which paper you should attempt next (always the one with the lowest score).

Section 1	1st Attempt	2nd Attempt	3rd Attempt
Specimen			
2016			
2017			
Mock Paper A			
Mock Paper B			

Section 2	1st Attempt	2nd Attempt	3rd Attempt
Specimen			
2016			
2017			
Mock Paper A			
Mock Paper B			

SPECIMEN

Section 1

Question 1: A
Take each side of the square as length a. The removed semicircle has radius a/2. The area of the square is a^2 and the area removed is $\pi a^2/8$. Therefore,

$$100 = a^2 \left(1 - \frac{\pi}{8}\right)$$

Rearranging this gives:

$$\Rightarrow a^2 = 800/(8 - \pi)$$

Taking the square-root gives:

$$\therefore a = 20\sqrt{\frac{2}{8-\pi}}.$$

Question 2: C
The net force is 900 – 600 = 300N upwards. The parachutist has mass 60kg. Therefore, the acceleration in the upwards direction is given by:

$$F = ma \Rightarrow a = \frac{F}{m} = 5\,\text{ms}^{-2}.$$

Question 3: E
The ratio 1:2 indicates RQ is half the length of PQ, i.e. 10cm.

Label the point on the perpendicular and the hypotenuse S.

The length of PR is $10\sqrt{5}$cm using Pythagoras theorem. The triangle PQR and PSQ are similar. The ratio of RQ:PR is equivalent to that of SQ:PQ. Therefore SQ = $20 \times 10/(10\sqrt{5})$. Simplifying this gives $4\sqrt{5}$.

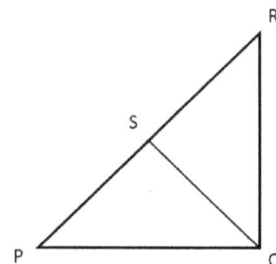

Question 4: C
Use the equation $c = f\lambda$. From the graph, the wave does one whole oscillation in 2 seconds. Therefore:

$$\Rightarrow f = \frac{1}{T} = 0.5\,\text{Hz}.$$

The wavelength of the wave is 1.5cm. Multiplying the two gives $c = 0.75\text{cms}^{-1}$.

Question 5: A

The length of the dotted line is x. Firstly, consider the distance from the centre of the base to one corner of the base. This is given by $\sqrt{(0.5)^2 + (0.5)^2}$. Now consider the triangle that has its hypotenuse along the dotted line. The other two sides are the vertical edge and the line from the centre of the base to the corner of the base. The vertical has length 1.

Now use Pythagoras to find the length:

$$\Rightarrow x^2 = 1^2 + 0.5^2 + 0.5^2 = 1.5.$$

Taking the square-root gives $\sqrt{3/2}$.

Question 6: B

The kinetic energy is $E = mv^2/2$. The momentum is given by $p = mv$. Divide the first equation by the second giving $5 = v/2$. Therefore $v = 10 ms^{-1}$. Substituting this result into the momentum equation gives 3 kg.

Question 7: D

The 12 hours span 360 degrees. Therefore, each hour is 30 degrees. At 9:45, the hour hand has moved ¾ of the way to 10. The answer is given by $30 \times 0.75 = 22.5°$.

Question 8: E

Consider the equation $P = IV$. Power has the unit Watt and Current has the unit Amp. Therefore voltage, or potential difference, is given by Watts per Amp.

Question 9: B

The area of a triangle is given by:

$$\Rightarrow A = \frac{1}{2}bh = \frac{1}{2}\left(4 + \sqrt{2}\right)\left(2 - \sqrt{2}\right)$$

Expanding the brackets gives:

$\frac{1}{2}\left(8 - 2\sqrt{2} - 2\right) = 3 - \sqrt{2}$.

Question 10: D

In 24 hours, source X has undergone 24/4.8 = 5 half-lives, and source Y has undergone 24/8 = 3 half-lives.

Therefore, the activity of X has decreased by a factor of $\left(\frac{1}{2}\right)^5$. This gives $320 \times \frac{1}{32} = 10$ Bq.

The activity of Y is 1/8th of its initial value, i.e. 60 Bq. Adding these two contributions gives 70 Bq.

Question 11: D

The radius of the cylinder is r and the height of the cylinder is 2r. It has a volume of $\pi r^2 h = 2\pi r^3$. Dividing the area of the sphere by this gives 2/3.

Question 12: E

The potential energy lost is given by:

$$\Rightarrow E = mgh = 100 \times 10 \times 100 = 100,000 \text{ J}.$$

The distance travelled down the slope is 10 times the vertical distance dropped, i.e. 1000 m. For constant speed, all the energy must be dissipated by resistive forces. The work done by the resistive force is $W = Fd$.

Therefore:

$$\Rightarrow F = W/d = 100000/1000 = 100N.$$

Question 13: A

For $0 < x < 1$: x^2 and $\sqrt{x}$ are both less than one. For the other three options, we want the denominator to be as small as possible to give the largest number. $1 + x < x$ rules out option C. In this range of x, $x < \sqrt{x}$. The largest option is therefore $1/x$.

Question 14: A

Assuming all the initial kinetic energy is converted into GPE:

$$\Rightarrow KE = \frac{1}{2}mv^2 = mgh = GPE$$

Rearranging gives:

$$\Rightarrow h = v^2/(2g) = 144/(20) = 7.2 \text{ m}.$$

Question 15: C

The ratio AB:BC is equal to the ratio AD:DE. AD has length x cm. Therefore:

$$\frac{4}{x} = \frac{x}{x+3} \Rightarrow 4(x+3) = x^2 \Rightarrow x^2 - 4x - 12 = 0.$$

Factorising this gives: $(x - 6)(x + 2) = 0$.

The two possible solutions are $x = 6$ cm and $x = -2$cm. A negative length is unphysical, and this solution can be discarded. The length of DE is $x + 3 = 9$cm.

Question 16: C

To bring the lorry to rest, all of the kinetic energy must be dissipated by the resistive forces. The initial energy is given by $\frac{1}{2}mv^2$. The resistive force must do work given by $W = Fd$ which is equal to the initial energy. Thus:

$$Fd = \frac{1}{2}mv^2 \Rightarrow d = mv^2/(2F).$$

Question 17: C

As Q is 1.4 times its original value, Q^2 is $1.4 \times 1.4 = 1.96$ times its original value. Therefore, P is $1/1.96$ its original value. This is close to $1/2$ i.e. a 50% decrease. However, $1/1.96$ is larger than $1/2$. Therefore, it has decreased by a factor of close to but less than 50%. The answer is 49%.

Question 18: B

An alpha particle consists of 2 neutrons and two protons; so, alpha decay would decrease the value of the mass number by 4 and the proton number by 2. A beta decay ejects an electron from the nucleus by converting a neutron to a proton. This doesn't affect the mass number but increases the proton number by one.

From X to Y, an alpha decay must occur, so P is equal to N - 4. From Y to Z, the mass number is unchanged, so it must be a beta decay. Thus, Q equals R-1.

Question 19: D

$x \propto z^2$ and $y \propto 1/z^3$. This means $x^3 \propto z^6$ and $y^2 \propto 1/z^6$

Inverting the second equation yields $z^6 \propto 1/y^2$. Combining this with the first equation gives $x^3 \propto 1/y^2$.

Question 20: B

For the pulse to travel to the foetus and back, it must travel 20 cm. It travels at a speed of 500 ms^{-1}. The time taken is given by $t = 0.2/500 = 0.4 \text{ ms}$.

Question 21: C

Label the distance QX x. Therefore, the distance PX is $6x$ and the distance XR is $\frac{3}{2}x$.

The distance PR is $\frac{15}{2}x$. As M is the midpoint of PR the distance PM is $\frac{15}{4}x$. The distance MX is:

$$\Rightarrow PX - MX = 6x - \frac{15}{4}x = \frac{9}{4}x \quad \Rightarrow \frac{QX}{MX} = \frac{x}{(9x)/4} = \frac{4}{9}.$$

Question 22: B

Constant acceleration gives a straight line on a velocity-time graph, so we can discard R and S. Graph P has an acceleration of $10/24 \text{ ms}^{-2} \neq 2.4 \text{ ms}^{-2}$. The final graph Q reaches a velocity of 58 ms^{-1}. The change in velocity is therefore 48 ms^{-1} over 20 seconds, i.e. an acceleration of 2.4 ms^{-2}.

Question 23: D

Move all the terms to the left-hand side:

$$\Rightarrow x^2 + 2x - 8 \geq 0.$$

Factorising this gives:

$$\Rightarrow (x + 4)(x - 2) \geq 0.$$

Therefore, $x^2 + 2x - 8$ changes sign at $x = -4$ and $x = 2$. Note that the x^2 term is positive, so the graph is U-shaped. It is therefore greater than or equal to 0 when $x \geq 2$ or when $x \leq -4$.

Question 24: D

Fission is when a nucleus splits into two parts - not gamma decay. A half-life is the time taken for half of the substance to decay but each nuclei decays instantaneously at a random time - B is false. The number of neutrons is the mass number – the atomic number - C is False.

Under beta decay, a neutron in the nucleus becomes a proton and an electron. The electron is ejected so the number of particles in the nucleus is conserved. An alpha particle is made up of two protons and two neutrons which are ejected from the nucleus. No neutrons are converted to protons. D is the true statement.

Question 25: A

The area of a cylinder is given by $\pi r^2 h$. The surface area is $2\pi r^2 + 2\pi rh$

$\Rightarrow \pi r^2 h = 2\pi r^2 + 2\pi rh$

$\Rightarrow rh = 2r + 2h$

$\Rightarrow h(r - 2) = 2r$

$\Rightarrow h = \frac{2r}{r-2}$.

Question 26: D

The two resistances in series add. The total resistance of the circuit is $R_1 + R_2$.

Using Ohm's Law gives the current through the circuit as $I = \frac{V}{R} = \frac{V}{R_1+R_2}$.

This is the current that flows through each resistor. Thus, the power dissipated by the first resistor is given by:

$\Rightarrow P = IV = I^2R = \frac{V^2 R_1}{(R_1+R_2)^2}$.

Question 27: B

From 1 to 2, the square has been rotated by 90 degrees clockwise. It has then been reflected in $y = x$. To transform 3 to 1, the square must be reflected in the y-axis.

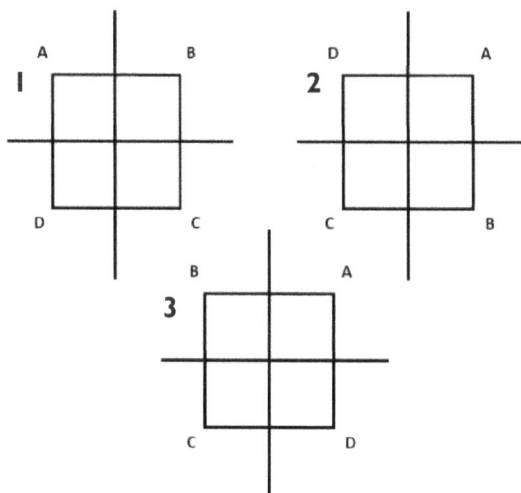

Question 28: B

The speed of sound is around 340 ms^{-1} so P is not true. The distance between the point X and Y corresponds to twice the amplitude i.e. the peak to peak distance. Therefore, the amplitude is 2.5mm. The wavelength cannot be deduced from the information given as the speed is not given. The frequency is $f = \frac{1}{T} = 5 \text{Hz}$. The only statement that can be deduced is S.

Question 29: D

$$\frac{x^2 - 4}{x^2 - 2x} = \frac{(x + 2)(x - 2)}{x(x - 2)} = \frac{x + 2}{x}$$

Question 30: A

The acceleration of the car is uniform, so the force is constant. Use $F = ma$ with an acceleration of $a = \frac{10}{5} = 2 \text{ ms}^{-2}$. This gives a net force of 2 kN. The resistive force must therefore be $f = 3 - 2 = 1$ kN.

Question 31: F

$a^x b^{2x} c^{3x} = (ab^2 c^3)^x = 2$

$\log 2 = x \log(ab^2 c^3)$

$\Rightarrow x = \frac{log(2)}{log(ab^2 c^3)}.$

Question 32: C

Consider P initially moving from left to right and Q from right to left. In the collision momentum is conserved. The total initial momentum is given by:

$\Rightarrow p = 2 \times 3 - 5r = 6 - 5r.$

The final momentum is given by $\frac{5r}{2} - 2$. Equating these gives $6 - 5r = \frac{5r}{2} - 2$. Rearrangement gives $r = \frac{16}{15}.$

Question 33: D

A gives $\tan\left(\frac{3\pi}{4}\right) = -1$.

B is $\log_{10} 100 = 2 \log_{10} 10 = 2$.

C is $\sin(\pi/2) = 1$ and $1^{10} = 1$.

D gives $\log_2 10 > \log_2 8 = 3$. Therefore, D is larger than 3. $\sqrt{2} - 1 < 1$, so, the tenth power of this is very small and certainly less than 1. Thus, D is the largest.

Question 34: A

Initially, the parachutist should experience an increasing air resistance until they reach terminal velocity when it becomes constant. When the parachute is opened, the air resistance will increase substantially. Then, as the parachutist decelerates, the air resistance will decrease until the new terminal velocity is reached.

As the parachutist experiences no net force at terminal velocity, the air resistance must be the same as in the previous period of terminal velocity. The only graph that satisfies this is A.

Question 35: E

Let $y = 2^x$. Then substitute this into the original equation, giving $y^2 - 8y + 15 = 0$. Factorising this gives $(y - 5)(y - 3) = 0$. This yields two solutions: $2^x = 5$ and $2^x = 3$. Taking logs of these gives solutions:

$$\Rightarrow x_1 = \frac{log_{10}5}{log_{10}2} \text{ and } x_2 = \frac{log_{10}3}{log_{10}2}$$

Summing these and factoring out the denominator gives:

$$x_1 + x_2 = \frac{1}{\log_{10}2}(\log_{10}5 + \log_{10}3) = \frac{\log_{10}15}{\log_{10}2}.$$

Question 36: F

The first graph decreases at a decreasing rate. This could be either the acceleration as it reaches terminal velocity or the resultant force on the car. The gravitational energy does not change as the car travels horizontally. The second graph increases to a steady state. This could by air resistance or velocity which are both constant at terminal velocity. Finally, Z can only be the weight of the car which remains constant throughout. The kinetic energy increases as the car accelerates. The only option which satisfies all three is option F.

Question 37: E

The first inequality can be shown to be true by subtracting both a and b from both sides.

For the second inequality, move the 2ab to the left-hand side $a^2 + b^2 - 2ab = (a - b)^2 \geq 0$. This is true assuming both a and b and real numbers.

The third inequality is violated for a negative value of c. Therefore, the answer is E.

Question 38: C

Initially, the frictional force balances the horizontal force applied. However, the frictional force cannot exceed μR so, after a time, the block will accelerate. As the horizontal force continues to increase and the frictional force remains constant, the block will increasingly accelerate.

Question 39: A

The term $(-1)^n$ is equal to -1 when n is odd and +1 when n is even. This means the first six terms of the sequence are: $2, 1, 2, 1, 2, 1$. Therefore, when n is odd, $a_n = 2$ and for n even, $a_n = 1$. The summation to n = 100 can be written as:

$$\Rightarrow 2 \times 50 + 1 \times 50 = 150.$$

Question 40: C

Require momentum to be conserved during the collision. The balls all have mass m. The initial momentum is 3m. Let the final velocity of the red ball be v. Then the final momentum is $m + mv = 3m$. Divide by m and this yields $v = 2ms^{-1}$.

Question 41: C

Roots are found where the graph $y = x^4 - 4x^3 + 4x^2 - 10 = 0$. To find how many points satisfy this, consider the location of the stationary points.

$$\Rightarrow \frac{dy}{dx} = 4x^3 - 12x^2 + 8x = 4x(x^2 - 3x + 2) = 4x(x - 1)(x - 2).$$

Stationary points are found where this equals zero i.e. $x = 0, x = 1, x = 2$. Now we want to find whether these stationary points are located above or below the x axis.

$$\Rightarrow y(0) = -1, \ y(1) = -9, \ y(2) = -10.$$

All 3 of these are located beneath the x-axis so there are no roots between $x = 0$ and $x = 2$. However, when x is a large negative or positive value the x^4 term dominates so y is a large positive value. Therefore, it must intersect the x axis at 2 points, so it has two roots.

Question 42: B

All of the kinetic energy of the rock is converted into gravitational potential energy, so we can write the equation $\frac{mv^2}{2} = mgh$. Cancelling the masses and inserting the velocity of 20 ms^{-1} gives $gh = 200$. The only option which satisfies this equation is option B.

Question 43: F

Consider a point in the centre of the base PQRS. Find the distance of this point from point P using Pythagoras's theorem:

$$x^2 = 5^2 + 5^2 \Rightarrow x = 5\sqrt{2}\text{cm}.$$

Now form a right-angled triangle using the point at the centre of the base, P and T. The desired angle is the one at point P. We have the adjacent, with length $5\sqrt{2}$cm, and the hypotenuse, 12cm. Using $\cos\theta = \frac{A}{H} = \frac{5\sqrt{2}}{12}$ gives the answer as F.

Question 44: A

The ball initially only has gravitational potential energy which, as it falls, is all converted into kinetic energy. During the bounce, half of the energy is dissipated. The remaining energy is then all converted into gravitational potential energy as the ball rises. As $GPE = mgh$, a 50% decrease in the energy causes the height of the bounce to decrease by 50% as the gravitational field strength and mass are constant. After the first bounce, the ball reaches 8m. After the second 4m. After the 3rd, 2m and, after the 4th, 1m. Therefore, the answer is A.

Question 45: D

For variables x and y, a straight line has the form $y = mx + c$. For a graph of $\log y$ and $\log x$, the relation should look like $\log y = m \log x + c$ with m and c both constants. The first gives $\log y = \frac{x \log a}{6}$.

B gives $\log y = x \log b + \log a$. C gives $2 \log y = \log(a + x^b)$. D gives $\log y = b \log x + \log a$. E gives $\log y = \frac{a}{x} \log b$ The only one with the correct form is D.

Question 46: E

If the elevator was moving with constant velocity, the motion would have no effect on the scale reading. If the elevator was moving downward with a decreasing speed, the scales would have to exert a net force upwards on the man. This would cause the scale reading to be greater than the weight of the man. If the elevator were moving upwards with increasing speed, the scale would have to exert a force great enough to accelerate the man upwards, so the reading would be larger than the man's weight. If the elevator is moving upwards with decreasing speed, the resultant force is downwards so the normal force of the scales is smaller.

Question 47: D

For real distinct roots in a quadratic equation, we require the discriminant $b^2 - 4ac > 0$. For the given equation, this is $(a - 2)^2 + 8a > 0$. Expanding the brackets yields $a^2 + 4a + 4 = (a + 2)^2 > 0$. If a is a real number, this is satisfied when $a \neq -2$.

Question 48: B

Constant acceleration of $a = -1.5 \text{ ms}^{-1}$. We can use SUVAT equations in this case. The known variables are a, u and v and we need to find s. Choose the equation $v^2 = u^2 + 2as$.

Rearranging this gives: $s = \frac{1}{2a}(v^2 - u^2)$. Insert $u = 12 \text{ ms}^{-1}$ and $v = 0 \text{ ms}^{-1}$. This gives $s = 48 \text{ m}$.

Question 49: C

To be able to construct a triangle, the longest side must be shorter than the sum of the other two, so the longest side must be less than 6cm long. Therefore, there are only 3 possible constructions with sides of integer length: 5cm, 4cm and 3cm; 5cm, 5cm and 2cm; and 4cm, 4cm and 4cm.

Question 50: C

The vertical component of the tension in the angled rope must balance the weight of the particle. Resolving vertically gives $T_1 \cos(60) = T_1/2 = 5\text{N}$. This gives $T_1 = 10\text{N}$.

The forces must also balance horizontally. The horizontal component of the tension in the angled rope is $T_1 \sin(60) = 5\sqrt{3}\text{N}$. This is directly balanced by the horizontal rope, so the answer is C.

Question 51: F

If the angle at R is 90 degrees, the hypotenuse must be the opposite side i.e. PQ. Therefore, the length QR is given by $QR^2 = PQ^2 - PR^2$. To minimise QR, we must minimise PQ whilst maximising PR. The smallest possible value of PQ is 3.5cm and the maximum of PR is 2.5cm. Inserting these gives $QR^2 = 6$. Thus, the minimum value of QR is $\sqrt{6}$.

Question 52: D

Power is given by energy per unit time. The change in gravitational energy per unit time is given by $mg\frac{\Delta h}{\Delta t}$. Take the gradient of the graph when h = 10 metres.

$$\Rightarrow \frac{\Delta h}{\Delta t} = 15/30 = 0.5 \Rightarrow P = 20 \times 10 \times 0.5 = 100\text{W}.$$

Question 53: B

In the range $0 \leq x \leq \pi$, $-1 \leq \tan x \leq 1$ is satisfied when $x \leq \pi/4$ or $x \geq \frac{3\pi}{4}$.

The inequality $\sin y \geq 0.5$ is satisfied for $\pi/6 \leq y \leq 5\pi/6$. Substituting in $y = 2x$ gives $\frac{\pi}{12} \leq x \leq \frac{5\pi}{12}$.

These are both satisfied in the range $\frac{\pi}{12} \leq x \leq \frac{\pi}{4}$. Therefore, the interval has length $\pi/6$.

Question 54: A

All the carriages must accelerate at the same rate. The total mass of the system is 30,000 kg.

$$\Rightarrow a = F/m = 0.5 \text{ ms}^{-1}.$$

Consider now only the forces acting on the final carriage. There is only the tension T which causes an acceleration of 0.5ms^{-1}. This gives $T = ma = 5000 \times 0.5 = 2500\text{N}$.

Section 2

Question 1: C

This problem can be solved using the period and speed of the wave. Re-arrange the equation wave speed (v) = frequency (f) × wavelength (λ) to give:

$$\lambda = v/f$$

The period of a wave is the time taken for a particle of the medium through which the wave is travelling to perform one complete oscillation.

$$f = 1/period \quad \Rightarrow \quad \lambda = v \times period$$

The time taken to move from equilibrium position to maximum displacement is 0.2 s, and this is one quarter of the period. The wave speed is 60 cms⁻¹. $\lambda = 60 \times 4 \times 0.2 = 48$ cm

Question 2: D

This problem can be solved by considering the change in height and the motion along the slope. Both blocks fall through the same vertical height h so they transfer the same amount of gravitational potential energy to kinetic energy. Therefore, change in gravitational potential energy equates to change in kinetic energy

$$mgh = \frac{1}{2}mv^2$$

where m is the mass, g is the gravitational field strength and v is the final velocity, since both blocks start from rest. Both blocks, therefore, reach the same final velocity: $v_x = v_y$. As both blocks also accelerate down the slope at a constant rate, their average velocities are also the same. Therefore, $\frac{v_x}{2} = \frac{v_y}{2}$.

However, block X travels a longer distance along the slope than block Y so takes a longer time to travel through the same vertical height, meaning that $t_x > t_y$. Therefore, $t_x > t_y$ and $v_x = v_y$.

Question 3: G

This problem can be solved by considering the conservation of momentum and the conservation of energy. The probe is travelling through the vacuum of deep space, its engine is switched off and there are no gravitational forces acting on it. Therefore, it is an isolated system and the total momentum before the collision will equal the total momentum afterwards. The fuel explodes and so its chemical energy decreases but since energy must be conserved and the system is isolated, some or all of the energy is transferred to kinetic energy and so increases the total kinetic energy.

Therefore, the correct answer is the total momentum after the explosion is the same as before, but the total kinetic energy has increased.

Question 4: A

This problem can be solved by using conservation of energy and work done. The kinetic energy (KE) lost by the object is equal to the gain in gravitational potential energy (GPE) plus the work done against the frictional force (W).

Gain in GPE $= mgh = 5 \times g \times 3 = 150$ J and W = Fd

Where F is the frictional force and d is the distance travelled by the object.

Considering the that length PQ is equal to distance d, from the diagram we get that $\sin \alpha = \frac{3}{d}$ so $PQ = d = \frac{3}{\sin \alpha}$. Since $\tan \alpha = \frac{3}{4}$, this is a **3, 4, 5** triangle and so:

$$\sin \alpha = \frac{3}{5} = 0.6 \quad \therefore \quad d = \frac{3}{0.6} = 5 \text{ m} \quad \therefore \quad W = 5F$$

Since the change in KE is equal to the GPE gain plus the work done we have:

$$210 = 150 + 5F \quad \therefore \quad F = \frac{210 - 150}{5} \quad \therefore \quad F = 12N$$

Question 5: C
This problem can be solved by considering the vertical and horizontal motion separately and using the equations of motion, since the acceleration is constant.

Vertically:
$$s = ut + \frac{1}{2}at^2 \quad \therefore \quad 4 = \frac{1}{2}gt^2 \quad \therefore \quad t^2 = \frac{4}{5} \quad \therefore \quad t = \frac{2}{\sqrt{5}} \text{ s}$$

Horizontally
$$\text{distance} = \text{speed} \times \text{time} \quad \therefore \quad \frac{6\sqrt{5}}{5} = \frac{2\sqrt{5}}{5}v \quad \therefore \quad v = \frac{5 \times 6\sqrt{5}}{5 \times 2\sqrt{5}} \quad \therefore \quad v = 3 \text{ ms}^{-1}$$

Question 6: A
This problem can be solved by using Own's Law V=IR and
$$\text{resistivity} = \text{resistance} \times \frac{\text{cross-sectional area}}{\text{length}}, \rho = \frac{RA}{l}$$
The current in the resistor is given by:
$$I = \frac{V}{R} = \frac{\text{pd across resistor}}{\text{resistance of resistor}} = \frac{1.0}{1.0 \times 10^3} = 1.0 \times 10^{-3} \text{ A}$$

The wire and the resistor are connected in series so the current is the same in each of them, where both act as a potential divider. The 1.2V across the arrangement is shared between the wire and the resistor. There is 1.0V across the resistor so the pd across the wire is: (1.2 − 1.0) = 0.20 V

Therefore, the resistance of the wine is given by:
$$R = \frac{V}{I} = \frac{\text{pd across wire}}{\text{current in wire}} = \frac{0.2}{1.0 \times 10^{-3}} = 2.0 \times 10^2 \Omega$$

The resistance is then:
$$\rho = \frac{RA}{l} = 1.0 \times 10^{-6} \Omega m$$

Question 7: B

This problem can be solved by applying Newton's second law to each block. The forces acting on the 6 kg block are its weight 60N vertically downwards and the tension TN in the string, vertically upwards. The forces acting on the 4 kg block are the tension in the string, upwards along the plane, a component of the weight, 40sin30° downwards along the plane and the frictional force F. Given the masses of the blocks, the acceleration of the 4 kg block must be upwards along the plane, and F must act downwards along the plane. The system moves with constant acceleration, a, upwards along the plane. The acceleration of the 4kg block is the same as the acceleration of the 6kg block.

Applying Newton's second law to each block:

$60 - T = 6a$ and $T - 15 - 40\sin30° = 4a$

This gives two simultaneous equations in T and a:

$60 - T = 6a$

$T - 35 = 4a$

Adding these gives:

$10a = 25 \quad \therefore \quad a = 2.5$

The acceleration of each of the blocks is 2.5 ms^{-2}

Question 8: A

This problem can be solved by considering each set of resistors and the internal resistance separately. The combined resistance of the two 10 Ω resistors in parallel is given by:

$\frac{1}{R_{total}} = \frac{1}{10} + \frac{1}{10} + \frac{1}{10} \quad \therefore \quad R_{total} = 5.0\ \Omega$

The total resistance in the circuit, not including the internal resistance of the battery will be therefore

5.0Ω + 3.0Ω = 8.0Ω

The total resistance in the circuit including the internal resistance of the battery is:

$\frac{Emf}{I} = \frac{20V}{2.0A} = 10\Omega$

The internal resistance of the battery is 10Ω − 8.0Ω = 2.0Ω

Question 9: C

This problem can be solved by taking moments about the support. The forces on the telescope are the weights of each of the three tubes, each acting at the midpoint of the tubes, because the tubes are uniform. Each tube has a length of 20 cm.

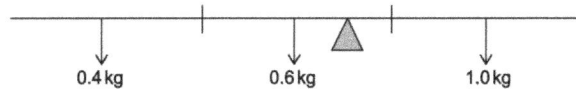

There is a normal contact force at the support, vertically upwards, but this calculation is not required, and the problem can be solved by taking moments about the support. For the system to be in equilibrium, the support must be closer to the centre of the 1.0 kg tube than the centre of the 0.4 kg tube, as shown in the diagram. The support is at an unknown distance x cm from the centre of the 0.6 kg tube, weight = 6 N

The moment of the 4N weight is 4 × (20 + x) anticlockwise

The moment of the 6N weight is 6x anticlockwise

The moment of the 10N weight is 10 × (20 − x) clockwise.

For the telescope to remain horizontal, anticlockwise and clockwise moments must balance:

$4(20 + x) + 6x = 10(20 + x) \qquad \therefore \qquad x = 6\,cm$

The support is 20 cm + 10 cm + 6 cm = 36 cm from the lighter end.

Question 10: C

This problem can be solved by first considering one cable individually. Calculate the force applied by one of the cables by combining the equation for Young modulus, $E = \sigma/\varepsilon$, and the equation defining stress, $\sigma = F/A$, to give, $F = \varepsilon E A = 0.0025 \times 2.0 \times 10^{11} \times 2.0 \times 10^{-4} = 1.0 \times 10^{5}N$

The four cables are connected in parallel so the total force on the boulder is 4.0×10^5 N = 400 kN. The boulder is moving at constant velocity so the resultant force is zero and the frictional force is equal to the total pulling force. Therefore, the magnitude of the frictional force is also 400 kN.

Power transfer is given by P = F_v = 400000 N × 0.20 ms^{-1} = 80000 W = 80 kW

Question 11: E

This problem can be solved by considering the phase difference as a fraction of a full cycle of vibration. One cycle of vibration causes a phase change of 2π. The phase difference between the vibrations at the two buildings is π/3, which is equivalent to 1/6 of a cycle.

Therefore the 1.0 km distance between the buildings must be 1/6 of a wavelength and so the wavelength of the seismic wave is 6.0 km. The frequency of the seismic wave $f = \frac{1}{period} = \frac{1}{2.0}$. The speed of the seismic wave is given by: $v = f\lambda = \frac{6.0}{2.0} = 3.0\,kms^{-1}$

Question 12: B

This problem can be solved by considering the situation when the object is falling at an acceleration of 0.5 g with velocity v_0.

Weight of object = mg

Drag force acting on object $F = kv_0^n$

Resultant force on object downwards = $mg - kv_0^n = ma = 0.5\,mg$

Rearranging: $\frac{1}{2}mg = kv_0^n$ $\therefore$ $k = \frac{mg}{2v_0^n}$

Now consider the object falling at terminal speed v_T.

$kv_t^n = mg$ $\therefore$ $\frac{mg}{2v_0^n}v_T^n = mg$ $\therefore$ $2v_0^n = v_T^n$

Raising both sides to the power (1/n)

$2^{\frac{1}{n}}v_0 = v_T$

Question 13: E

This problem can be solved by considering the ratio of the forces on and extensions of the springs. The weight of each mass is 0.10 kg ×10 Nkg −1 = 1.0 N. The weight of the springs can be ignored because their masses are negligible. The force on the upper spring is 2.0 N, and the force on the lower spring is 1.0 N. As they are subjected to different forces, the two springs will have different extensions.

Force, F, and extension, x, are related by F = kx and the spring constant, k, is the same for both springs.

Therefore, since the ratio of the forces on the upper and lower springs is 2:1, their extensions must also be in the ratio 2:1. The sum of the extensions of the two springs is 30.0 cm − (2 × 12.0 cm) = 6.0 cm.

Splitting this in the ratio 2:1 gives 4.0 cm: 2.0 cm for the extensions of the upper and lower springs, respectively.

The spring constant kk can be calculated by considering either of the springs:

Upper spring:

$k = \frac{F}{x} = \frac{2.0\ \text{N}}{4.0\ \text{cm}} = 0.50\ \text{Ncm}^{-1}$

Lower spring:

$k = \frac{F}{x} = \frac{1.0\ \text{N}}{2.0\ \text{cm}} = 0.50\ \text{Ncm}^{-1}$

Question 14: H

This problem can be solved by using trigonometry to find the angle θ (shown in the diagram) and then using the law of refraction to find the refractive index n of the block.

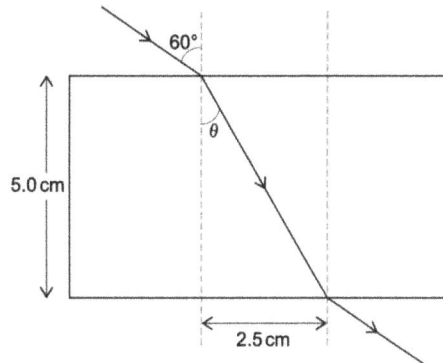

Using Pythagoras, $\sin\theta = \dfrac{2.5}{\sqrt{2.5^2+5.0^2}} = \dfrac{2.5}{\sqrt{\frac{25}{4}+25}} = \dfrac{2.5}{\sqrt{125/4}} = \dfrac{1}{\sqrt{5}}$

Since $\sin 60° = \dfrac{\sqrt{3}}{2}$ $\therefore$ $n = \dfrac{\sin 60°}{\sin\theta} = \dfrac{\sqrt{15}}{2}$

Question 15: B

This problem can be solved by considering the change in kinetic energy and the conservation of momentum. Let the speeds (in ms⁻¹) of the spheres after the collision be u and v respectively, both in the direction of motion of the 3 kg sphere before the collision.

Using the kinetic energy $\mathrm{KE} = \dfrac{1}{2}mv^2$, the total kinetic energy of the spheres are:

After the collision: $\mathrm{KE} = \dfrac{1}{2}3u^2 + \dfrac{1}{2}1v^2$

Before the collision: $\mathrm{KE} = \dfrac{1}{2}3 \times 2^2 + \dfrac{1}{2}1 \times 6^2$

The KE after the collision is 75% of the KE before, so: $\dfrac{1}{2}3u^2 + \dfrac{1}{2}1v^2 = \dfrac{3}{4}(\dfrac{1}{2}3 \times 2^2 + \dfrac{1}{2}1 \times 6^2)$

This simplifies to: $3u^2 + v^2 = \dfrac{3}{4} \times 48 = 36$

In the collision, linear momentum must be conserved, so taking into account the directions:

$3u + 1v = 2 \times 3 - 1 \times 6$ $\therefore$ $3u + v = 0$ $\therefore$ $v = -3u$

Substituting into the equation $3u^2 + v^2 = 36$ $\therefore$ $3u^2 + (-3u)^2 = 36$ $\therefore$ $v = -3\sqrt{3}$ or $3\sqrt{3}$

The speed of the 1kg sphere after the collision is $3\sqrt{3}$ ms⁻¹

Question 16: B

This problem can be solved by considering the displacements of point mass. Assume that positive vector quantities are in the direction from P towards Q time t = 0. The point of zero displacement is taken to be the position of P at time t = 0.

At time t, mass P has displacement: $s = \frac{1}{2}at^2 = \frac{1}{2} \times 6 \times t^2 = 3t^2$

The initial displacement of Q is +60m. At time t, mass Q has displacement:

$$s = 60 + ut + \frac{1}{2}at^2 = 60 - 14t + t^2$$

Masses P and Q meet when they have the same displacement at the same time, therefore:

$$3t^2 = 60 - 14t + t^2 \quad \therefore \quad 2t^2 + 14t - 60 = 0 \quad \therefore \quad (t-3)(t+10) = 0$$

Therefore, P and Q will meet when t – 3 = 0, so t = 3.0 s

Question 17: B

This problem can be solved by trigonometry and taking moments. sin(angle SPO) = 2/4, so angle SPO = 30°

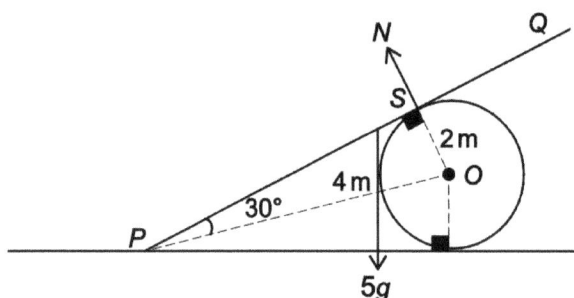

Using Pythagoras' Theorem, $PS = \sqrt{16-4} = \sqrt{12} = 2\sqrt{3}$

Taking moments about P: $2\sqrt{3}N = 2\cos 60° \times 5g = 5g \quad \therefore \quad N = \frac{5g}{2\sqrt{3}} = \frac{5\sqrt{3}g}{6}$

Question 18: A

This problem can be solved using Kirchhoff's laws and solving the resulting simultaneous equations. Label currents I_1, I_2 and I_3 as shown in the diagram.

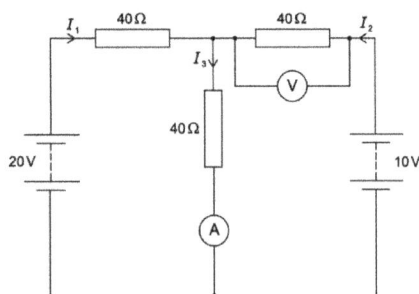

Using Kirchhoff's laws, we can derive the following equations:

$I_3 = I_1 + I_2$ (1) $\quad \therefore \quad 20 = 40I_1 = 40I_2$ (2) $\quad \therefore$

$10 = 40I_2 + 40I_3$ (3)

From (1) $I_1 = I_3 - I_2$, so substitute for I_1 into (2):

$20 = 80I_3 - 40I_2$

152

Add this to (3): $30 = 120I_3$, giving the ammeter reading I_3: $I_3 = \frac{30}{120} = 0.25A$

Substitute for I_3 into (3): $10 = 40I_2 + 10$ ∴ $I_2 = 0A$

The voltmeter is therefore connected across a resistor that has no current, so the voltage across it is zero, and the voltmeter reading is 0 V.

Question 19: E

This problem can be solved by considering both the hydrostatic pressure and the pressure due to the force of the water falling. The water falls 45 m. Its velocity just before hitting the rock can be calculated from conservation of energy:

change in kinetic energy = change in gravitational potential energy

$$\tfrac{1}{2}mv^2 = mgh \quad \therefore \quad v = \sqrt{2gh} = \sqrt{2 \times 10 \times 45} = \sqrt{900} = 30\,\mathrm{ms}^{-1}$$

The pressure on the rock surface arises from two things. Firstly, the hydrostatic pressure from water resting on the surface, and secondly the pressure exerted on the surface as the water stops, equal to its rate of change of momentum divided by the area of the rock.

hydrostatic pressure $P_1 = \rho gh = 1000 \times 10 \times 0.050 = 500$ Pa

pressure $P_2 = \frac{F}{A} = \frac{\Delta(mv)}{A\Delta t} = \frac{40 \times 30}{2.0} = 600$ Pa

Total pressure exerted on the rock is P = P_1 + P_2 = 500 + 600 = 1100 Pa.

Question 20: A

This problem can be solved by considering the circuit as a potential divider circuit and considering possible values of the voltages.

Component X and the 200 Ω resistor are in parallel. For the pd across component X to be greater than 2.0 V, the pd across the 200 Ω resistor must be greater than 2.0 V.

The battery supplies 12 V, so the pd across the thermistor must be less than 10 V.

Using this information and the equation $\frac{V_1}{V_2} = \frac{R_1}{R_2}$, the ratio of the resistances of the thermistor and the 200 Ω resistor must be less than the ratio of the voltages across them:

$$\frac{R}{200} < \frac{10}{2} \quad \therefore \quad R < 5 \times 200 \quad \therefore \quad R < 1000\,\Omega\,.$$

Therefore, $R_0 b^{-\mu T} < 1000$ ∴ $b^{-\mu T} < \frac{1000}{R_0 1000}$ ∴ $-\mu T < \log_b\left(\frac{1000}{R_0}\right)$ ∴ $T > (\tfrac{1}{\mu})(\log_b R_0 - \log_b 1000)$

END OF PAPER

2016

Section 1

Question 1: G

Rearranging the expression gives $\frac{x}{2} < 14$. Multiply by 2 to get $x < 28$.

Question 2: D

An alpha decay (emission of two protons and two neutrons) reduces the mass number by four and the proton number by two. Beta decay leaves the mass number unchanged and increases the proton number by one as a neutron is converted to a proton and an electron. Therefore, there has been 1 alpha decay and 2 beta decays.

Question 3: B

Expand the expression given:

$$\Rightarrow \left(\sqrt{3} - \sqrt{2}\right)\left(\sqrt{3} - \sqrt{2}\right) = 3 - 2\sqrt{6} + 2 = 5 - 2\sqrt{6}.$$

This is the same as option B.

Question 4: F

The kinetic energy of an object is proportional to the square of the velocity, so it is not a straight line. The potential energy of an object being lifted increases linearly with height as $\text{GPE} = mgh = 20\text{kg} \times 10\text{ms}^{-2} \times 2\text{m} = 400\text{J}$. Therefore, the graph cannot represent option 2.

If the resultant force is constant, the acceleration of the object is constant, so the velocity increases linearly with time. $a = \frac{F}{m} = 100/20 = 5\text{ms}^{-2}$. Work done is given by $W = Fd$. For a constant speed, d increases linearly with time. $W = 5\text{N} \times 2\text{m} = 10\text{J}$ so, the graph could represent number 4 as well.

Question 5: C

It is given that $\frac{Q}{R} = \frac{5}{2}$ and $\frac{R}{S} = \frac{3}{10}$. Multiply the first equation by $\frac{R}{S}$. This gives $\frac{Q}{S} = 15/20 = 3/4$.

Therefore, $Q:S = 3:4$.

Question 6: C

We require the total number of nucleons to be conserved. In the first diagram, there are initially 236 nucleons: 235 in the U nucleus and 1 proton. After fission, there are $144 + 89 + 2 = 235$ – therefore, nucleon number hasn't been conserved. In diagram 2, there are 236 nucleons initially and $137 + 96 + 3 = 236$ nucleons so this process is possible. In diagram 3, there are 236 at first and $145 + 87 + 3 = 235$ afterwards so this process can be discounted.

Question 7: E

If the mean age of 20 people is 28, the sum of all their ages is $20 \times 28 = 560$. The new mean age is 30 for 22 people, so the sum of the ages after the members join is $22 \times 30 = 660$. The two new members have added 100 years between them, so the mean age of the two is 50 years.

Question 8: D

The current through the circuit is given by $I = V/R = 24/(5 + R_V)$, where R_V is the resistance of the variable resistor. The power dissipated in the 5Ω resistor is $P = I^2R = \frac{24^2 \times 5}{(5 + R_V)^2}$.

To maximise the power P, we want to minimise the denominator and minimise R_V. Therefore, we choose $R_V = 3\Omega$ and insert this giving $P = \frac{24^2 \times 5}{8^2} = 3^2 \times 5 = 45W$.

Question 9: C

If the scanner decreases by 20% each year, its new value is 80% of its previous value. After two year its new value is $15000 \times 0.8 \times 0.8 = £9600$. Therefore, the value has decreased by $15000 - 9600 = £5400$.

Question 10: D

As $P = kR^2T^4$, an increase of R by a factor of 100 will increase P by a factor of $100^2 = 10^4$. The temperature decreasing by 50% will cause the power to become $\left(\frac{1}{2}\right)^4 = \frac{1}{16}$ of its previous value. Therefore, the new power will be $\frac{4.0}{16} \times 10^{26} \times 10^4 = 0.25 \times 10^{30} = 2.5 \times 10^{29}$ W.

Question 11: B

The interior angle at B is 30 degrees and the one at A is 60 degrees. This means the angle at C must be a right angle as $180 - 60 - 30 = 90$. Therefore, we can use $\sin\theta = \frac{O}{H} = \frac{BC}{AB} = \sin 60 = \sqrt{3}/2$. Rearranging this gives $BC = 2\sqrt{3}$km.

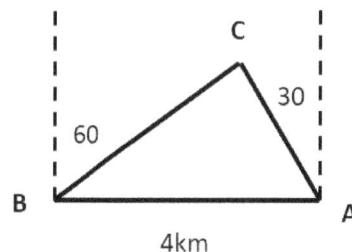

Question 12: E

The wave is transverse, so the particles of the medium oscillate perpendicular to the motion of the wave. In one minute, there are $5 \times 60 = 300$ oscillations. The amplitude measures from the equilibrium position to a maximum. In one full oscillation, a particle travels 4 times the amplitude i.e. 12 cm. In one minute, a particle of the medium will travel $300 \times 12 = 3600$ cm.

Question 13: F

For $x \propto \frac{1}{\sqrt{y}}$, $x = \frac{k}{\sqrt{y}}$. Insert $x = 8$, $y = 9$ which gives $k = 24$. Rearranging the equation gives $\sqrt{y} = k/x = 4$. Therefore, $y = 16$.

Question 14: H

The tension is constant throughout the rope. Consider first the forces acting on the mass:

$\Rightarrow F = ma = 5 \times 0.8 = 4N = T - mg$.

Therefore, the tension is $T = 54N$. The tension acts downwards on both sides of the pulley so the force on the coupling is $2T = 108$ N.

Question 15: E

Split the trapezium into a rectangle and a triangle with areas of $x(x-1)$ and $\frac{1}{2}x \times 6 = 3x$. i.e. $120 = x(x-1) + 3x = x^2 + 2x$.

Rearranging gives $x^2 + 2x - 120 = 0$. The factors of 120 that differ by 2 are 12 and 10 so $(x + 12)(x - 10) = 0$. Discard the negative solution as negative lengths are not possible. Therefore $x = 10$cm and RS has length 15cm.

Question 16: F

The current through the circuit is given by $I = V/R = 6/15 = 0.4$A. The power dissipated by the heater is $180/(3 \times 60) = 1$ W. Now use $P = IV$ to find $V = 2.5$V. Current is defined as the charge per unit time so the charge passing through the heater in 3 minutes is $0.4 \times 180 = 72$ C.

Question 17: B

Multiply both sides by the denominator of the fraction to give:

$$(3b^2 - 1)a = b^2 + 2.$$

Expanding the brackets:

$$\Rightarrow 3b^2 a - a = b^2 + 2$$

Then move all terms containing b to the LHS:

$$\Rightarrow b^2(3a - 1) = 2 + a$$

$$b^2 = \frac{2 + a}{3a - 1}$$

$$b = \pm\sqrt{\frac{2+a}{3a-1}}.$$

Question 18: F

The weight of the block is 30 N, so the mass is $m = 30/10 = 3$ kg. The volume of the cube without the hole is 10^3cm^3. The volume removed is $5^2 \times 10 = 250 \text{cm}^3$. The total volume is $1000 - 250 = 750 \text{cm}^3$. The density is $\rho = m/V = 3000/750 = 4 \text{ gcm}^{-3}$.

Question 19: C

The volume of a cylinder is $V = \pi r^2 h$. First, we need to find the radius. The circumference of the circle is 5m so $r = \frac{5}{2\pi}$. Using this expression in the volume gives:

$$\Rightarrow V = \pi \times \left(\frac{5}{2\pi}\right)^2 \times 10 = \frac{250}{4\pi} = \frac{125}{2\pi}.$$

Question 20: D

The balls R and S are at the same temperature as their surrounding so cannot lose heat by convection. P and Q are hotter than their surroundings, so both lose heat via convection. Dull surfaces are better at absorbing and emitting radiation than shiny surfaces. Also, hotter objects emit more thermal radiation than cooler objects, so S emits the most radiation.

Question 21: A

$$4 + \frac{4 - x^2}{x^2 - 2x} = \frac{4x^2 - 8x + 4 - x^2}{x^2 - 2x}$$

$$\Rightarrow \frac{3x^2 - 8x + 4}{x(x - 2)} = \frac{(3x - 2)(x - 2)}{x(x - 2)} = \frac{3x - 2}{x}$$

Therefore, the answer is $3 - \frac{2}{x}$.

Question 22: A

The distance travelled is given by the area under the velocity-time graph. In the first 20 seconds, the object travels $\frac{1}{2} \times 8 \times 20 = 80$ m. In the following 10 seconds, the object travels $\frac{1}{2} \times 2 \times 10 = 10$ m in the opposite direction. The total distance travelled is 90 m and the object is 70 m from its starting position after 30 s. The average speed does not depend on the direction of travel as speed is a scalar quantity. Therefore, the average speed is $90/30 = 3 \text{ ms}^{-1}$.

Question 23: D

There are 120 students: 46 girls and 74 boys. $\frac{2}{3} \times 36 = 24$ boys chose tennis and 12 girls chose tennis. 25 girls chose to swim and the $46 - 25 - 12 = 9$ remaining girls did archery. 27 students did archery - 9 girls and 18 boys. The remaining 32 boys all chose swimming. Therefore, if a boy is chosen at random, the chance he swims is $\frac{32}{74} = \frac{16}{37}$.

Question 24: G

The mass of each component is given by $m = \rho V$, so the total mass per unit volume is $0.9X + 0.1Y$. The fraction of the mass that is tin is $\frac{0.1Y}{0.9X + 0.1Y} = \frac{Y}{9X + Y}$. Therefore, the percentage mass of tin is $\frac{Y}{9X + Y} \times 100$.

Question 25: D

$$\Rightarrow \frac{9^{2n+1} \times 3^{4-3n}}{27^{2-n}} = 3^{4n+2} \times 3^{4-3n} \times 3^{3n-6}$$

This equals $3^{6+n+3n-6} = 3^{4n}$.

Question 26: D

The initial and final momentum must be the same i.e. 0. The mass of the alpha particle is $\frac{4}{234}$ that of the thorium. Therefore $v_\alpha = \frac{234}{4} v_{Th}$. The kinetic energy of the Thorium is given by $E_{Th} = \frac{1}{2} m_{Th} v_{Th}^2$. The kinetic energy of the alpha particle is $E_\alpha = \frac{1}{2} \times \frac{4m_{Th}}{234} \times \left(\frac{234 v_{Th}}{4}\right)^2 = \frac{234}{4} E_{Th}$. Summing these gives $E = E_\alpha + \frac{4}{234} E_\alpha$. This yields $E_\alpha = \frac{234E}{238}$.

Question 27: E

The external angle of a regular polygon with n sides is $\frac{360}{n}$. This corresponds to the angle RQT. As the triangle is isosceles, the angle RTQ is the same as the angle RQT. Therefore, the angle $x = 180 - \frac{2 \times 360}{n}$. Rearranging this gives $\frac{720}{n} = 180 - x$. Further manipulation yields $n = \frac{720}{180 - x}$.

Question 28: G

For the sound to reflect from the left building and back to the student, the wave must travel 96 m. This takes 0.3 s. For the sound wave to reflect from the right building, the sound wave would have to travel 160 m, taking 0.5 s. The first click occurs at 0 seconds, then another must occur 0.3 s later and 0.2 s after that. The lowest frequency that would produce clicks with both of these intervals is 10Hz i.e. 1 click every 0.1 s.

Question 29: C

Substitute in x = 2: $8 + 4p + 2q + p^2 = 0$.

With x = 1: $1 + p + q + p^2 = -3.5$.

Multiply the second equation by 2 and subtract it from the first:

$\Rightarrow 6 + 2p - p^2 = 7$

$\Rightarrow p^2 - 2p + 1 = 0$

$\Rightarrow (p - 1)^2 = 0$.

Therefore, the only solution is $p = 1$.

Question 30: C

Newton's third law: For every action there is an equal and opposite reaction. This only applies to the internal forces in this diagram as the particles of the air have not been considered. The only internal forces are N and P which are equal and opposite. The answer is 4 only.

Question 31: C

The square has a side length 6. The centre of the square is $(-2, 3)$. The diameter of the circle is also 6 so it has a radius 3. The equation for a circle radius, r, and centred at (x_0, y_0), has the equation $(x - x_0)^2 + (y - y_0)^2 = r^2$. Inserting the values gives:

$\Rightarrow (x + 2)^2 + (y - 3)^2 = 9$

$\Rightarrow x^2 + 4x + y^2 - 6y + 4 = 0$.

Question 32: C

The forces acting on the mass are its weight downwards and the tension upwards: $F = T - mg = ma = 1600N$. Rearranging gives $T = 1600 + 8000 = 9600N$.

Question 33: C

For the series: $a_1 = 8$, $a_5 = 2$. A geometric series means consecutive terms differ by the same factor. $8 \times r^4 = 2$. $r = \pm (1/4)^{1/4} = \pm \frac{1}{\sqrt{2}}$. As the sixth term is real and positive, we can discard the imaginary and negative roots. So $r = \frac{1}{\sqrt{2}}$. Using this gives the series sum as $\frac{8}{1 - \frac{1}{\sqrt{2}}} = \frac{8\sqrt{2}}{\sqrt{2} - 1} = \frac{8\sqrt{2}(\sqrt{2} + 1)}{2 - 1} = 8(2 + \sqrt{2})$.

Question 34: A

The vertical component of the force applied to the trolley is $50 \sin(37) = 50 \times 0.6 = 30$N. The horizontal component is $50 \cos(37) = 40$N. The total force downwards is the sum of the vertical component of the force applied by the shopper and the trolley's weight i.e. $30 + 350 = 380$ N. This is balanced by the reaction force from the surface. The work done by the pushing force is: $W = Fd = 40$ N $\times 15$ m $= 600$ J.

Question 35: A

Tangents to a circle are at 90 degrees to the radius of the circle at that point. Thereby we can form two right-angled triangles with points at P, C and where the two tangents touch the circle. The hypotenuse of these triangles is 20cm long and the radii are 10cm. Using Pythagoras's theorem, the distance from P to the point on the circle is $\sqrt{20^2 - 10^2} = 10\sqrt{3}$cm. Now consider the angle formed by the line PC and from C to the tangent line: $\cos \theta = \frac{10}{20} = 0.5$. Therefore, $\theta = 60°$. The total area of the two right angled triangles is $2 \times \frac{1}{2} \times 10\sqrt{3} \times 10 = 100\sqrt{3}$cm^2. The removed area is $\frac{1}{3}$ of the area of the circle i.e. $\frac{100\pi}{3}$. Subtracting this area gives $\frac{100}{3}\left(3\sqrt{3} - \pi\right)$.

Question 36: H

The moments of each object are given by $F \times d$. The total moment must be zero to remain in balance. The resultant moment of the boy and the woman is $g(1.2 \times 35 - 0.8 \times 60) = -60$ Nm with positive moments causing a clockwise rotation. The moment of the plank must be 60 Nm to balance this so $d = \frac{60 \text{Nm}}{15 g} = 0.4$m to the right of the pivot. The total force between the pivot and the plank is the sum of the weights i.e. $F = (35 + 15 + 60)g = 1100$N.

Question 37: D

Use $\tan \theta = \frac{\cos \theta}{\sin \theta}$ and multiply through by $\cos \theta$. This gives $7 \cos^2 \theta - 3 \sin^2 \theta = \cos \theta$. Now use the identity $\sin^2 \theta = 1 - \cos^2 \theta$. This gives $10 \cos^2 \theta - \cos \theta - 3 = 0$. Substitute $x = \cos \theta$ and factorise to find $(5x - 3)(2x + 1) = 0$. The two solutions are $\cos \theta = \frac{3}{5}$ and $\cos \theta = -\frac{1}{2}$.

Question 38: B

The initial energy is $E = \frac{1}{2}mv^2 + mgh = 100 \times 25 + 16000 = 18{,}500$ J. At point Y, the new energy is $E_Y = 100 \times 81 + 4000 = 12{,}100$J. The difference between these two must be the work done against resistive forces. This is 6400J.

Question 39: G

Rearrange this equation to get $3x^2 - (a + 2)x + 3 = 0$. To find the number of distinct real roots, the discriminant is required: $b^2 - 4ac$. For two real roots, this must be greater than zero. $(a + 2)^2 - 36 > 0$. This gives $(a + 2)^2 > 36$. This happens when $a + 2 > 6$ or when $a + 2 < -6$. Therefore, we require $a > 4$ or $a < -8$.

Question 40: B

The first must be untrue. The weight of the block acts vertically downwards so has no horizontal component. For the block to be stationary, the forces must balance in all directions so frictional forces are required to balance the horizontal force P. The second is true as this allows the frictional force to balance P horizontally. The third is not true as the horizontal forces could not balance. The moment of force P around the edge in contact is $F \times l$ where l is the perpendicular distance. The distance between P and the table surface is $\frac{1}{2}(1 + \sqrt{3})d$ so the moment is greater than $P \times d$.

Question 41: D

A line perpendicular to one of gradient m has a gradient $\frac{-1}{m}$. At $y = 0$, the first line satisfies $x = \frac{-3}{m}$. The second line crosses the x axis at $x = \frac{-2}{p} = 2m$. Therefore, $5 = 2m + \frac{3}{m}$ which gives $2m^2 - 5m + 3 = 0$. Factorising this gives $(2m - 3)(m - 1) = 0$. Two possible solution $m = 1$ and $m = \frac{3}{2}$. Discard the first solution as $m > 1$. This gives $p = -2/3$. $m + p = \frac{3}{2} - \frac{2}{3} = \frac{5}{6}$.

Question 42: A

The gravitational potential energy is given by mgh, with h the height of the centre of mass of the block. The first arrangement has GPE $= mg\left(\frac{b}{2} + \frac{3b}{2}\right) = 2mgb$. In the second arrangement, GPE $= mg\left(\frac{a}{2} + \frac{3a}{2}\right) = 2mga$. Therefore, the expression is $2mg(a - b)$.

Question 43: G

$f(x)$ is an increasing function when its derivative with respect to x is greater than zero. $\frac{df}{dx} = 3x^2 - a^2 > 0$. Rearranging gives $x^2 > \frac{a^2}{3}$. This is true when $x < \frac{a}{\sqrt{3}}$ or when $x < -\frac{a}{\sqrt{3}}$.

Question 44: G

Initially, the object has kinetic energy $\frac{1}{2}m \times 8^2 = 32m$. When it has $v = 2ms^{-1}$, the kinetic energy is $\frac{1}{2}m \times 2^2 = 2m$. Therefore, the object has gained $30m = mgh$ of gravitational potential energy. This gives $h = 3$. Use $s = ut + \frac{1}{2}at^2 = 8t - \frac{1}{2}gt^2$. Inserting $s = 3m$ gives $5t^2 - 8t + 3 = 0$. Factorising leads to $(5t - 3)(t - 1) = 0$. Therefore, the two possible solutions are $t = 0.6$ s and $t = 1$ s. The smaller value of t is as the object is on the way up and the larger value as the object is falling. Therefore, G is the correct answer.

Question 45: E

Translation by the vector (4, 3) gives the new equation: $y - 3 = (x - 4)^2$.

To reflect in $y = -1$, isolate $y + 1$ on the LHS i.e. $y + 1 = 4 + (x - 4)^2$. A reflection in $y = -1$, causes the sign of the RHS to be changed, giving $y + 1 = -4 - (x - 4)^2$. Rearranging this yields: $y = -5 - (x - 4)^2$.

Question 46: E

Momentum must be conserved in the collision. The initial momentum is $4 \times 10 = 40 \text{ kgms}^{-1}$. If P has velocity v after the collision, the final momentum is $4v + 20$. Therefore, $4v = 20$. $v = 5\text{ms}^{-1}$. This is to the right. The initial energy is $\frac{1}{2} \times 4 \times 100 = 200$ J. The kinetic energy after the collision is $\frac{1}{2} \times 2 \times 100 + \frac{1}{2} \times 4 \times 25 = 100 + 50 = 150$ J. The energy lost is 50 J.

Question 47: C

$\Rightarrow 2x^4 - 9x^2 + 4 > 0$

Substitute in $y = x^2$ into the expression. This gives $2y^2 - 9y + 4 > 0$. Factorise this to get $(2y - 1)(y - 4) > 0$. Therefore, the two roots are at $y = \frac{1}{2}$ and $y = 4$. For y large, positive or negative, the quadratic will be greater than zero. Therefore, $x^2 < \frac{1}{2}$ or $x^2 > 4$ satisifies the original inequality.

For $x^2 < \frac{1}{2}$, require $-\frac{1}{\sqrt{2}} < x < \frac{1}{\sqrt{2}}$. For $x^2 > 4$, require $x < -2$ or $x > 2$.

Question 48: A

Add the two vectors together at 90 degrees to each other. The magnitude of the resultant force is $\sqrt{9^2 + 12^2} = 15$N. The maximum friction is $F = \mu R = 0.25 \times 20 = 5$N. As the object moves, the net force is $F = 10$ N $= 2 \times a$. This gives $a = 5 \text{ ms}^{-2}$.

Question 49: B

The distance parallel to the y-axis between the lines $y = 10$ and $y = 4x^3 - 12x^2 - 36x - 15$ is $f(x) = 25 - 4x^3 + 12x^2 + 36x$. We want to find the minimum of f(x) for negative values of x.

$$\frac{df}{dx} = -12x^2 + 24x + 36 = -12(x^2 - 2x - 3) = -12(x - 3)(x + 1)$$

There are two stationary points: $x = 3$ and $x = -1$. The highest point of the cubic, for negative x, is at $x = -1$, where $f = 25 + 4 + 12 - 36 = 5$.

Question 50: C

Consider first the 30 kg mass. This is accelerating at 2.5 ms^{-2} so the net force is $F = 75\text{N} = mg - T = 300 - T$. This gives $T = 225$N. Now consider the 20 kg mass. The forces acting parallel to the plane are the tension up the plane, $mg \sin \theta$ down the plane and friction down the plane. $225 - 100 - F_{\text{Friction}} = ma = 50$. Rearranging gives $F_{\text{Friction}} = 75$N.

Question 51: A

Expand the expression as follows:

$$\Rightarrow x = 3(y - 1)^2 + 4 = 3(y^2 - 2y + 1) + 4 = 3y^2 - 6y + 7.$$

The two lines intersect at $7 = 3y^2 - 6y + 7$ i.e. $3y^2 - 6y = 3y(y - 2) = 0$. Integrate, from $y = 0$ to $y = 2$, the difference $f(y) = 3y^2 - 6y$. $\int_0^2 dy(3y^2 - 6y) = [y^3 - 3y^2]_0^2 = 8 - 12 = -4$. We only need the magnitude of the area, so we can discard the negative.

Question 52: E

Initially, the momentum is 4000×7425. Let the mass of ejected fuel be m. After the fuel is ejected, momentum conservation requires:

$$\Rightarrow 4000 \times 7425 = 7500(4000 - m) + 6000m$$

$$4 \times 7425 = 7.5(4000 - m) + 6m$$

$$4 \times 7425 = 30000 - 1.5m$$

$$\Rightarrow 1.5m = 30000 - 29700 = 300$$

$$\Rightarrow m = 200 \text{ kg}.$$

Question 53: C

To find the number of roots of the equation consider the locations of the turning points:

$$\Rightarrow \frac{dy}{dx} = 12x^3 - 12x^2 - 24x = 12x(x^2 - x - 2) = 12x(x - 2)(x + 1)$$

Therefore, stationary points are at $x = -1, 0, 2$. Define $f = 3x^4 - 4x^3 - 12x^2 + 20 - k$.

$$f(-1) = 3 + 4 - 12 + 20 - k = 15 - k.$$

$$f(0) = 20 - k.$$

$$f(2) = 48 - 32 - 48 + 20 - k = -12 - k.$$

For 4 distinct roots, we require stationary points to alternate above and below the x-axis. Need $15 - k < 0$, $20 - k > 0$ and $-12 - k < 0$. i.e. $k > 15$, $k < 20$ and $k > -12$. The required conditions are $15 < k < 20$.

Question 54: B

For the object to be stationary, the forces must cancel in all directions. The string must be at an angle such that the horizontal component of the tension is 30 N. Let the angle the string makes with the vertical be θ. The vertical component of the tension must balance the weight of the object i.e. 40 N. The magnitude of the tension in the string is $\sqrt{30^2 + 40^2} = 50$N. This gives $\sin\theta = 0.6$ and $\cos\theta = 0.8$. The vertical distance from the ceiling to the block is now $0.35\cos\theta = 0.28$m. Therefore, the change in height is 0.07m. Therefore, the change in GPE is $mg\Delta h = 40 \times 0.07 = 2.8$ J.

END OF SECTION

Section 2

Question 1a: A

The relation between the tension in the wire and its extension is $F = kx$ where k is the stiffness.

This means that the largest value of $k = \frac{F}{x}$ will correspond to the largest gradient on the graph i.e. S1.

Question 1b: B

The strain of a wire is defined as $\epsilon = \frac{x}{L}$, where x is the extension of the wire and L is its natural length. The samples have natural length 100 mm; so at 2% strain, the extension is 2 mm. A sample that obeys Hooke's Law has a tension that is directly proportional to the extension giving a straight line through the origin. Samples S1, S3, S4 and S5 obey this well beyond 2 mm so the answer is S2.

Question 1c: A

Young's Modulus is $E = \frac{\sigma}{\epsilon} = \frac{FL}{xA}$. The ratio $\frac{F}{x}$ is the stiffness which is the gradient of the graph:

$$\Rightarrow k = \frac{250}{5 \times 10^{-3}} = 5 \times 10^4 \, \mathrm{Nm^{-1}} \cdot \frac{L}{A} = \frac{100 \times 10^{-3}}{\left(5 \times 10^{-3}\right)^2} = 4 \times 10^3 \, \mathrm{m^{-1}}.$$

Multiplying these gives: $E = 20 \times 10^7 \, \mathrm{Pa} = 200 \mathrm{MPa}$.

Question 1d: A

The work done against the tension T is $W = \int dx \, T$. This gives $W = \frac{ax^2}{2} - \frac{bx^3}{3} + c$. Set $c = 0$ as no work has been done when the extension is zero. At $x = 10\mathrm{mm} = 10^{-2}\mathrm{m}$, the work done is $W = \frac{a}{2} \times 10^{-4} - \frac{b}{3} \times 10^{-6} \, \mathrm{Nm}$.

Question 2a: F

Currents are equal for components that are in series whereas voltages are equal across components that are in parallel. Therefore, the voltages across R_2 and R_3 are the same.

Question 2b: A

To find the current, we need to find the total resistance of the circuit. First consider the two resistors in parallel. The total resistance of the resistors in parallel is $\frac{1}{R_T} = \frac{1}{R_2} + \frac{1}{R_3}$. This yields $R_T = \frac{R_2 R_3}{R_2 + R_3}$. The total resistance of the circuit is then the sum of the two

resistances in series: $R_{Tot} = R_1 + R_T = \frac{R_1(R_2 + R_3) + R_2 R_3}{R_2 + R_3} = \frac{R_1 R_2 + R_1 R_3 + R_2 R_3}{R_2 + R_3}$. This gives the current through the voltage source as $I = \frac{V}{R_{Tot}} = \frac{V(R_2 + R_3)}{R_1 R_2 + R_1 R_3 + R_2 R_3}$.

Question 2c: D

We know the current through the first resistor from the previous part which is now $I = \frac{V(R_1+R_3)}{R_1^2+2R_1R_3}$. The voltage across the first resistor is then $V_1 = IR_1 = \frac{V(R_1+R_3)}{R_1+2R_3}$. Subtracting this from the total voltage gives the voltage

across each of the resistors in parallel: $V_3 = \frac{V(R_1+2R_3-R_1-R_3)}{R_1+2R_3} = \frac{VR_3}{R_1+2R_3}$. The power dissipated across R_3 is then $P = \frac{V_3^2}{R_3} = \frac{V^2R_3}{(R_1+2R_3)^2}$.

Question 2d: B

Consider $\frac{1}{P} = \frac{(R_1+2R_3)^2}{V^2R_3}$ and take the derivative with respect to R_3.

$$\Rightarrow \frac{d}{dR_3}\left(\frac{1}{P}\right) = \frac{2\times2\times R_3(R_1+2R_3)-(R_1+2R_3)^2}{R_3^2} = \frac{R_1+2R_3}{R_3^2} \times (4R_3-R_1-2R_3).$$

The derivative is set to zero for a stationary point which only occurs when $(4R_3-R_1-2R_3) = (2R_3-R_1) = 0$. This is when $R_3 = \frac{1}{2}R_1$.

Question 3a: C

1 ns $= 10^{-9}$ s. The speed of light is $c = 3 \times 10^8$ ms^{-1}. This gives a distance of $3 \times 10^8 \times 10^{-9} = 0.3$ m.

Question 3b: F

The speed of light in a material with refractive index n is $v = \frac{c}{n}$. For $n = 1.5, v = 2 \times 10^8$ ms^{-1}. Therefore, the time taken to travel 9 km is:

$$\Rightarrow T = \frac{9\times10^3}{2\times10^8} = 4.5 \times 10^{-5} s = 45 \text{ μs}.$$

Question 3c: A

A larger refractive index leads to a smaller speed of propagation, so the speeds obey the inequality $v_{nom} < v_{blue} < v_{red}$. The faster the light travels, the shorter the propagation time will be,

so the answer is $T_{red} < T_{blue} < T_{nom}$.

Question 4a: E

At point B, the cyclist has kinetic energy $E +$ Mgh sourced from the work done by the cyclist and from gravitational

potential energy. This gives the cyclist a speed of v where $\frac{1}{2}Mv^2 = E + Mgh$. This gives $v = \sqrt{\frac{2E}{M} + 2gh}$. The component down the slope is $\sqrt{2\left(\frac{E}{M} + gh\right)} \times \cos\theta$.

Question 4b: B

Consider the motion of the cyclist perpendicular to the plane. The initial velocity in this direction is $V \sin \theta$ and the component of the gravitational acceleration is $g \cos \theta$.

Now use $s = ut + \frac{1}{2}at^2$ and set $s =$

0 as this is when the bike is in contact with the plane. This gives $0 = Vt \sin \theta -$

$\frac{1}{2}gt^2 \cos \theta$. Factorise this to get $0 = t\left(V \sin \theta - \frac{gt}{2}\cos \theta\right)$. The solution $t =$

0 corresponds to the cyclist leaving the ground. The solution $t = \frac{2V}{g} \tan \theta$ is when the cyclist lands at

C.

Question 4c: C

Consider the components parallel to the slope. There is an acceleration of $g \sin \theta$ down the slope and an initial velocity of $V \cos \theta$. Insert these

values, and the time from the previous part, into $s = ut + \frac{1}{2}at^2$ which gives:

$$\Rightarrow L = V \cos \theta \times \frac{2V}{g} \tan \theta + \frac{1}{2}g \sin \theta \left(\frac{2V}{g} \tan \theta\right)^2 = \frac{2V^2 \sin \theta}{g}\left(1 + tan^2 \theta\right).$$

Use $1 + \tan^2 \theta = 1/\cos^2 \theta$. This gives $L = \frac{2V^2 \sin \theta}{g \cos^2 \theta}$.

Question 4d: D

The maximum perpendicular distance of the cyclist from the plane occurs at the time halfway between it leaving the plane and landing i.e. $t = \frac{V}{g} \tan \theta$. Insert this into $s = Vt \cos \theta + \frac{1}{2}gt^2 \sin \theta$. This gives:

$$\Rightarrow s = V \cos \theta \times \frac{V}{g} \tan \theta + \frac{1}{2}g \sin \theta \times \left(\frac{V}{g} \tan \theta\right)^2 = \frac{V^2}{g} sin \theta + \frac{V^2}{2g} sin \theta \, tan^2 \theta = \frac{V^2}{g} \sin \theta \left(1 + \frac{1}{2}\tan^2 \theta\right).$$

END OF PAPER

2017

Section 1

Question 1: F

The surd $\sqrt{12} = 2\sqrt{3}$. Inserting this into the expression gives $\frac{\left(3\sqrt{3}\right)^2}{\left(\sqrt{3}\right)^2} = 3^2$

$= 9$.

Question 2: B

The car decelerates between $t = 110\ seconds$ and $t = 130\ seconds$. The distance travelled is the area under the graph between these times. This is equal to $20 \times 20 + \frac{1}{2} \times 20 \times 10 = 500$ m.

Question 3: E

Move all terms to the LHS to get $2x^2 + x - 15 \geq 0$. Factorise this equation to get $(2x - 5)(x + 3) \geq 0$. The equation $2x^2 + x - 15 = 0$ has roots at $x = 2.5$ and $x = -3$. When x is a large positive or negative number, x^2 dominates so the inequality is satisfied when x tends to $\pm\infty$. Therefore, $x \leq -3$ or $x \geq 2.5$ solves the inequality.

Question 4: G

When the water is heated its density decreases as its volume increases. This means that the mass in a fixed volume decrease. Statements 2 and 3 are true.

Question 5: B

Add 5 to both sides and divide by 3: $\frac{y+5}{3} = \left(\frac{x}{2} - 1\right)^2$. Square roots give $\pm\sqrt{\frac{y+5}{3}} = \frac{x}{2} - 1$. Add 1 and multiply by 2 to get $x = 2 \pm 2\sqrt{\frac{y+5}{3}}$.

Question 6: D

The car gains potential energy: $mgh = 1200 \times 10 \times 1 = 12$ kJ. All the other energy inputted is lost to the surroundings so the total energy lost is $28 - 12 = 16$ kJ.

Question 7: G

Sam gives $2x + 5y = P$. Lesley: $3x + 2y = Q$. Multiply the first equation by 3 and the second by 2: $6x + 15y = 3P$ and $6x + 4y = 2Q$. Subtract the second equation from the first to eliminate x: $11y = 3P - 2Q$. Rearranging this gives $y = \frac{3P - 2Q}{11}$.

Question 8: D

The time difference corresponds to a difference in distance travelled of $d = 3.0 \times 10^8 \times 4.0 \times 10^{-10} = 1.2 \times 10^{-1}$m. If the source is a distance x to the left of Q the gamma ray must travel $3 - x$ to the left detector and a distance $3 + x$ to the right detector. So, the source is $\frac{d}{2} = 6$ cm away from the midpoint Q.

Question 9: E

$P = kQ^2$. Insert $P = 2$, $Q = 4$ to get $P = Q^2/8$.

Secondly, $Q = \frac{k_2}{R}$. Insert $Q = 2$, $R = 5$ to find $k_2 = 10$. Substitute $Q = \frac{10}{R}$ into the equation for P to get $P = \frac{100}{8R^2} = \frac{25}{2R^2}$.

Question 10: B

In fission, the total number of nucleons and the number of protons must be conserved. These conditions give equations $240 = w + y + z$ and $94 = 54 + x$. Rearranging the first equation gives:

$\Rightarrow z = 240 - (w + y)$.

Question 11: D

Factorising the numerator gives $2 - \frac{(3x-2)(3x+2)}{x(2-3x)} = 2 + \frac{3x+2}{x} = 5 + \frac{2}{x}$.

Question 12: F

The power supplied is $P = IV$. The work done lifting the mass is $mgh = 20 \times 10 \times 6 = 1200$ J. If the system is 80% efficient, then the total input energy is $\frac{1200}{0.8} = 1500$ J. The power is $P = \frac{W}{\Delta t} = \frac{1500}{5} = 300$ W. The current is therefore $I = \frac{300}{12} = 25$A.

Question 13: D

Rewrite the expression using $4 = 2^2$, $8 = 2^3$ and $\sqrt{2} = 2^{1/2}$:

$\Rightarrow 2^{3+2x}2^{2x}2^{-3x} = 2^{5/2}$.

Take logarithms to find:

$3 + 2x + 2x - 3x = 5/2 \Rightarrow x = -0.5$.

Question 14: A

Emission of an alpha particle would cause the mass number to decrease by 4 and the proton number to decrease by 2. Emission of a beta particle leaves the mass number unchanged but increases the proton number by 1. If the alpha particle were emitted first, the mass number would be $P - 4$ and proton number $Q - 2$. Emission of 3 beta particles would then increase the atomic number to $Q - 1$, Q and $Q + 1$. If the beta particles were emitted first, the atomic mass would be P and the atomic number would increase to $Q + 1$, $Q + 2$, $Q + 3$. The only one that does not feature is A.

Question 15: F

There are $3X$ girls and 100 students in total. The number of girls studying Spanish is $35 - Y$. The number of girls studying German is $3X - X - (35 - Y) = 2X + Y - 35$ i.e. Total number of girls minus girls studying French and Spanish. Adding the number of boys doing German gives $2X + 3Y - 35$.

Question 16: A

Density is given by $\rho = M/V$, where the volume $V \propto r^3$. Therefore, $\rho \propto r^{-3}$. The mass can be assumed to be constant as it is dominated by the nucleus of the atom. As the radius of the iron atom is 3×10^4 times that of the nucleus, the density of the atom is $(3.0 \times 10^4)^{-3}$ times that of the nucleus.

Question 17: C

The exterior angle of an n-sided polygon is $\frac{360}{n}$.

Therefore, we can write $\frac{360}{n} - \frac{360}{n+3} = 4$. Multiply by $n(n+3)$ to get:

$\Rightarrow 360(n+3) - 360n = 4n(n+3)$.

Divide by 4 and expand to get: $n^2 + 3n - 270 = 0$. Factorise to find $(n-15)(n+18) = 0$. This gives $n = 15$ or $n = -18$. We can discard the negative value, so the answer is C.

Question 18: E

The wave equation is $v = f\lambda$. Between time t_1 and t_2 there is 3/2 full waves so the period is $\frac{2}{3}(t_2 - t_1)$ and the frequency is $\frac{3}{2(t_2-t_1)}$. The distance between x_2 and x_1 is half a wavelength. Combining these gives $v = \frac{3(x_2-x_1)}{t_2-t_1}$.

Question 19: B

The angle SLR is 40 degrees. The angle LRC is the same as SLR so is 40 degrees. The lengths CL and RC are the same, so the triangle is isosceles. Therefore, the angles LRC and CLR are the same. These are both 40 degrees so the angle LCR is $180 - 80 = 100$. The bearing of L from C is given by $180 - 100 = 80$ degrees.

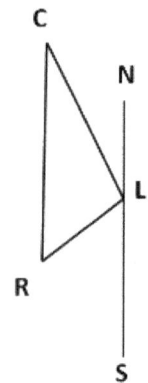

Question 20: C

The power is $P = IV$. This gives a current $I = 150/12 = 50/4 = 12.5A$. Current is defined as charge per unit time so the charge passing through the element is $12.5 \times 20 \times 60 = 250 \times 60 = 15,000C$.

Question 21: B

At 4:40, the minute-hand points towards 8 and the hour hand has moved $\frac{2}{3}$ of the way towards 5. Each hour subtends an angle of 30 degrees. Between 8 and 5 there are 90 degrees. Between 5 and the hour hand is $\frac{1}{3} \times 30 = 10$. Therefore, the angle between the two hands is $90 + 10 = 100$ degrees.

Question 22: C

Momentum must be conserved in the collision. $M_f \times 2 = M_p \times 5$, where M_f is the mass of the freight train and M_p the mass of the passenger train. The mass of the passenger train is $M_p = \frac{2M_f}{5}$ and $M_f = 390 + 210 = 600$. This gives $M_p = 240$. Therefore, there are $240 - 140 = 100$ tonnes of passenger carriages. This is 10 carriages.

Question 23: C

There are $4 + x$ rabbits. The chance that the first rabbit is male is $\frac{x}{x+4}$. Now that one male rabbit has been removed, the chance the second rabbit is male is $\frac{x-1}{x+3}$. This gives $\frac{x(x-1)}{(x+4)(x+3)} = \frac{1}{3}$. Rearranging this gives, $3x(x-1) = (x+4)(x+3)$.

$3x^2 - 3x = x^2 + 7x + 12$

$\Rightarrow 2x^2 - 10x - 12 = 0$

Factorising gives $2(x-6)(x+1) = 0$. Therefore $x = 6$ or $x = -1$. Take the positive value.

Question 24: C

The parachutist has kinetic energy $\frac{1}{2}mv^2 = 72 \times 5^2/2 = 900$ J. The gravitational potential energy lost is $mg\Delta h$. In one second, the parachutist loses 5m of height so the rate of change of GPE is $72 \times 10 \times 5 = 3600$ J s^{-1}. The air resistance's third law pair is the force that acts downwards on the air particles not the gravitational force.

Question 25: F

Firstly, calculate the radius of the circle by considering the distance from the point O to where the corner of the square and the circle meet. This distance is $\sqrt{x^2 + \left(\frac{x}{2}\right)^2} = \frac{x\sqrt{5}}{2}$. The shaded area is half the area of the circle with the area of the square subtracted i.e. $\frac{5\pi x^2}{8} - x^2 = \left(\frac{5\pi-8}{8}\right)x^2$.

Question 26: D

In six hours, X has undergone 2 half-lives and Y has undergone 3 half-lives. Originally, X was $\frac{1}{2}$ of the mixture but it is now $\frac{1}{2} \times \frac{1}{4} = \frac{1}{8}$. Similarly, Y now makes up $\frac{1}{16}$ of the mixture. The remaining is Z i.e. $\frac{13}{16}$.

Question 27: G

The volume of a cylinder is $\pi r^2 h$. The volume of the pipe is $16\pi(5^2 - 4^2) = 16 \times 9\pi \text{cm}^3$. Multiply this by the density to get 1152π g.

Question 28: E

Kinetic energy is given by $\frac{1}{2}mv^2$. The mass of car Y is M and its velocity is v. Y has kinetic energy $\frac{1}{2}Mv^2$. Car X has energy $\frac{1}{2} \times \frac{4}{5}M \times \left(\frac{3}{2}v\right)^2 = \frac{1}{2}Mv^2 \times \frac{9}{5}$. Therefore, X has 1.8 times the kinetic energy of car Y.

Question 29: E

$\Rightarrow 1 - \left(\frac{3+\sqrt{3}}{6-2\sqrt{3}}\right)^2 = 1 - \frac{9+3+6\sqrt{3}}{36+12-24\sqrt{3}} = 1 - \frac{2+\sqrt{3}}{4(2-\sqrt{3})} = 1 - \frac{\left(2+\sqrt{3}\right)^2}{4(2-\sqrt{3})(2+\sqrt{3})} = 1 - \frac{7+4\sqrt{3}}{4} = \frac{-3}{4} - \sqrt{3}.$

Question 30: B

The moment of a force about the point P is given by $F \times d$, with d the perpendicular distance. If the load is moved to the left by 5 m then the moment around P has increased by $4000 \times 5 = 20000\text{Nm}$. To balance this, the counterweight must move such that its moment increases by the same amount but in the opposite direction. Thus, it must move to the right by $\frac{20000}{2000\times10} = 1$ m.

Question 31: C

The first equation can be rearranged to give $\sin x = -\frac{1}{2}$. Similarly, the second equation gives $\cos 2x = \frac{1}{2}$. $\cos 2x = 1 - 2\sin^2 x$ — so, this can be simplified to $\sin^2 x = \frac{1}{4}$ i.e. $\sin x = \pm\frac{1}{2}$. Consider the graph of $\sin x$. There are two points where $\sin x = \frac{1}{2}$ for values of x smaller than that where $\sin x = -\frac{1}{2}$. Therefore, including the point at k, there are 3 values of x that satisfy at least one of these equations.

Question 32: A

Initially the velocity is positive and then decreases as kinetic energy is converted into gravitational potential energy. When it reaches the top of it motion the velocity is zero and then the velocity will become negative. When the ball is caught the velocity should have the same magnitude as when it was thrown as no energy has been lost (neglecting air resistance). The acceleration of the ball is constant, so the velocity will change linearly with time, so the graph should be a straight line.

Question 33: B

Rewrite the equation $3(3^x)^2 - 6(3^x) = 0$. Then divide by 3 and substitute $y = 3^x$ to get $y^2 - 2y = 0$. This has solutions $y(y - 2) = 0$. Therefore, the two possible solutions are $3^x = 2$ and $3^x = 0$. The second requires $x = -\infty$ so can be ignored. Now take the logarithm with base 3 to get $x = \log_3 2$

Question 34: C

If the aircraft is travelling at a constant speed, then the resultant force on the aircraft must be zero.

Question 35: C

First find the radius of the circle. If the angle QOR is $\pi/6$ then the remaining angle is $2\pi - \pi/6 = 11\pi/6$. The long arc has length $11\pi r/6 = 22\pi$. This gives $r = 12$mm. Therefore, the distance SO and OP are both 30mm. The area of the triangle is $\frac{1}{2}ab \sin\theta = 450 \times \frac{1}{2} = 225$mm^2. The remaining area of the circle is $\frac{11}{12}\pi r^2 = 132\pi$. Adding these contributions gives $132\pi + 225$.

Question 36: E

The moments around the pivot must balance for the bar to be in equilibrium. It is subject to 2 moments: one arising from its weight and other from F. The moment due to its weight is $Mg \times d = 60 \times 10 \times 2 = 1200$ Nm. The moment due to F is only due to the vertical component of F i.e. $F \sin 60$. The moment is $F \sin 60 \times 4$. Equating the two gives $F = \frac{300}{\sin 60}$.

Question 37: F

The equations given can be rewritten $y = 2^{3p}$ and $z = 2^{-2q}$. Insert these into $\frac{y^3}{z^2} = 2^{9p+4q}$. Taking the logarithm, of base 2, of this gives $\log_2 \frac{y^3}{z^2} = 9p + 4q$.

Question 38: D

Use $v^2 = u^2 + 2as$ with $u = 40, a = -14.4, s = 20$. This gives $v^2 = 1600 - 4 \times 144 = 1024$. Taking the square root gives $v = 32$ms^{-1}. It is easiest to do this by noticing $1024 = 2^{10}$.

Question 39: D

Take the derivative to find $\frac{dy}{dx} = 3x^2 + 2px + q$. At stationary points, this equals zero. Inserting the two values of x gives $0 = 12 + 4p + q$ and $0 = 48 + 8p + q$. Subtracting the first equation from the second gives $4p + 36 = 0$ so $p = -9$. Using this in either of the first two equations gives $q = 24$.

Question 40: F

The force meter will measure the tension in the string. The tension is constant throughout, as the system is in equilibrium. Consider the hanging mass. Here the tension must balance the weight of the object i.e. $T = 10 \times 1 = 10N$. This will be the reading on the meter.

Question 41: D

The area of the triangle in given by:

$$\Rightarrow \frac{1}{2}PQ \times QR \times \sin 60 = 2x(8 - 3x) \times \frac{\sqrt{3}}{2} = \sqrt{3}x(8 - 3x).$$

Take the derivative of this with respect to x and set this equal to zero giving $\sqrt{3}(8 - 6x) = 0$. This gives a maximum at $x = 4/3$. Substitute this into the equation for the area to get $A = \sqrt{3} \times \frac{4}{3} \times 4 = \frac{16\sqrt{3}}{3}$.

Question 42: A

As the apple falls, it converts GPE into kinetic energy. However, some energy is dissipated by doing work against resistive forces. The initial GPE is $mgh = 0.1 \times 10 \times 4 = 4\,J$. When the apple hits the ground, it has kinetic energy of $\frac{1}{2}mv^2 = 0.5 \times 0.1 \times 64 = 3.2\,J$. Therefore, $0.8\,J$ of work has been done against resistive forces.

Question 43: E

Take the derivative using the chain rule to get $\frac{dy}{dx} = 6 \times 3 \times (2 + 3x)^5$. Using the binomial expansion, the x^3 term is: $\Rightarrow 18 \times \frac{5!}{3!2!} \times (3x)^3 \times 2^2 = 18 \times 27 \times 4 \times \frac{5 \times 4}{2} = 19440$.

Question 44: A

When the stone reaches the top of the cliff on the way down, it will have the same kinetic energy as when it is initially released. Therefore, its velocity is $13\,ms^{-1}$ downwards. Use $s = ut + \frac{1}{2}at^2$ with $a = 10\,ms^{-2}$ from gravity and $u = 13\,ms^{-1}$, where we have chosen downwards to be positive. This gives $5t^2 + 13t - 6 = 0$. Factorise to find $(5t - 2)(t + 3) = 0$. Can neglect the negative time so the solution is $t = 2/5 = 0.4s$

Question 45: A

The sum to infinity of a geometric series is $\frac{a}{1-r} = \frac{4}{3}$. As $a = 1$, rearrangement gives $r = \frac{1}{4}$. Therefore, we require solutions to $\sin 2x = \frac{1}{2}$. The solutions to this are $2x = \frac{\pi}{6} + 2n\pi, \frac{5\pi}{6} + 2n\pi$ for integers n. Divide by 2 for $x = \frac{\pi}{12} + n\pi$ and $x = \frac{5\pi}{12} + n\pi$. Choose $n = 1$ for the desired range to get the answers $x = \frac{13}{12}\pi$ and $x = \frac{17\pi}{12}$.

Question 46: B

The area under the graph is $\frac{1}{2}bh = 0.2 \times 192 = 38.4$ J. This is all converted into kinetic energy at the point where the arrow leaves the bow. At the maximum height the arrow has converted all its kinetic energy into gravitational potential energy so $38.4 = mgh = 0.024 \times 10 \times h$. This gives $h = 38.4/0.24 = 160$ m. The answer is B.

Question 47: C

The second term is $u_2 = 2p + 3$, the third term is $u_3 = 2p^2 + 3p + 3$ and $u_4 = 2p^3 + 3p^2 + 3p + 3 = -7$. This gives a cubic $2p^3 + 3p^2 + 3p + 10 = 0$. A solution of this equation is $p = -2$. Factorising the cubic gives $(p + 2)(p^2 - p + 5) = 0$. The quadratic factor does not have any real roots, so the only solution is $p = -2$. This gives $u_2 = -1$ and $u_3 = 5$. Summing the four terms gives $2 - 1 + 5 - 7 = -1$.

Question 48: H

Resolve the gravitational force parallel and perpendicular to the plane at 20 degrees. The gravitational force down the plane is $mg\sin 20$ and the force into the plane is $mg\cos 20$. The gravitational force is balanced by the resistive force $F = \mu N$ with μ a constant coefficient of friction. As the forces are balanced in the first scenario, $\mu mg\cos 20 = mg\sin 20$. This gives $\mu = \tan 20$.

Now consider the forces when the surface is at 25 degrees. The gravitational force down the plane is now $mg\sin 25$ and the force into the plane is $mg\cos 25$. The net force down the plane is $mg\sin 25 - \mu mg\cos 25$. Using $F = ma$, the acceleration down the plane is $g(\sin 25 - \cos 25\tan 20)$.

Question 49: A

The numerator is zero when $x = 1$, so it can be factorised to give $(x - 1)(x^2 - 5x + 4x) = (x - 1)(x - 4)(x - 1)$. The inequality can now be written $\frac{(x-1)^2(x-4)}{x} > 0$. The graph $\frac{(x-1)^2(x-4)}{x} = 0$ crosses the x axis at $x = 4$ but does not at $x = 1$. This is because it is a repeated root.

When x is a large positive or negative number, the numerator is dominated by the x^3 term so the inequality become $\frac{x^3}{x} = x^2 > 0$ which is true. There is also an asymptote at $x = 0$. Near this point, the numerator is dominated by the constant term, so it reduces to $\frac{-4}{x} > 0$. This is true when x is negative and false for positive x. Summarising this the inequality is satisfied when x is negative; between 0 and 1 the function is negative and at $x = 1$ the function is zero. Between 1 and 4 the function is negative and beyond 4 it becomes positive again. Therefore, A is the correct answer.

Question 50: B

As the suitcase is not on the verge of slipping, we can only use an inequality to describe the resistive force in terms of the coefficient of friction $F > \mu R$. But, as the suitcase is travelling at a constant speed, the net force on it must be zero. The frictional force must balance the component of the gravitational force that acts down the plane. This is $mg\sin\theta$.

Question 51: H

To stretch a graph by scale factor $\frac{1}{2}$, replace all occurrences of x with 2x. To then shift by $-\pi/4$ in the x-direction replace x with $x + \frac{\pi}{4}$. This gives:

$$\Rightarrow y = \sin 2\left(x + \tfrac{\pi}{4}\right) = y = \sin\left(2x + \tfrac{\pi}{2}\right).$$

Question 52: B

The initial momentum of the ball is $-mu$. The change in momentum is given by the area under the Force-Time graph which is $F(t_2 - t_1)$. The final momentum is $F(t_2 - t_1) - mu$.

Question 53: B

For a straight line, $y = mx + c$, another line described by $y = gx + d$ is perpendicular if $m = \frac{-1}{g}$. Using the gradients from the two lines gives $-1 = (2p^2 - p)(p - 2)$. Expand this to get $2p^3 - 4p^2 - p^2 + 2p = -1$. Collecting the terms gives $2p^3 - 5p^2 + 2p + 1 = 0$. This has a root at $p = 1$. It can be factorised to $(p - 1)(2p^2 - 3p - 1) = 0$. The quadratic factor has solutions $p = \frac{3 \pm \sqrt{17}}{4}$. Take the positive root for the larger solution and take $\sqrt{17} \approx 4$. This gives $p \approx 7/4 = 1.75$.

Question 54: A

After hitting the ground for the first time, the ball has momentum $P = mv$. It takes 0.8 s for the ball to reach the ground again at which point it has an equal velocity but in the opposite direction and momentum $P = -mv$. Use $\Delta P = F\Delta t$ with $F = mg$. This gives $v = \frac{mg \times 0.8}{2m} = 4 \text{ ms}^{-1}$. The kinetic energy of the ball after the first bounce is $\frac{1}{2}mv^2 = 8m = mgh$ as it is all converted into gravitational energy. This gives $h = \frac{8m}{10m} = 0.8$.

Section 2

Question 1a: B
The acceleration of the ball is $a = 0.4t$. The velocity is the integral of this: $v = 0.2t^2$. At $t = 0.5$ seconds, the velocity should be $v = 0.05 \text{ ms}^{-1}$ so the correct answer is B.

Question 1b: A
The ball is initially at a distance d. Take downwards as positive and the initial displacement of the ball as $-d$. The displacement is the integral of the velocity with respect to time which is $s = 0.2t^3/3 - d$, using the condition at $t = 0$. The ball hits the floor when $s = 0$ giving $t^3 = 15d$. Therefore, $t = (15d)^{1/3}$.

Question 1c: A
As the gravitational force continues to increase, the ball will require more energy to reach the same height. However, this does not happen, and the ball will bounce less high each time meaning only P is correct.

Question 1d: C
The unit of mass is kg and the unit of acceleration is ms^{-2}. Therefore, multiplying these gives a unit of force: kgms^{-2}.

Question 1e: E
D is a force so must have units kgms^{-2}. The units of v^2 are m^2s^{-2}. The units of area and density are m^2 and kgm^{-3} respectively. Multiplying these together gives $\text{kgm}^2\text{m}^2\text{m}^{-3}\text{s}^{-2} = \text{kgms}^{-2}$. This has the same units as a force so X does not have units.

Question 2a: C
A resistor obeys Ohm's law: $V = IR$, so current and voltage are directly proportional to each other. This means the resistor must correspond to the straight line through the origin which is device Y.

Question 2b: B
The filament lamp is 9W at 6V. Use the equation $P = IV$ to find which device has the correct current at this voltage. The current at 6V should be $I = 9/6 = 1.5A$. Therefore, the filament lamp is device X.

Question 2c: D
The resistance of the filament lamp at 6V is: $R = V/I = 4\Omega$. The resistor has a resistance of 8Ω. As the resistances are in parallel, they add in reciprocal: $\frac{1}{R_T} = \frac{1}{R_1} + \frac{1}{R_2}$. This gives $R_T = \frac{R_1 R_2}{R_1 + R_2} = 32/12 = 8/3 \ \Omega$. The current drawn from the supply is then $I = V/R = 18/8 = 2.25A$.

Question 2d: C
The same current flows through two components that are in series, so C must be correct.

Question 2e: B
The currents through W and Y are the same and the voltages across the components sum to give 6. Choose a current at which $V_W + V_Y = 6$. This is when $I = 0.5A$, $V_W = 2V$ and $V_Y = 4V$. The power dissipated by W is $P = IV = 1$ W.

Question 3a: D

Hooke's law behaviour gives a straight line through the origin which is observed up to point P. The strain is given by $\frac{x}{L}$, where L is the natural length of the cord. At the point Q, where the cord fractures, this is $\frac{0.05}{0.5} = 0.1$.

Question 3b: C

The work done is equal to the area under the graph. Each box of 10 N by 0.01 m is 0.1J of energy. There are approximately six of these under the graph, so the work done is 0.6 J.

Question 3c: B

As the cord is half the length, it can only be stretched half as much before it fractures. This means the maximum work that can be done on the cord is $\frac{U}{2}$. When the catapult is fired, this energy is converted into kinetic energy, giving:

$\Rightarrow \frac{1}{2}mv^2 = U/2$. Therefore, $V_{max} = \sqrt{U/m}$.

Question 3d: D

Now that there are two cords, twice as much work can be done as in the previous part so $\frac{1}{2}mv^2 = U$. This gives $v = \sqrt{2U/m} = \sqrt{2}V_{max}$.

Question 4a: B

The rays from the two slits will positively interfere when the distances travelled differ by a whole number of wavelengths. The rays will destructively interfere when the path lengths differ by a half-integer number of wavelengths. Positive interference corresponds to the brightest parts whereas destructive interference gives a minimum. For a diffraction pattern, the light from the two slits must have the same frequency i.e. they must be coherent.

Question 4b: D

The introduction of the material will cause the phase of the light from one slit to be shifted. This means the points where the two rays are in phase will move causing the diffraction pattern to shift in the y direction.

Question 4c: E

Radio waves are an EM wave, so they travel at the speed of light. Use $c = f\lambda$ to find the wavelength of the signal:

$\Rightarrow \lambda = c/f = \frac{3\times10^8}{6\times10^8} = 0.5$ m.

This means the separation of the two transmitters is of comparable size to the wavelength, so diffraction will be significant. Therefore, E is the correct answer.

END OF PAPER

2018

Section 1

Question 1: E

It is easiest to present the information from the question in a table:

	Men	**Women**	**Total**
Passed 1st attempt			167
Failed 1st attempt		143	
Total	300	200	

The number of women who passed their first attempt can be calculated as:

$\Rightarrow$ Total women – Women who failed $= 200 - 143 = 57$.

The number of men who passed their first attempt is:

$\Rightarrow$ Total people who passed 1st attempt – Women who passed 1st attempt $= 167 - 57 = 110$.

Question 2: B

Alpha particles: atomic number (Z) = 2, mass number (A) = 4.

Beta particles: atomic number (Z) = -1, mass number (A) = 0.

Particles released: $5\alpha + 2\beta$: $\Delta Z = 5(2) + 2(-1) = 8$.

Question 3: B

It may help to draw a diagram:

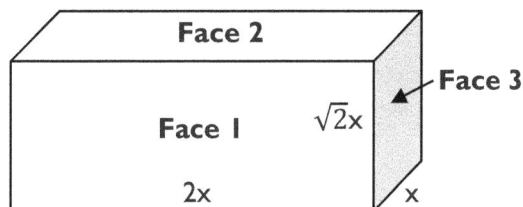

$\Rightarrow$ Volume = $(2x)\left(\sqrt{2}x\right)(x) = 2\sqrt{2}x^3$

There are 6 faces in total, but 3 different proportions, so surface area is:

$\Rightarrow 2(2x)\left(\sqrt{2}x\right) + 2(2x)(x) + 2\left(\sqrt{2}x\right)(x) = 2x^2\left(2\sqrt{2} + 2 + \sqrt{2}\right)$

Volume = 2 × S.A:

$\Rightarrow 2\sqrt{2}x^3 = 4x^2\left(3\sqrt{2} + 2\right)$

$2\sqrt{2}x^3 - 12\sqrt{2}x^2 - 8x^2 = 0$ (Factorise out x^2)

It is possible to disregard the $x^2 = 0$ solution in this case because this is the solution for the cuboid having no dimensions, no volume and no surface area. It is not always possible to disregard this solution – take care not to 'divide' by x without good reason, as it may cause solutions to be 'lost'.

$\Rightarrow 2\sqrt{2}x - 12\sqrt{2} - 8 = 0$

Rearranging for x gives: $x = \frac{12\sqrt{2}+8}{2\sqrt{2}} = 6 + \frac{8}{2\sqrt{2}} = 6 + 2\sqrt{2}$.

Question 4: B

Combining resistors in series using: $R_{total} = R_1 + R_2 + R_3 = 30\Omega$.

Calculate terminal voltage using: $V = IR$. $V = 0.2(30) = 6V$.

After removal of resistor R₃: $R_{total} = 27\Omega$.

Calculate new current using: $I = \frac{V}{R}$. $I = \frac{6}{27} = \frac{2}{9} \approx 0.22$.

Question 5: H

Find gradient of line joining the two points in terms of p using $m = \frac{\Delta y}{\Delta x}$:

$$\Rightarrow m = \frac{2p - (p - 1)}{(1 - p) - p} = \frac{p + 1}{1 - 2p}$$

Rearrange the equation of the line to $y = mx + c$ form, to find m:

$$y = -\frac{2}{3}x - \frac{1}{3}, \quad m = -\frac{2}{3}$$

'Parallel' means the two gradients are equal: $\frac{p+1}{1-2p} = -\frac{2}{3}$.

Rearranging for p gives:

$$\Rightarrow 3(p + 1) = -2(1 - 2p) \Rightarrow 3p + 3 = -2 + 4p \Rightarrow p = 5.$$

Question 6: G

UV has higher frequency and lower wavelength than visible light.

Convert minimum wavelength of visible light to frequency using $f = \frac{c}{\lambda}$. $f = \frac{3 \times 10^8}{400 \times 10^{-9}} = 7.5 \times 10^{14}$ Hz.

This is the minimum frequency of UV, as it is the boundary between UV and visible light.

Question 7: D

Let $QR = x$. Draw a labelled diagram:

Use area to find 'x': $96 = (2x)(x)$.

$$\Rightarrow x = \sqrt{\frac{96}{2}} = \sqrt{48} = 4\sqrt{3}.$$

The Pythagorean theorem gives

$$\Rightarrow (2r)^2 = (2x)^2 + (x)^2$$

$$\Rightarrow 4r^2 = \left(8\sqrt{3}\right)^2 + \left(4\sqrt{3}\right)^2 = 240.$$

Rearranging for 'r' gives: $r = \sqrt{60} = 2\sqrt{15}$.

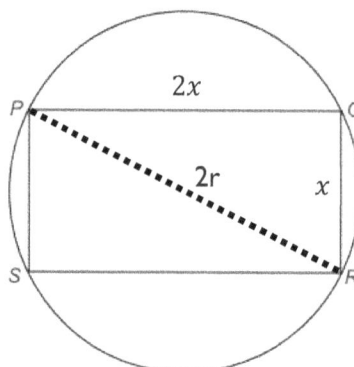

Question 8: E

Energy transferred (J) = Power (W) × Time (s):

$100 \times (10 \times 60) = 60,000$ J is the total energy transferred.

Efficiency of 5% means 95% of the energy is wasted.

Energy wasted = $0.95 \times 60,000 = 57,000$J.

Question 9: E

If the ratio of the heights is $4:5$, the ratio of volumes is $4^3:5^3$.

Multiplying 320 by a ratio of $64:125$ gives $\frac{320}{64}(125) = 625$ cm³.

Question 10: C

Electrical energy transferred (J) = Current (A) x Voltage (V) x Time (s) $= 1250 \times 400 \times 4 = 2,000,000$ J.

Efficiency of 45% means 45% of the energy input is converted to useful energy (kinetic) output.

Kinetic energy $= 0.45 \times 2,000,000 = 900,000$ J.

Question 11: C

Draw a diagram - remember to label the y-intercepts!

Find P by substituting y=3x - 23 into 5x + 2y = 20:

$\Rightarrow 5x + 2(3x - 23) = 20 \Rightarrow x = 6.$

Use $Area = \frac{1}{2}base \times height$ with base equal to the distance along the

y-axis and height equal to the x-coordinate of P.

$\Rightarrow Area = \frac{1}{2}(33)(6) = 99.$

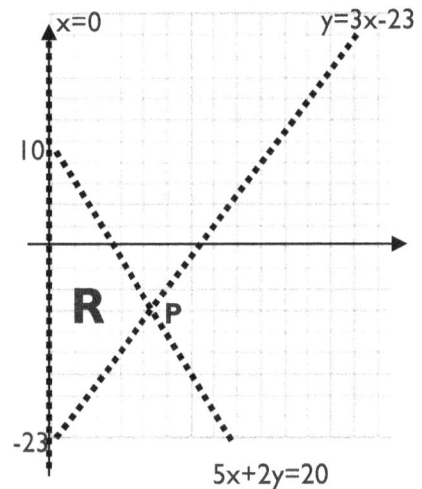

Question 12: A

Momentum = Mass $\times$ Velocity and $K.E. = \frac{1}{2} \times$ Mass $\times$ Velocity^2.

Divide KE by momentum to find velocity:

$\Rightarrow \frac{K.E.}{momentum} = \frac{\frac{1}{2}mv^2}{mv} = \frac{1}{2}v.$

$\Rightarrow \frac{1}{2}v = \frac{96}{24} \Rightarrow v = 8$ m/s.

Substitute back into momentum or KE equation to find mass: $m = \frac{24}{8} = 3$ kg.

Question 13: E

Length scale of $1:40$ means a volume scale of $1^3:40^3$.

Convert mass of full-sized pillar into cm³: $12\pi \times 10^6$.

Volume of model is therefore: $\frac{Full-sized\ pillar}{40^3} = \frac{12\pi \times 10^6}{64 \times 10^3} = \frac{3\pi \times 10^3}{16}.$

Mass = Density x Volume $= \frac{4}{3} \cdot \frac{3\pi \times 10^3}{16} = \frac{1000\pi}{4} = 250\pi.$

Question 14: A
The background count-rate is the value the count tends towards once the isotope has decayed – in this case, 20 cpm.

The half-life is calculated as the time taken for the isotope to fall to half the previous value, but an adjustment must be made to remove the background count rate. Therefore, the initial reading of 120 on the graph corresponds to an activity of 100cpm for the sample, and so the half-life should be measured to a sample activity of 50cpm (70 on the graph). This gives 40 seconds.

Question 15: D
The internal angles in a regular pentagon are all 108°. RSU is equilateral so ∠RSU=60°. Therefore ∠UST is 48°. As RSU is equilateral, SU = RS. All sides are the same length in a regular pentagon so SU = ST and therefore STU is isosceles. This means ∠STU can be found by:

$$\Rightarrow \angle STU = \tfrac{1}{2}(180 - 48) = 66°.$$

Question 16: C
Resultant force downwards = Weight – Air resistance = Mass x Acceleration.

$$\Rightarrow 10m - 12 = 2m \Rightarrow m = 1.5 \text{ kg}.$$

Question 17: C
The original price is p. An increase of 125% gives a price of 2.25p. A decrease of 40% gives a price of (0.6)(2.25p) = 1.35p = q.

Question 18: E
A frequency of 10 Hz means a particle in the rope completes 10 full cycles per second, or 200 full cycles in 20 seconds.

An amplitude of 4 means the particle travels a distance of 16 cm for each full cycle (rest-top-rest-bottom-rest is a full cycle).

The total distance travelled is therefore 20 x 16 = 3200 cm. The speed is unnecessary and is included to obscure the solution!

Question 19: E
Draw a diagram. Arrows show North (N). Using 'C-angles' (interior angles).

$$\Rightarrow \angle PQN=180-65=115°$$

$$\Rightarrow \angle QRN=180-155=25°$$

At point Q the angle inside the triangle is:

$$\Rightarrow 360 - 155 - 115 = 90°.$$

Because PQR is isosceles, once the interior angle at Q is known, the angles QPR and QRP can be calculated as $\tfrac{1}{2}(180 - 90) = 45°$.

Finally, the bearing at R can be calculated as 360 – 25 - 45=290°.

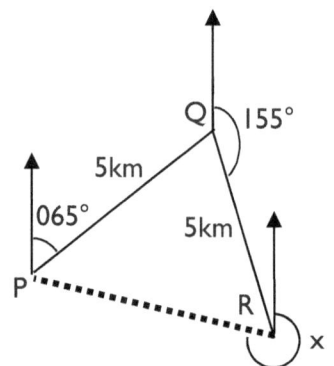

Question 20: C

Overall equation: $Mass\ balance\ reading = m_{cyl} + \rho_{fluid}V$.

Forming equation for addition of water: $290 = m_{cyl} + V$.

Form equation for addition of oil: $270 = m_{cyl} + 0.9V$.

Solving as simultaneous equations gives: $20 = 0.1V$ so $V = 200\text{cm}^3$.

Substituting this back into either equation gives: $m_{cyl} = 90\text{g}$.

Question 21: C

The alternate segment theorem means that $\angle$PST = $\angle$PQS = 75°.

Because the angle between a tangent and a radius is a right angle, $\angle$PSO = 90 - 75=15°.

As PQS is isosceles, $\angle$QSP = $\frac{1}{2}(180 - 75) = 52.5°$.

$\angle$QSO is therefore 52.5 – 15 = 37.5°.

Question 22: B

At terminal velocity (the straight-line/constant-gradient section of the graph), weight downwards = air resistance upwards, giving:

$$\Rightarrow 1000 = kv^2 \text{ so } k = \frac{1000}{v^2}$$

The velocity is the gradient of the distance-time graph, and is found to be 50 m/s. Therefore $k = \frac{1000}{50^2} = 0.4$.

Question 23: C

From the question:

$$h = b + 3 \Rightarrow Area = \frac{hb}{2} = 14.$$

Combine as simultaneous equations to get: $\frac{b(b+3)}{2} = 14$.

This gives the quadratic $b^2 + 3b - 28 = 0$, the only appropriate solution of which is $b = 4$cm. Therefore $h = 7$cm.

Use Pythagoras's theorem on half the triangle:

$$\Rightarrow s^2 = h^2 + \left(\frac{b}{2}\right)^2 = (b+3)^2 + \left(\frac{b}{2}\right)^2.$$

$$\Rightarrow s^2 = b^2 + 6b + 9 + \frac{1}{4}b^2 = \frac{5}{4}\left(4^2\right) + 6(4) + 9$$

Therefore, $s^2 = 53$ and so s must lie between 7 and 8.

Question 24: H

Nucleons in = 235 (Uranium) + 1 (thermal neutron) = 236.

Nucleons out = 236 = 88 (Bromine) + 145 (Lanthanum) + x (neutrons).

Rearrange to find the number of neutrons out as: 236 − 88 - 145 = 3.

Protons In = 92 (Uranium).

Protons Out = 92 = 35 (Bromine) + y (Lanthanum).

Rearrange to find atomic number of Lanthanum is 57.

Beta decay involves a neutron turning into a proton and an electron, therefore when Lanthanum decays the proton number goes from 57 to 58.

Question 25: G

For the 1st term (n=1): $2 = p(1^2) + q = p + q$.

For the 2nd term (n=2): $17 = p(2^2) + q = 4p + q$.

Solve as simultaneous equations (subtract) to give:

$\Rightarrow 3p = 15$ so $p = 5$ and $q = -3$.

$\Rightarrow \frac{p-q}{p+q} = \frac{5-(-3)}{5+(-3)} = \frac{8}{2} = 4$.

Question 26: C

Let the resistance of X, Y and Z = R.

Create an expression for the total resistance of X and Y: $\frac{1}{R_{total}} = \frac{1}{R} + \frac{1}{R} \Rightarrow R_{total} = \frac{R}{2}$.

The circuit can therefore be considered as a resistor, Z, of resistance R, in series with a resistor, 'XY', of resistance R/2.

The power dissipated is proportional to the resistance, and therefore two-thirds of the power (12W) will be dissipated through Z, with the other third (6W) dissipated in the combination of X and Y. The power will split between X and Y equally (3W each).

Question 27: B

The number of red sweets taken will always be exactly 1, because at the point that the child has taken one red sweet, they stop taking any more. Therefore, the options which give more green sweets than red are:

2, 3, 4, 5, or 6 green sweets followed by a red.

Considering the opposite scenarios (fewer green than red) reduces the number of calculations required:

The chance of the child picking a red sweet on their first pick is $\frac{1}{2}$.

The chance of the child picking one green and then red is $\left(\frac{1}{2}\right)\left(\frac{6}{11}\right) = \frac{3}{11}$.

All remaining scenarios give more green than red, so the probability can be worked out as $1 - \frac{1}{2} - \frac{3}{11} = \frac{5}{22}$.

Question 28: A

Distance = Speed × Time – so, as the wave is travelling at the same speed in all directions, and the time taken to detection is the same for X and Y, the distance between the epicentre and X must equal the distance between the epicentre and Y. This doesn't mean that '1' must be true, however, as the epicentre could be equidistant but lie to one side of XY, rather than on the line.

'2' isn't necessarily true, as Z needn't be equidistant from X and Y. For example, the scenario below would give the detection pattern seen:

'3' also needn't be true in all cases. The distance travelled between detection by X or Y, and Z is:

$$\Rightarrow Distance = speed \times time = 4(60) = 240 \text{km}.$$

However, as shown in the diagram, while Z must be no more than 240km from one of X or Y – it could be considerably further from the other (if it was on the opposite side of the epicentre).

Question 29: A

Draw a diagram.

The area between C and L is equal to the area below C (between the two points of intersection) minus the rectangular area below L.

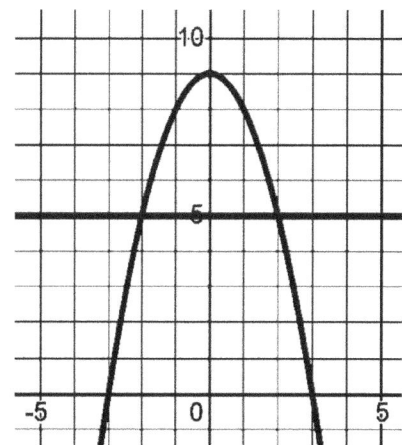

Find the points of intersection:

$$9 - x^2 = 5 \Rightarrow x = \pm 2$$

Find the area below C by integrating:

$$\Rightarrow \int_{-2}^{2} 9 - x^2 dx = \left[9x - \frac{x^3}{3}\right]_{-2}^{2} = \left[9(2) - \frac{2^3}{3}\right] - \left[9(-2) - \frac{-2^3}{3}\right] = \frac{92}{3}.$$

The area below L is base × height = 4 × 5 = 20.

The area between C and L is therefore $\frac{92}{3} - 20 = \frac{32}{3}$.

Question 30: E

SUVAT: $s = 1,600$ m, $u = 0$ m/s, $v = 80$ m/s, $a = ?$. Use $v^2 = u^2 + 2as$:

$\Rightarrow 80^2 = 2(1600)a \Rightarrow a = \frac{6400}{3200} = 2$ m/s².

Question 31: D

Rearranging gives: $2 \sin^3 \theta - \sin \theta = 0$.

Factorise (dividing would lose solutions):

$\Rightarrow sin\theta(2 \sin^2 \theta - 1) = 0 \Rightarrow sin\theta = 0$ or $sin\theta = \pm\frac{1}{\sqrt{2}}$.

Considering the graph of $y = sin\theta$ (overlaid with $y = 0$, $y = \frac{1}{\sqrt{2}}$, and $y = -\frac{1}{\sqrt{2}}$) shows 5 solutions in the interval $-\frac{\pi}{2} \le \theta \le \pi$:

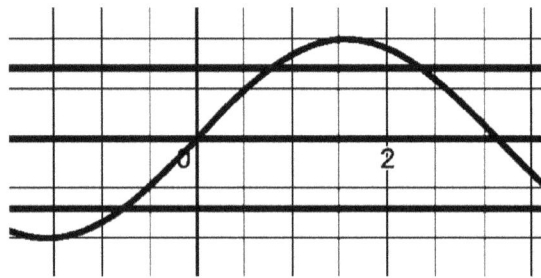

Question 32: F

A: Not necessarily true – balanced overall doesn't mean 'equal'. B: There will likely be more than 3: normal reaction, friction, weight, driving force. C: Not true – the driving force up the slope will be balanced by both friction and the component of the weight parallel to the slope. D: Not true – mass is straight down, contact is perpendicular to plane. E – Not enough information to know. F – This must be correct. Constant speed means no overall resultant force (Newton's First Law).

Question 33: C

The tangent to the curve has gradient found by differentiation:

$\Rightarrow \frac{dy}{dx} = 6x - 2$.

The gradient of the curve is equal to the gradient of the tangent, so:

$\Rightarrow 6x - 2 = 1 \Rightarrow x = \frac{1}{2}$.

The y-coordinate is found by substituting x back into the equation of the curve:

$\Rightarrow y = 3\left(\frac{1}{2}\right)^2 - 2\left(\frac{1}{2}\right) + 1 = \frac{3}{4}$.

The value of k can then be found as $\frac{3}{4} = \frac{1}{2} + k$, giving $k = \frac{1}{4}$.

Question 34: D

All blocks have the same acceleration, calculated using Newton's 2nd Law (N2L): $30 = (3 + 4 + 6 + 2)a$ giving $a = 2$ m/s².

Considering N2L on 'Z' alone: $30 - T_1 = 2(2)$ gives $T_1 = 26$N.

Then considering N2L on 'Y' alone: $T_1 - T_2 = 6(2)$.

Therefore $26 - T_2 = 12$, so $T_2 = 14$N.

Question 35: D

Area of a sector $= \frac{1}{2}r^2\theta$ and Arc length $= r\theta$.

$$\Rightarrow Area\ (S) = \frac{1}{2}r^2\theta = 10\pi \text{ and } Area\ (T) = \frac{1}{2}r^2\left(\theta + \frac{\pi}{20}\right) = \frac{25}{2}\pi$$

Solving simultaneously (subtracting Area (S) from Area (T)) gives:

$$\Rightarrow \frac{1}{2}r^2\left(\frac{\pi}{20}\right) = \frac{5}{2}\pi \Rightarrow r = 10\text{cm}.$$

Substituting back into the expression for Area (S) gives $\theta = \frac{\pi}{5}$. Arc length (T) = $10\left(\frac{\pi}{5} + \frac{\pi}{20}\right) = \frac{10\pi}{4} = \frac{5}{2}\pi$.

Question 36: F

Let contact force at X = F_x. Take moments around Y: $10g(2) + 40g(3) = F_x(4)$. Solve for $F_x = 35g = 350$N.

Question 37: E

nth term of A.P = $a + (n-1)d$. 13th term = 6 x 1st term: $a + 12d = 6[a]$. 11th term = 2 x 5th term − 1: $a + 10d = 2[a + 4d] - 1$.

$$\Rightarrow 5a = 12d \text{ and } a - 1 = 2d$$

Solve simultaneously: $5 = 2d \Rightarrow d = \frac{5}{2}$ and $a = 6$.

3rd term = $a + 2d = 6 + 2\left(\frac{5}{2}\right) = 11$.

Question 38: B

Initial GPE = Final KE + Work done against friction

$$mgl\sin\theta = KE + kmgl\sin\theta \Rightarrow KE = mgl\sin\theta(l - k)$$

As a proportion of the initial GPE, $mgl\sin\theta$:

$$\Rightarrow \frac{KE}{GPE} = \frac{mgl\sin\theta(l-k)}{mgl\sin\theta} = 1 - k.$$

Question 39: B

n^{th} term of G.P = ar^{n-1} where r is a constant ratio, therefore:

1st term: $a = p - 2$

2nd term: $r = 2p + 2$ so $r = \frac{2p+2}{p-2}$

3rd term: $ar^2 = 5p + 4$ so $r = \frac{5p+14}{2p+2}$

Solving simultaneously by setting the two expressions for r equal to each other gives p = 8 or p = -4. 'p' must equal 8 as the question states the terms are all greater than 0. 'r' can be calculated as $\frac{2(8)+2}{8-2} = 3$ and 'a' as 8 – 2 = 6.

The 5th term is therefore $ar^4 = 6(3^4) = 486$.

Question 40: D

Change in momentum = Impulse = Force x Time.

Change in momentum of X when force is applied = 5 x 3 = 15 kg m/s.

$\Rightarrow \Delta mv = m(v_f - v_1) \Rightarrow 15 = 2(v_f - 4.5) \Rightarrow v_f = 12$ m/s

Conservation of momentum: Momentum of X+Y before = Momentum of XY after .

$2(12) + 3(0) = (3+2)v_{combined}$ which can be rearranged to give the speed of the combined 'XY' as $\frac{24}{5}$ m/s.

Question 41: C

In general $\log(a) + \log(b) = \log(ab)$.

$$\Rightarrow \log_2\left(\frac{5}{4}\right) + \log_2\left(\frac{6}{7}\right) + \cdots + \log_2\left(\frac{64}{63}\right) = \log_2\left(\frac{(5)(6)\ldots(64)}{(4)(5)\ldots(63)}\right)$$

As an aside: $n! = n \times (n-1) \times (n-2) \ldots \times 3 \times 2 \times 1$.

In this case, not all terms are included, so instead of 64! the expression for the numerator is $\frac{64!}{4!}$. Similarly, instead of 63! the expression for the denominator is $\frac{63!}{3!}$.

The expression is therefore equal to $\log_2\left(\frac{64!}{4!} \cdot \frac{3!}{63!}\right)$ which simplifies to $\log_2\left(\frac{64}{4}\right) = \log_2(16) = 4$.

Question 42: G

Initial energy = KE + GPE = $\frac{1}{2}(0.2)(4^2) + 0.2(10)(0.45) = 2.5$ J.

Energy after bounce = KE = $\frac{1}{2}(0.2)(2^2) = 0.4$ J.

Energy lost in bounce = 2.5 – 0.4 = 2.1 J.

Question 43: C

Draw a diagram.

Consider the simplest case, when the tangent is horizontal (or vertical).

The co-ordinates of P would therefore be:
$x = -3 - 5\sqrt{3}$ and $y = 2 + \sqrt{5}$.

The shortest distance between the point P and the circle can then be generalised as the length of the line marked 'L' minus the radius.

$Shortest\ distance = \sqrt{(x_p - x_c)^2 + (y_p - y_c)^2} - r$ where

the point P has coordinates (x_p, y_p) and the circle has centre (x_c, y_c).

Substituting in the values above gives $shortest\ distance = 3\sqrt{5}$.

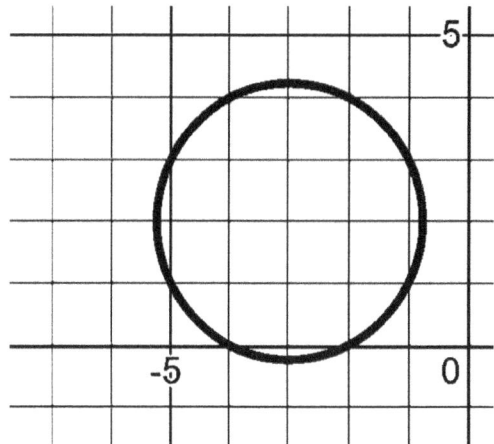

Question 44: F

(Momentum of X + Momentum of Y) before = Momentum of 'XY' after.

After: XY both have velocity '-v', and therefore momentum = -3mv.

Before: X has momentum 'mv'.

Rearrange to find the momentum of Y before = -3mv – mv = -4mv. Speed of Y before is therefore -2v.

Question 45: D

Distance AC = $\sqrt{(u-2)^2 + (-1-(-3))^2} = \sqrt{u^2 - 4u + 8}$.

Distance AB = $\sqrt{(3-2)^2 + (1-(-3))^2} = \sqrt{1 + 16} = \sqrt{17}$.

⇒ AC = 2AB: $\sqrt{u^2 - 4u + 8} = 2\sqrt{17}$.

⇒ Square both sides and rearrange: $u^2 - 4u - 60 = 0$.

⇒ Factorising gives: $(u - 10)(u + 6) = 0$ so $u = 10$ or $u = -6$.

Question 46: A

Resolve horizontally: $P = T sin30$ (where $sin(30) = \frac{1}{2}$).

Question 47: D

'Brute force' method:

Binomial expansion: $(a + x)^n = a^n + na^{n-1}x + \frac{n(n-1)}{2!}a^{n-2}x^2 + \cdots$

Expansion of $(1 - 2x)^5 = 1 + 5(-2x) + \frac{5(4)}{2}(-2x)^2 + \frac{5(4)(3)}{(3)(2)}(-2x)^3 = 1 - 10x + 40x^2 - 80x^3 + \cdots$

Expansion of $(1 + 2x)^5 = 1 + 5(2x) + \frac{5(4)}{2}(2x)^2 + \frac{5(4)(3)}{(3)(2)}(2x)^3 + \cdots = 1 + 10x + 40x^2 + 80x^3 + \cdots$

When multiplied together, the x^3 terms would be:

$1(80x^3) - 10x(40x^2) + 40x^2(10x) - 80x^3(1) = 0$.

Alternative 'clever' method:

$(1 - 2x)^5(1 + 2x)^5 = [(1 - 2x)(1 + 2x)]^5 = [1 - 4x^2]^5$

Because the bracket to be expanded contains only an x^2 term, there will be no terms in the expansion with x raised to an odd power.

Question 48: A

Although this looks like SHM, the angle is not small enough for the equations to be appropriate.

Find the vertical distance between 'pivot' and pendulum at the 'amplitude position' using Pythagoras: $\sqrt{50^2 - 30^2} = 40$cm.

This means between amplitude and equilibrium, the bob loses 10 cm of GPE: $0.01(10)(0.1) = 0.01$.

It will gain KE = GPE lost, and therefore $0.01 = \frac{1}{2}mv^2$ can be rearranged to give $v = \sqrt{\frac{0.02}{0.01}} = \sqrt{2}$m/s.

Question 49: E

$$\Rightarrow \int_0^2 x^m dx = \left[\frac{x^{m+1}}{m+1}\right]_0^2 = \left[\frac{2^{m+1}}{m+1}\right] = \frac{16\sqrt{2}}{7}$$

$$\Rightarrow \int_0^2 x^{m+1} dx = \left[\frac{x^{m+2}}{m+2}\right]_0^2 = \left[\frac{2^{m+2}}{m+2}\right] = \frac{32\sqrt{2}}{9}$$

$$\left[\frac{2^{m+2}}{m+2}\right] = \left[\frac{2^{m+1}}{m+1}\right]\left[\frac{2(m+1)}{m+2}\right] \Rightarrow \frac{32\sqrt{2}}{9} = \frac{16\sqrt{2}}{7}\left[\frac{2(m+1)}{m+2}\right]$$

Which can be rearranged for $\frac{1}{9} = \frac{m+1}{7(m+2)}$ giving $m = \frac{5}{2}$.

Question 50: E

Total momentum (to the right) before $= mu$.

Total momentum (to the right) after $= MU_M - mv$.

Setting before and after equal gives $MU_M - mv = mu$ which can be rearranged for $U_M = \frac{mu+mv}{M}$.

Question 51: A

$f'(x) = ax + g(x)$ can be integrated between 2 and 4 to give:

$$\Rightarrow \int_2^4 f'(x)dx = \int_2^4 ax + g(x)dx$$

$$\Rightarrow f(4) - f(2) = \left[\frac{ax^2}{2}\right]_2^4 + \int_2^4 g(x)dx$$

$18 = \left(\frac{4^2}{2} - \frac{2^2}{2}\right)a + 12$ giving $a = \frac{6}{6} = 1$.

Question 52: F

Initial GPE = Final KE + Work done against friction, where the work done against friction is half the change in GPE:

$$\Rightarrow 0.5\, mgh = \frac{1}{2}mv^2 \Rightarrow 0.5(10)h = \frac{1}{2}(10^2) \Rightarrow h = 10\text{m}.$$

Question 53: C

It may help to draw a diagram:

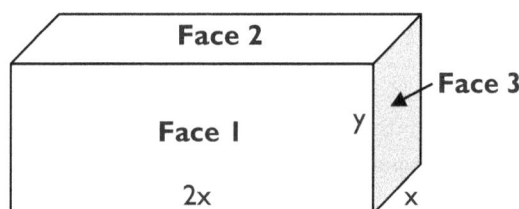

Volume of the cuboid = $2x^2y = 576$. This can be rearranged to give $y = \frac{288}{x^2}$.

S.A of the cuboid = $2(2xy + xy + 2x^2) = 4x^2 + 6xy$. Substitute to eliminate y: $S.A = 4x^2 + \frac{1728}{x}$.

If this is a maximum value, then the differential of the surface area with respect to x (or y) is equal to 0.

$\Rightarrow \frac{dSA}{dx} = 8x - \frac{1728}{x^2} = 0$, which can be rearranged to find $x = 6$.

When $x = 6$, $y = 8$ and the largest face therefore has surface area = 12 x 8 = 96 cm².

Question 54: B

SUVAT (1ˢᵗ object): $s = 0$ m, $a = -10$ m/s², $u = 40$ m/s, $t = T$.

Use $s = ut + \frac{1}{2}at^2$ for both objects.

For object 1 this gives: $0 = 40T - 5T^2$ which can be solved to give $T = 0$, which isn't a relevant solution, or $T = 8s$.

SUVAT (2ⁿᵈ object): $s = -h$ m, $a = -10$ m/s², $u = 0$ m/s, $t = 8 - 2s \Rightarrow -h = -5(6^2) \Rightarrow h = 180$m.

END OF SECTION

Section 2

Question 1: B

Man: Distance = Speed x Time = 9t.

Boy – use SUVAT: $a = 0.8$ m/s², $u = 5$ m, $t = ? \Rightarrow s = ut + \frac{1}{2}at^2 = 5t + 0.4t^2$

Set the distances travelled equal to each other: $9t = 5t + 0.4t^2$, which can be solved to give $t(0.4t - 4) = 0$.

This means the boy and man have the same displacement at t = 0 (given in the question) and at t = 10s.

Question 2: B

$\rho_{mixture} = \frac{mass_{mixture}}{volume_{mixture}} = \frac{\rho_p V_p + \rho_Q V_Q}{V_p + V_Q}$.

Question 3: C

The resistance of the combination (WY) would increase if W increased. Considering this as a potential divider would mean that the voltage dropped across WY would increase, and the voltage dropped across X would decrease. As a result, the power dissipated through WY would increase, and so the power dissipated in X would decrease.

The current flowing through Y would increase and so would the power dissipated through Y.

Question 4: A

As there are no external forces acting, the total momentum of the system remains 0 at all points. Therefore, following the collision, the magnets must have 0 velocity.

Question 5: F

In diagram 2: The buoyancy force of the water on the stone = 1 N. Therefore, the force of the stone on the water (and by extension the balance) = 1 N. The reading will have increased by 100g.

In diagram 3: The stone weighs 300g (as shown in diagram 1), therefore the mass balance reading will have increased by 300g.

Question 6: F

From diagram 1: The wavelength is twice the length of the tube = 1 m.

Use $f = \frac{c}{\lambda}$ to find frequency is 1000 Hz and $T = \frac{1}{f} = \frac{1}{1000} s$.

From diagram 2: X is 1.5 cycles from the origin, therefore it occurs at 0.0015s.

Question 7: A

The collision is elastic, so there is no loss in kinetic energy due to the collision. The mass of the tennis ball can be worked out from KE as $m = \frac{2KE}{v^2} = 0.06$ kg. Impulse applied = Force x time = Area under graph = $\frac{2}{1000} \times 1500 = 3$ Ns.

Change in momentum = Impulse = Mass x (Final velocity – Initial velocity).

$\Rightarrow 3 = 0.06(v - (-30))$ so $v = \frac{3}{0.06} - 30 = 20$ m/s.

Question 8: G
Let the reaction force at P be F_p, and the reaction force at Q be F_Q.

Take moments around P:

$$5g(0.2) + 74g(1) + 24g(1) = F_Q(1.5) \Rightarrow F_Q = \frac{1+74+24}{1.5}g = 660\text{N}.$$

Question 9: E
Consider the top 'branch' as a potential divider with total voltage 12V. The voltage dropped across the 2 and 3Ω resistors is $12\left(\frac{2+3}{1+2+3}\right)$ = 10V. Therefore the 'top' of the voltmeter is at 2V. Considering the bottom 'branch' similarly gives a voltage drop of 2V meaning the 'bottom' of the voltmeter is at 10V. The voltmeter reading will be the difference.

Question 10: D
Spring constants can be combined (with the opposite rules to resistors).

The springs in parallel therefore have a combined constant of 3k.

This can be combined with the spring in series as $\frac{1}{k_{total}} = \frac{1}{3k} + \frac{1}{k} \Rightarrow k_{total} = \frac{3k}{4}$.

Extension: $x = \frac{F}{k} = \frac{4mg}{3k}$, and EPE = $\frac{1}{2}Fx = \frac{1}{2}mg\left(\frac{4mg}{3k}\right) = \frac{2(mg)^2}{3k}$.

Question 11: B
GPE lost = KE gained + Work done against friction.

$$\Rightarrow 3.6(10)(1.5sin30) = \frac{1}{2}(3.6)\left(2^2\right) + W_f \Rightarrow W_f = 19.8\text{J}$$

Rate at which work is done:

R = $\frac{W_f}{time} = W_f.\frac{av.\ speed}{distance} = 19.8\left(\frac{1}{1.5}\right) = 13.2$ J/s.

Question 12: D
The total current in the circuit = Current through R = 20 + Current through R = 30.

$$\Rightarrow I_{total} = \frac{4.8}{20} + \frac{4.8}{30} = 0.24 + 0.16 = 0.4\text{A}$$

If 6V are produced by the battery, but there are only 4.8V over the resistors, the 'lost volts' dropped due to the internal resistance must be 1.2V.

The internal resistance can be calculated as lost volts divided by circuit current: $r = \frac{1.2}{0.4} = 3\Omega$.

Question 13: F
The stick dips into the water every 0.8 s, therefore T = 0.8s. In the 1 second which has passed, the wave has progressed by 1.25 wavelengths. Therefore, the difference between wave-crest Q at time t and the wave-crest at time t+1 corresponds to 1.25 wavelengths.

0.25 wavelengths = 1.5 cm, so the wavelength is 6 cm.

Question 14: B
Though counterintuitive, because the surfaces of P and Q are smooth, Q is not being held in place by friction. This means that the acceleration of Q due to the tension in the string is equal to the acceleration of P.

Resolve vertically on R: $m_R g = T$. Resolve horizontally on Q: $T = m_Q a$. This gives: $\frac{m_r g}{m_q} = a$.

Question 15: G
Surface area of a cube $= 6l^2 = 96 \Rightarrow l = 4$cm.

$F = kx$ finds the weight of the cube as $2 \times 10^4 \times 1.6 \times 10^{-4} = 3.2$ N.

$\Rightarrow$ Pressure $= \frac{Force}{Area} = \frac{3.2}{(4\times10^{-2})^2} = 2000$ N/m².

$\Rightarrow$ Density $= \frac{mass}{volume} = \frac{\frac{W}{g}}{l^3} = \frac{0.32}{(4\times10^{-2})^3} = 5{,}000$ kg/m³.

Question 16: A
$\Rightarrow M = density_{copper}(A)(L) + density_{aluminium}(6A)(L) = 9ALd$ or $A = \frac{M}{9Ld}$

Find the resistance of copper cylinder as $\frac{\rho L}{A}$. $R_c = \frac{2\rho L}{A}$.

Find the resistance of a single aluminium cylinder as $R_a = \frac{3\rho L}{A}$.

Combining as resistors in parallel:

$\Rightarrow \frac{1}{R_{total}} = \frac{A}{2\rho L} + \frac{6A}{3\rho L}$ gives $R_{total} = \frac{2\rho L}{5A}$.

Substituting A gives $R_{total} = \frac{2\rho L}{5}\left(\frac{9Ld}{M}\right) = \frac{18\rho L^2 d}{5M}$.

Question 17: D
Mass flow rate $= \rho Av = 2{,}400$ kg/s.

Mass flow rate must be conserved. This means the velocity of the fluid in the smaller cross-section is $\frac{2400}{800\times0.25} = 12$ m/s.

$\Rightarrow Force = \frac{\Delta mv}{t} = \frac{m}{t}.\Delta v = 2400(12 - 5) = 16{,}800$ N.

Question 18: B

Consider between release and Q: GPE lost = Work done against friction

Work done = Force x distance (in this case the proportion of the circumference).

$\Rightarrow mgr\cos 45 = 2\pi r \left(\frac{135}{360}\right) F$, where 'F' is the constant magnitude of the friction.

$\Rightarrow F = \frac{2\sqrt{2}mg}{3\pi}$.

Consider between release and P: GPE lost = KE gained + Work done.

$\Rightarrow mgr = KE + 2\pi r \left(\frac{90}{360}\right) F$ gives $KE = mgr - \frac{\pi r F}{2}$.

The expression for F can then be substituted in.

END OF PAPER

2019

Section 1A

Question 1: F

Expand each of the brackets out:

$$\Rightarrow \left(\sqrt{7}+\sqrt{3}\right)^2 - \left(\sqrt{7}-\sqrt{3}\right)^2 = \left(\left(\sqrt{7}\right)^2 + 2\sqrt{3}\sqrt{7} + \left(\sqrt{3}\right)^2\right) - \left(\left(\sqrt{7}\right)^2 - 2\sqrt{3}\sqrt{7} + \left(\sqrt{3}\right)^2\right)$$

The $\left(\sqrt{7}\right)^2$ and $\left(\sqrt{3}\right)^2$ in each bracket cancel to leave:

$$\Rightarrow 2\sqrt{3}\sqrt{7} - \left(-2\sqrt{3}\sqrt{7}\right) = 2\sqrt{21} + 2\sqrt{21} = 4\sqrt{21}.$$

Question 2: F

At a current of 8.0 mA, the graph shows that the diode has a voltage of 1.2 V across it. Since the battery provides a voltage of 6.0 V, there is a remaining voltage of $6.0 - 1.2 = 4.8$ V across the resistor. Use Ohm's law on the resistor to calculate the resistance:

$$V = IR \Rightarrow 4.8 = 8 \times 10^{-3} \times R$$

$$\Rightarrow R = \frac{4.8 \times 10^3}{8} = 0.6 \times 10^3 = 600\ \Omega.$$

Question 3: E

$$y = 3 - 4\left(1 - \frac{x}{2}\right)^2 \Rightarrow 4\left(1-\frac{x}{2}\right)^2 = 3 - y \Rightarrow \left(1-\frac{x}{2}\right)^2 = \frac{3-y}{4} \Rightarrow 1 - \frac{x}{2} = \pm\sqrt{\frac{3-y}{4}}$$

Note that even when moved to the other side of the equation the $\pm$ remains the same:

$$1 \pm \sqrt{\frac{3-y}{4}} = \frac{x}{2} \Rightarrow x = 2 \pm 2\sqrt{\frac{3-y}{4}}.$$

Question 4: A

When in the same medium, all electromagnetic waves travel at the same speed, so:

$$\frac{Speed\ of\ P}{Speed\ of\ Q} = 1.0.$$

Use the equation for wave speed to calculate the ratio of frequencies:

$$v = f\lambda \Rightarrow f = \frac{v}{\lambda}$$

Since speed is constant, frequency is inversely proportional to wavelength:

$$\frac{frequency\ of\ P}{frequency\ of\ Q} = 1 \Big/ \frac{wavelength\ of\ P}{wavelength\ of\ Q}$$

$$\Rightarrow \frac{1}{1.0 \times 10^8} = 1.0 \times 10^{-8}.$$

Only answers A and B are now possible, meaning that P must be a microwave. Since the wavelength of Q is much shorter than the wavelength of P, Q must be an X-ray, because the relative length of wavelengths of these waves are: X-rays < microwaves < radio waves.

Question 5: C

The resistance to the motion of the car can be written as $F = kv^2$, where v is the speed and k is a constant. Increasing the speed by 20% leads to v becoming 1.2v and thus:

$$F = k(1.2v)^2$$

$$F = 1.44kv^2$$

This is therefore an increase of 44%.

Question 6: A

Since air is an ideal gas at a constant temperature, $p \times V$ remains constant:

$$p_1V_1 = p_2V_2 \Rightarrow \left(1.0 \times 10^5\right) \times \left(2.0 \times 10^{-3}\right) = p_2 \times \left(4.0 \times 10^{-4}\right)$$

$$\Rightarrow p_2 = \frac{2 \times 10^2}{4 \times 10^{-4}} = 0.5 \times 10^6 = 5 \times 10^5\ \text{Pa}$$

The pressure exerted on the gas is due to the hydrostatic pressure plus the constant atmospheric pressure ($p = \rho g h + p_{atm}$), where ρ is the density of water. Using $\rho = 1000$ kgm-3, $p_{atm} = 1 \times 10^5\ Pa$ and $g = 10\ Nkg^{-1}$:

$$5 \times 10^5 = 1000 \times 10 \times h + 1 \times 10^5 \Rightarrow h = \frac{4 \times 10^5}{10^4}$$

$$\Rightarrow h = 40\ \text{m}.$$

Question 7: E

Substitute the first point (2, 6) into the equation:

$$6 = p(2)^2 + 2q \Rightarrow 6 = 4p + 2q \Rightarrow 2p + q = 3 \dots (1)$$

Substitute the second point (4, -4) into the equation:

$$-4 = p(4)^2 + 4q \Rightarrow -4 = 8p + 4q \Rightarrow 4p + q = -1 \dots (2)$$

Eq. (2) – (1):

$$4p + q - (2p + q) = -1 - (3) \Rightarrow 2p = -4 \Rightarrow p = -2.$$

Substitute p back into Equation (1):

$$2(-2) + q = 3 \Rightarrow -4 + q = 3 \Rightarrow q = 7$$

$$\therefore q - p = 7 - (-2) \Rightarrow q - p = 9.$$

Question 8: A

The voltage ratio is given by:

$$\Rightarrow \frac{V_p}{V_s} = \frac{n_p}{n_s}$$

Since the transformer is 100% efficient, $V_p I_p = V_s I_s$. This means that:

$$\frac{I_s}{I_p} = \frac{n_p}{n_s} \Rightarrow \frac{I_s}{4} = \frac{240}{4800} \Rightarrow I_s = \frac{4 \times 24}{480}$$

$$\Rightarrow I_s = \frac{4}{20} = 0.2 \text{ A}$$

The power produced as heat is given by $P = I^2 R$:

$$P = 0.2^2 \times 1500 = 0.04 \times 1500$$

$$\Rightarrow P = 4 \times 15 = 60 \text{ W}.$$

Question 9: D

First, divide the top and bottom of the fraction by x and factorise the bottom bracket:

$$4 - \frac{x(3x + 1)}{x^2(3x^2 - 2x - 1)} = 4 - \frac{3x + 1}{x(3x + 1)(x - 1)}$$

Cancel $(3x + 1)$ from the top and bottom and include 4 in the fraction:

$$4 - \frac{1}{x(x-1)} = \frac{4x(x-1)-1}{x(x-1)} = \frac{4x^2-4x-1}{x(x-1)}.$$

Question 10: F

The rate of conduction is increased most under the following conditions: large temperature difference, large surafce area and short path.

The lowest rate of thermal conduction therefore occurs when there is a small temperature difference (θ close to 90°C), small area (low d) and long path (large l). The option best satisfying these criteria is therefore option F.

Question 11: B

The volume of the ball is given by $V = \frac{4}{3}\pi r^3 = 192$ cm³. For the children's game, the radius is r_{child} and is $\frac{3}{4}$ the size of the radius of the adult ball, r, thus $r_{child} = \frac{3}{4}r$:

$$V_{child} = \frac{4}{3}\pi\left(\frac{3}{4}r\right)^3 \Rightarrow V_{child} = \left(\frac{3}{4}\right)^3 \times \frac{4}{3}\pi\ r^3$$

As $\frac{4}{3}\pi r^3 = 192$, we obtain:

$$V_{child} = \frac{27}{64} \times 192 \Rightarrow V_{child} = 27 \times 3 = 81 \text{ cm}^3.$$

Question 12: C

An alpha particle consists of 2 protons and 2 neutrons, and a beta particle consists of 1 electron. When an alpha particle is emitted, the atomic number decreases by 2 and the mass number decreases by 4. When a beta particle is emitted, the atomic number increases by 1 but the mass number remains constant.

There are more alpha particles than beta particles emitted, which means that at least four alpha particles must be emitted. The mass number must therefore fall by at least 16, excluding options A and B.

If four alpha particles and three beta particles are emitted, the mass number would fall by 16 and the atomic number would fall by $4 \times 2 + (3 \times -1) = 5$. Option E is therefore incorrect.

If five alpha particles and two beta particles are emitted, the mass number would fall by 20 and the atomic number would fall by $5 \times 2 + (2 \times -1) = 8$. Option C is therefore correct.

Repeating the method for six alpha particles and one beta particle, the mass number falls by 24 and the atomic number should fall by $6 \times 2 + (1 \times -1) = 11$, shows that option D is incorrect.

Question 13: F

First, use Pythagoras' theorem to calculate x:

$$(3x)^2 = (x+4)^2 + (2x+2)^2$$

$$\Rightarrow 9x^2 = x^2 + 8x + 16 + 4x^2 + 8x + 4 \Rightarrow 4x^2 - 16x - 20 = 0$$

$$\Rightarrow x^2 - 4x - 5 = 0 \Rightarrow (x-5)(x+1) = 0.$$

x must be positive as the length $3x$ can't be negative, so $x = 5$.

The sides of the triangle can now be calculated as $x + 4 = 9$ cm, $2x + 2 = 12$ cm and $3x = 15$ cm. Use the formula for the area of a right-angled triangle:

$$A = \frac{1}{2} \times base \times height \Rightarrow A = \frac{1}{2} \times 12 \times 9$$

$$\Rightarrow A = 6 \times 9 = 54 \text{ cm}^2.$$

Question 14: C

Kinetic energy is given by $KE = \frac{1}{2}mv^2$. Initially,

$$32 = \frac{1}{2} \times 4 \times v_1^2 \Rightarrow v_1 = \sqrt{16} = 4 \text{ ms}^{-1}.$$

After three seconds:

$$\Rightarrow 200 = \frac{1}{2} \times 4 \times v_2^2 \Rightarrow v_2 = \sqrt{100} = 10 \text{ ms}^{-1}$$

The acceleration during this time can therefore be calculated as:

$$\Rightarrow a = \frac{v_2 - v_1}{t} = \frac{10 - 4}{3} = 2 \text{ ms}^{-2}$$

Finally, use Newton's second law to calculate the resultant force:

$$F = ma \Rightarrow F = 4 \times 2 = 8 \text{ N}.$$

Question 15: D

The area of a rhombus is given by $A = \frac{1}{2}ab$, where a and b are the lengths of the two diagonals. Therefore:

$$\frac{1}{2}(3x+2)(8-2x) = 11 \Rightarrow 24x + 16 - 6x^2 - 4x = 22$$

$$6x^2 - 20x + 6 = 0 \Rightarrow 3x^2 - 10x + 3 = 0$$

$$(x-3)(3x-1) = 0 \Rightarrow x = 3 \text{ or } x = \frac{1}{3}$$

Since $PR = 3x + 2$, $PR = 3 \times (3) + 2 = 11$ or $PR = 3 \times \left(\frac{1}{3}\right) + 2 = 3$.

The difference between the two possible values is $11 - 3 = 8$ cm.

Question 16: E

The voltage across the two resistors in series is 2 V, as they have the same resistance. The voltage across the whole two branches of the resistors in parallel is therefore also 2 V, since parallel components have the same voltage. The total resistance of these two branches can be calculated using:

$$\frac{1}{R_p} = \frac{1}{R} + \frac{1}{2R} \Rightarrow \frac{1}{R_p} = \frac{2+1}{2R} \Rightarrow \frac{1}{R_p} = \frac{3}{2R} \Rightarrow R_p = \frac{2R}{3}.$$

Using Ohm's law, the current in the circuit is given by: $2 = I \times \frac{2R}{3} \Rightarrow I = \frac{3}{R}$ A.

The total resistance of the circuit is given by: $R_T = R + \frac{2R}{3} + R = \frac{8R}{3}.$

The voltage across the battery is therefore:

$$V_{battery} = \frac{3}{R} \times \frac{8R}{3} \Rightarrow V_{battery} = 8.0 \text{ V}.$$

Question 17: E

Using Pythagoras' theorem, $PR^2 = RT^2 = 4^2 + 3^2 = 25 \Rightarrow PR = RT = 5$ cm.

Now, let angle $QPR = \theta$:

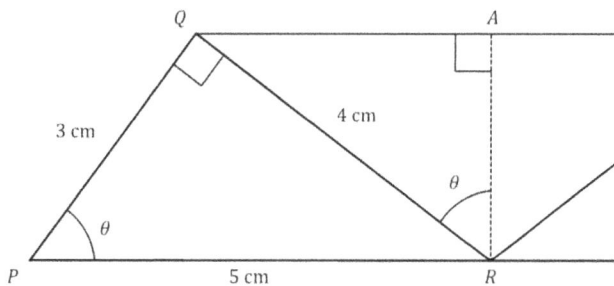

Angle QRP must be $90 - \theta$, and so angle QRA must be θ. This means that triangles PQR and AQR are similar. In triangle PQR the hypotenuse, PR, is 5 cm and in triangle AQR the hypotenuse, QR, is 4 cm, giving a scaling factor of $\frac{4}{5}$. The desired length is QA, the opposite length to angle θ which can be found by scaling the opposite in triangle PQR:

$$\Rightarrow Opposite_{AQR} = \frac{4}{5} \times Opposite_{PQR}$$

$$QA = \frac{4}{5} \times QR \Rightarrow QA = \frac{4}{5} \times 4 = 3.2 \text{ cm}$$

Since $QS = 2 \times QA$, $QS = 6.4$ cm.

Question 18: B

The cross-sectional area of the block is:

$$A = 22 \times 22 - \pi \times \left(\frac{14}{2}\right)^2 = 22 \times 22 - \frac{22}{7} \times 7^2 = 22 \times 22 - 22 \times 7 = 22(22 - 7) = 22 \times 15$$

$$\Rightarrow A = 330 \text{ cm}^2.$$

The mass of the block can therefore be calculated using:

$$m = area \times length \times density$$

$$\Rightarrow m = 330 \times 100 \times 0.1 = 3300 \text{ g} = 3.30 \text{ kg}.$$

Question 19: E

Using the formula in the question, the volume of the pyramid is:

$$V = \frac{1}{3} \times 12^2 \times h \Rightarrow V = 48h \text{ cm}^3$$

The length of each triangle along the surface from the base to the top is shown by l in the diagrams below.

Using Pythagoras' theorem,

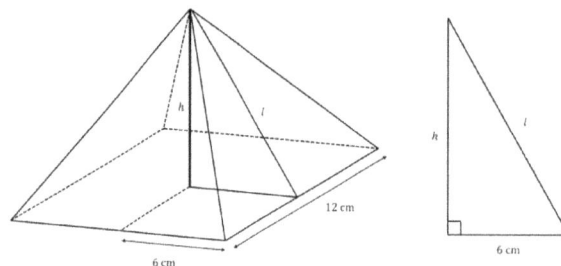

$$l^2 = \left(\frac{12}{2}\right)^2 + h^2 \Rightarrow l = \sqrt{36 + h^2}$$

The area of each triangle is therefore:

$$A_t = \frac{1}{2} \times 12 \times \sqrt{36 + h^2} \Rightarrow A_t = 6\sqrt{36 + h^2}.$$

Including the area of the base, the total surface area of the pyramid is:

$$A = 12^2 + 4 \times 6\sqrt{36 + h^2} = 144 + 24\sqrt{36 + h^2} \text{ cm}^2$$

Equate the area and volume:

$$\Rightarrow 144 + 24\sqrt{36 + h^2} = 48h \Rightarrow 6 + \sqrt{36 + h^2} = 2h \Rightarrow \sqrt{36 + h^2} = 2h - 6$$

$$36 + h^2 = (2h - 6)^2 \Rightarrow 36 + h^2 = 4h^2 - 24h + 36 \Rightarrow 3h^2 - 24h = 0$$

$$h^2 - 8h = 0 \Rightarrow h(h - 8) = 0.$$

Since h can't be zero, $h = 8$.

Question 20: B

Let there initially be a atoms of X and a atoms of Y. After 3 years, $\frac{a}{2}$ X atoms will have decayed, so there will be $a - \frac{a}{2} = \frac{a}{2}$ X atoms and $a + \frac{a}{2} = \frac{3a}{2}$ Y atoms.

After 6 years, a further $\frac{a}{4}$ X atoms will have decayed, so there will be $\frac{a}{2} - \frac{a}{4} = \frac{a}{4}$ X atoms and $\frac{3a}{2} + \frac{a}{4} = \frac{7a}{4}$ Y atoms.

The ratio number of X atoms: number of Y atoms is therefore:

$$\Rightarrow \frac{a}{4} : \frac{7a}{4} = 1 : 7.$$

Section 1B

Question 21: C

Firstly, sketch the lines and identify the required area:

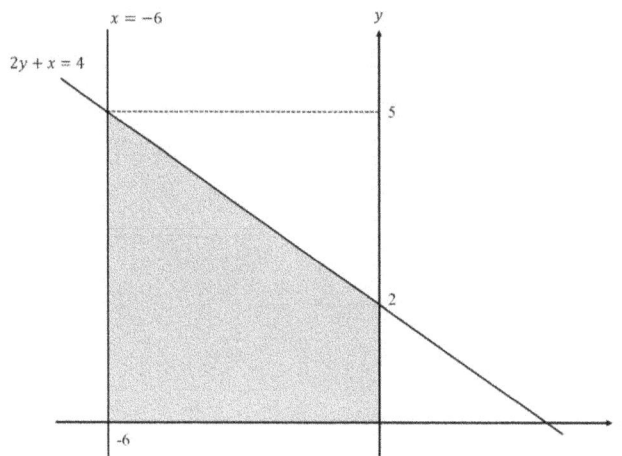

This shape is a trapezium, so its area can be found using the formula:

$A = \frac{a+b}{2}h \Rightarrow A = \frac{2+5}{2} \times 6 \Rightarrow A = \frac{7 \times 6}{2} \Rightarrow A = 21.$

Question 22: C

The gravitational potential energy of the water is converted to electrical energy with 90% efficiency. Let the volume of water passing through the turbine per minute be equal to V m^3. The loss of GPE per minute is then given by:

$\Rightarrow GPE_{loss} = V \times density \times g \times height = V \times 1000 \times 10 \times 150 = 1.5V \times 10^6$ J/min

The output power can be converted into output energy per minute, by converting megawatts to watts and multiplying by 60 seconds:

$\Rightarrow P_{out} = 1800 \text{ MW} = 1800 \times 10^6 \text{ } W \Rightarrow E_{out} = 1800 \times 10^6 \times 60$ J/min

Using an efficiency of 90%, the GPE loss and output energy can be equated:

$\Rightarrow 1800 \times 10^6 \times 60 = 0.9 \times 1.5 \times 10^6 V$

$V = \frac{1800 \times 10^6 \times 60}{0.9 \times 1.5 \times 10^6} = \frac{1800 \times 60}{0.9 \times 1.5} = \frac{18 \times 10^2 \times 60}{9 \times 10^{-1} \times 15 \times 10^{-1}}$

$\Rightarrow V = \frac{18}{9} \times \frac{60}{15} \times \frac{10^2}{10^{-2}} = 2 \times 4 \times 10^4 = 8 \times 10^4.$

Question 23: B

Differentiate the equation of the curve to find the turning points: $\frac{dy}{dx} = 3x^2 + 2px + q$.

At the maximum and minimum, $\frac{dy}{dx} = 0$.

When $x = -1$,

$3(-1)^2 + 2p(-1) + q = 0 \Rightarrow 3 - 2p + q = 0 \Rightarrow q = 2p - 3$

When $x = 3$,

$3(3^2) + 2p(3) + q = 0 \Rightarrow 27 + 6p + q = 0 \Rightarrow q = -6p - 27$

Equate the two equations for q:

$2p - 3 = -6p - 27 \Rightarrow 8p = -24 \Rightarrow p = -3$.

Question 24: E

Consider conservation of momentum: $Total\ momentum\ before = total\ momentum\ after$

$1000 \times 30 + 500 \times 20 = 1000 \times v_p + 500 \times 30$

Divide through by 1000:

$30 + 10 = v_p + 15 \Rightarrow v_p = 25\ \text{ms}^{-1}$.

The average force on Q can be calculated using: $Average\ force \times time = change\ in\ momentum$.

$F \times 0.2 = 500 \times 30 - 500 \times 20 \Rightarrow F \times \frac{1}{5} = 500(30 - 20)$

$F = 5 \times 500 \times 10 \Rightarrow F = 25000\ \text{N}$.

Question 25: B

Expand the brackets in the denominator:

$$\frac{1}{(1-\sqrt{2})^3} = \frac{1}{(1-\sqrt{2})(1-\sqrt{2})^2} = \frac{1}{(1-\sqrt{2})(1-2\sqrt{2}+2)} = \frac{1}{(1-\sqrt{2})(3-2\sqrt{2})}$$

$$\Rightarrow \frac{1}{3 - 3\sqrt{2} - 2\sqrt{2} + 2 \times 2} = \frac{1}{7 - 5\sqrt{2}}$$

Now rationalise the denominator:

$$\frac{1}{7-5\sqrt{2}} = \frac{1}{7-5\sqrt{2}} \times \frac{7+5\sqrt{2}}{7+5\sqrt{2}} = \frac{7+5\sqrt{2}}{(7-5\sqrt{2})(7+5\sqrt{2})}$$

$$= \frac{7+5\sqrt{2}}{49 - 25 \times 2} = \frac{-(7+5\sqrt{2})}{1} = -7 - 5\sqrt{2}.$$

Question 26: C

The resistance of the wire is given by:

$$R = \frac{Resistivity \times Length}{area} = \frac{1.6 \times 10^{-7} \times 0.5}{4 \times 10^{-7}} = \frac{0.8}{4} = 0.2 \ \Omega.$$

The potential difference is calculated using Ohm's law:

$$V = IR = 4 \times 0.2 = 0.8 \text{ V}.$$

Question 27: D

$7 \cos x + \tan x \sin x = 5$. Using $\tan x = \frac{\sin x}{\cos x}$ and multiplying through by $\cos x$:

$$\Rightarrow 7 \cos x + \frac{\sin^2 x}{\cos x} = 5 \Rightarrow 7 \cos^2 x + \sin^2 x = 5 \cos x.$$

Use the identity $\sin^2 x + \cos^2 x = 1$:

$$7 \cos^2 x + (1 - \cos^2 x) = 5 \cos x \Rightarrow 6 \cos^2 x - 5 \cos x + 1 = 0$$

$$(2 \cos x - 1)(3 \cos x - 1) = 0 \Rightarrow \cos x = \frac{1}{2} \text{ or } \cos x = \frac{1}{3}$$

Using the solution $\cos x = \frac{1}{2}$,

$$x = \cos^{-1} \frac{1}{2} \Rightarrow x = 60° \Rightarrow \tan x = \sqrt{3}$$

The only possible option is D.

To check further, the identity $\sin^2 x + \cos^2 x = 1$ can be divided through by $\cos^2 x$ to get:

$$\tan^2 x + 1 = \frac{1}{\cos^2 x} \Rightarrow \tan^2 x = \frac{1}{\cos^2 x} - 1$$

The results can be substituted in for the other solution $\left(\cos x = \frac{1}{3} \right)$ to confirm that option D is correct:

$$\tan^2 x = \frac{1}{\left(\frac{1}{3}\right)^2} - 1 \Rightarrow \tan^2 x = \frac{1}{\left(\frac{1}{9}\right)} - 1 \Rightarrow \tan^2 x = 9 - 1 = 8 \Rightarrow \tan x = 2\sqrt{2}.$$

Question 28: C

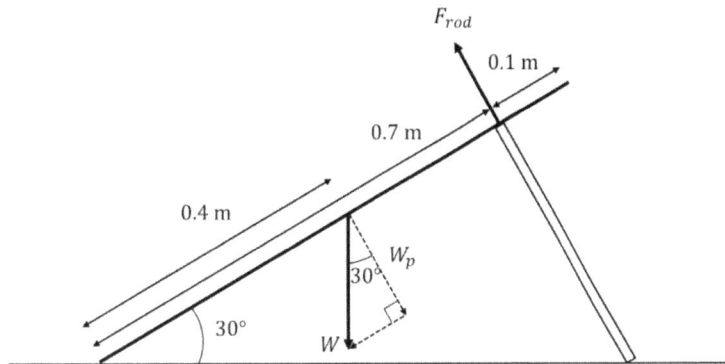

Equilibrium of moments about the hinge will need to be considered to find the required normal force. First, find the component of the weight acting perpendicular to the length of the trap door, W_P:

$$\cos 30 = \frac{W_p}{W}$$

Since $W = mg$,

$$W_p = mg \cos 30 \Rightarrow W_p = 14 \times 10 \times \frac{\sqrt{3}}{2} \Rightarrow W_p = 70\sqrt{3}$$

Now consider equilibrium of moments about the hinge, where the weight acts halfway along the rod, at a distance of 0.4 m:

$$F_{rod} \times 0.7 = 70\sqrt{3} \times 0.4 \Rightarrow F_{rod} = \frac{70\sqrt{3} \times 0.4}{0.7}$$

$$F_{rod} = 100\sqrt{3} \times 0.4 \Rightarrow F_{rod} = 40\sqrt{3} \ N.$$

Question 29: E

The height of the triangle can be calculated using Pythagoras' theorem:

$$h^2 = 8^2 - \left(\frac{8}{2}\right)^2 \Rightarrow h^2 = 64 - 16$$

$$h^2 = 48 \Rightarrow h = \sqrt{48} = 4\sqrt{3}.$$

The area of the triangle is therefore:

$$A_t = \frac{1}{2} \times base \times height$$

$$A_t = \frac{1}{2} \times 8 \times 4\sqrt{3} \Rightarrow A_t = 16\sqrt{3}$$

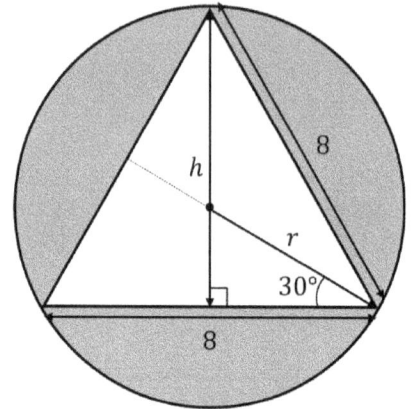

By drawing a line from one corner of the triangle to the middle of the opposite side (passing through the centre) and considering that angles in an equilateral triangle are 60°, the following triangle is formed:

Using trigonometric ratios,

$$\cos 30 = \frac{4}{r} \Rightarrow \frac{\sqrt{3}}{2} = \frac{4}{r}$$

$$r = \frac{4 \times 2}{\sqrt{3}} \Rightarrow r = \frac{8}{\sqrt{3}} \Rightarrow r = \frac{8\sqrt{3}}{3}.$$

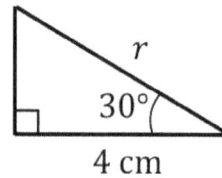

The area of the circle is therefore:

$$A_c = \pi r^2 \Rightarrow A_c = \pi \times \left(\frac{8\sqrt{3}}{3}\right)^2$$

$$A_c = \pi \times \frac{64 \times 3}{9} \Rightarrow A_c = \frac{64\pi}{3}$$

The total shaded area is therefore:

$$\Rightarrow A_c - A_t = \frac{64\pi}{3} - 16\sqrt{3} = \frac{16}{3}\left(4\pi - 3\sqrt{3}\right).$$

Question 30: C

The distance from the top speaker to the microphone is calculated using Pythagoras' theorem:

$$d^2 = 30^2 + 40^2 = 900 + 1600 = 2500 \Rightarrow d = 50 \text{ cm}$$

The wavelength of the sound can be calculated using $v = f\lambda$:

$$320 = 400\lambda \Rightarrow \lambda = \frac{320}{400} = \frac{4}{5} \text{ m} = 80 \text{ cm}$$

The phase difference is given by

$$\Rightarrow \Delta\phi = 360° \times \frac{path\ difference}{\lambda} = 360° \times \frac{50 - 40}{80} = 360° \times \frac{1}{8} = 45°.$$

Question 31: B

Rearrange the equation to form a quadratic in 5^x:

$$3 \times 5^{2x+1} - 5^x - 2 = 0 \Rightarrow 3 \times 5 \times 5^{2x} - 5^x - 2 = 0 \Rightarrow 15 \times \left(5^x\right)^2 - 5^x - 2 = 0$$

Use the quadratic equation:

$$5^x = \frac{1 \pm \sqrt{(-1)^2 - 4 \times 15 \times (-2)}}{2 \times 15} = \frac{1 \pm \sqrt{1 + 120}}{30} = \frac{1 \pm 11}{30} = \frac{12}{30} \text{ or } -\frac{10}{30}$$

5^x must be positive, so:

$$\Rightarrow 5^x = \frac{12}{30} = \frac{2}{5}$$

Now take logs and rearrange:

$$\log_5(5^x) = \log_5\left(\frac{2}{5}\right) \Rightarrow x = \log_5\left(\frac{2}{5}\right).$$

Question 32: D

Consider conservation of energy:

$$\Delta KE + \Delta GPE = 0 \Rightarrow (KE_{final} - KE_{initial}) + (GPE_{final} - GPE_{initial}) = 0$$

At maximum height, the ball is stationary thus $KE_{final} = 0$:

$$0 - \frac{1}{2}mv^2 + mgh - 0 = 0 \Rightarrow gh - \frac{1}{2}v^2 = 0$$

$$1.6 \times h = \frac{1}{2} \times 80^2 \Rightarrow h = \frac{3200}{1.6} \Rightarrow h = 2000 \text{ m}.$$

Question 33: A

Firstly, rearrange the second equation:

$$\int_{-2}^{2} 5f(x)dx - \int_{-2}^{4} f(x)dx = 7$$

$$\Rightarrow 5\int_{-2}^{2} f(x)dx - \left[\int_{2}^{4} f(x)dx + \int_{-2}^{2} f(x)dx\right] = 7$$

$$\Rightarrow 4\int_{-2}^{2} f(x)dx - \int_{2}^{4} f(x)dx = 7$$

Similarly, the first equation can be written as:

$$2\int_{-2}^{2} f(x)dx + \int_{2}^{4} f(x)dx = 4$$

These are now effectively two standard simultaneous equations. Denote $a = \int_{-2}^{2} f(x)dx$ and $b = \int_{2}^{4} f(x)dx$:

$$4a - b = 7 \ldots (1), \quad 2a + b = 4 \ldots (2)$$

Eq. $(1) + (2)$:

$$6a = 11 \Rightarrow a = \frac{11}{6}.$$

Substitute into (2):

$$2 \times \frac{11}{6} + b = 4 \Rightarrow b = 4 - \frac{11}{3} = \frac{12-11}{3} \Rightarrow b = \int_{2}^{4} f(x) \, dx = \frac{1}{3}.$$

Question 34: F

Rearrange Ohm's law to make I the subject, as on the graph:

$$V = IR \Rightarrow I = \frac{1}{R}V$$

The gradient of the line is therefore equal to $\frac{1}{R}$. For the bottom line,

$$\frac{1}{R_1} = \frac{20 \times 10^{-3}}{6} \Rightarrow R_1 = \frac{6}{20 \times 10^{-3}} \Rightarrow R_1 = 300 \ \Omega.$$

This corresponds to the case where only the 300 Ω resistor is connected. The top line must be therefore the case when both resistors are connected in parallel:

$$\frac{1}{R_T} = \frac{30 \times 10^{-3}}{6} \Rightarrow \frac{1}{R_T} = 5 \times 10^{-3} \Rightarrow \frac{1}{R_T} = \frac{1}{200}.$$

For the two resistors in parallel, where R_1 is the resistance of the first resistor and R_T is the resistance when both resistors are connected in parallel:

$$\frac{1}{R_T} = \frac{1}{R_1} + \frac{1}{R} \Rightarrow \frac{1}{200} = \frac{1}{R} + \frac{1}{300} \Rightarrow \frac{1}{R} = \frac{1}{600} \Rightarrow R = 600 \ \Omega.$$

Question 35: E

Use the binomial expansion on $(3 + 2t)^7$:

$$(3 + 2t)^7 = 3^7 + \binom{7}{1} \times 3^6 \times 2t + \binom{7}{2} \times 3^5 \times (2t)^2 + \binom{7}{3} \times 3^4 \times (2t^3)^3 + \cdots$$

After the integration, the required coefficient is of x^4, so before the integration the coefficient of t^3 is required:

$$\binom{7}{3} \times 3^4 \times (2t^3)^3 = \frac{7!}{3!\,(7-3)!} \times 3^4 \times 8t^3$$

$$= \frac{7!}{3!\,4!} \times 3^4 \times 8t^3 = \frac{7 \times 6 \times 5}{3 \times 2} \times 3^4 \times 8t^3$$

$$= \frac{7 \times 6 \times 5}{6} \times 3^4 \times 2^3 t^3 = 7 \times 5 \times 3^4 \times 2^3 t^3$$

$$\Rightarrow \int_0^x (7 \times 5 \times 3^4 \times 2^3) t^3 \, dt = 7 \times 5 \times 3^4 \times 2^3 \left[\frac{t^4}{4}\right]_0^x = 7 \times 5 \times 3^4 \times 2^3 \times \frac{1}{4} x^4 = 7 \times 5 \times 3^4 \times 2x^4$$

$$= 7 \times 81 \times 10x^4 = 567 \times 10x^4 = 5670x^4.$$

Question 36: D

Use the equations for Young modulus (E), stress (σ) and strain (ϵ):

$$E = \frac{\sigma}{\epsilon} \Rightarrow E = \frac{F}{A} \Big/ \frac{x}{L} \Rightarrow E = \frac{FL}{x}$$

Now rearrange and substitute values to find the extension, x:

$$x = \frac{FL}{AE} = \frac{mgL}{AE} = \frac{4 \times 10 \times 2.4}{2 \times 10^{-6} \times 1.2 \times 10^{11}}$$

$$= \frac{4}{2} \times \frac{2.4}{1.2} \times \frac{10}{10^5} = 4 \times 10^{-4} \text{ m}.$$

Now use the formula for strain energy:

$$U = \frac{1}{2}Fx \Rightarrow U = \frac{1}{2}mgx = \frac{1}{2} \times 4 \times 10 \times 4 \times 10^{-4} = 8 \times 10^{-3} \text{ J}.$$

Question 37: D

First, solve the simultaneous equations:

$$3\tan\alpha - 2\sin\beta = 2 \quad \text{... (1)}, \quad 5\tan\alpha + 6\sin\beta = 8 \text{ ... (2)}$$

$(1) \times 3 + (2)$:

$$9\tan\alpha + 5\tan\alpha = 6 + 8 \Rightarrow 14\tan\alpha = 14 \Rightarrow \tan\alpha = 1 \Rightarrow \alpha = 45°$$

Substitute into Eq. (1):

$$3 - 2\sin\beta = 2 \Rightarrow \sin\beta = \frac{1}{2} \Rightarrow \beta = 30°$$

Finally, internal angles of a triangle must sum to $180°$:

$$\theta + 45 + 30 = 180 \Rightarrow \theta = 105°.$$

Question 38: A

Note that α is not the angle of incidence; the angle of incidence is defined as the angle between the incident ray of light and the normal to the surface. The angle of incidence in both cases is therefore $90 - \alpha$ and the angle of refraction in the second case is $90 - \beta$, as shown in the diagram.

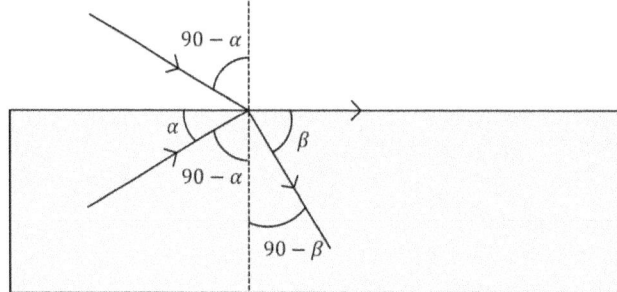

In the first case, the light must be incident at the critical angle and so the formula for the critical angle can be used:

$$\frac{n_2}{n_1} = \sin(90 - \alpha) = \cos\alpha$$

In the second case the light refracts according to:

$$\frac{\sin\theta_1}{\sin\theta_2} = \frac{n_2}{n_1} \Rightarrow \frac{\sin(90-\beta)}{\sin(90-\alpha)} = \frac{n_2}{n_1} = \cos\alpha$$

$$\frac{\cos\beta}{\cos\alpha} = \cos\alpha \Rightarrow \cos\beta = \cos^2\alpha.$$

Question 39: F

Substituting both $x = -1$ and $= 4$ into $x^3 - 2x^2 - 7x - 4$ gives zero, showing that these are both roots. The expression can now be factorised:

$$x^3 - 2x^2 - 7x - 4 = (x + 1)(x^2 - 3x - 4) = (x + 1)(x + 1)(x - 4) = (x + 1)^2(x - 4).$$

As there is a positive x^3 coefficient, the graph of $y = x^3 - 2x^2 - 7x - 4$ takes the form shown below.

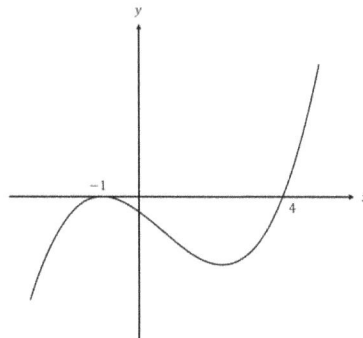

The expression is therefore positive for $x > 4$.

Question 40: B

At this stage, the person has a constant upwards acceleration of:

$$a = \frac{0-2}{25-20} = -\frac{2}{5} \text{ ms}^{-2}$$

Now use Newton's second law to find the contact force:

$$F - mg = ma$$

$$F - 80 \times 10 = 80 \times -\frac{2}{5} \Rightarrow F = 800 - 32 \Rightarrow F = 768 \text{ N.}$$

END OF SECTION

Section 2

Question 1: A

Use the formula for refraction:

$$\frac{\sin\theta_1}{\sin\theta_2} = \frac{n_2}{n_1}$$

Since the refractive index in a vacuum is equal to 1,

$$\frac{\sin x}{\sin y} = n_{medium}$$

When $x = 60°$ and $y = 45°$,

$$\Rightarrow n_{medium} = \frac{\sin 60}{\sin 45} = \frac{\frac{\sqrt{3}}{2}}{\frac{\sqrt{2}}{2}} = \frac{\sqrt{3}}{\sqrt{2}}.$$

When $x = 45°$,

$$\frac{\sin 45}{\sin y} = \frac{\sqrt{3}}{\sqrt{2}} \Rightarrow \sin y = \frac{\sqrt{2}}{\sqrt{3}} \times \sin 45 = \frac{\sqrt{2}}{\sqrt{3}} \times \frac{1}{\sqrt{2}} = \frac{1}{\sqrt{3}}.$$

Question 2: A

Using Ohm's law,

$$V = IR \Rightarrow I = \frac{V}{R}$$

Substitute this into $P = IV$:

$$P = \left(\frac{V}{R}\right)V \Rightarrow P = \frac{V^2}{R}$$

The order of increasing power is therefore equivalent to the order of decreasing resistance.

Let each resistor have resistance R. The total resistance of arrangement X is given by:

$$R_X = R + R = 2R$$

Two resistors in parallel have equivalent resistance of $\frac{R}{2}$ $\left(\frac{1}{R_T} = \frac{1}{R} + \frac{1}{R} = \frac{2}{R} \rightarrow R_T = \frac{R}{2}\right)$ and so the total resistance of arrangement Y is given by:

$$R_Y = R + \frac{R}{2} = \frac{3R}{2}$$

Finally, the total resistance of arrangement Z is given by:

$$\frac{1}{R_Z} = \frac{1}{R} + \frac{1}{2R} = \frac{2+1}{2R} = \frac{3}{2R} \Rightarrow R_Z = \frac{2R}{3}.$$

The order of decreasing resistance (and therefore increasing power) is X, Y, Z.

Question 3: C

First, find the component of the block's weight acting parallel to the slope:

$$\sin 30 = \frac{W_p}{W} \Rightarrow W_p = mg \sin 30$$

$$= 2 \times 10 \times \frac{1}{2} = 10 \text{ N}$$

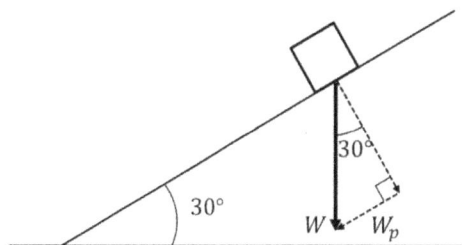

Since the load moves at a constant speed, there must be no resultant force:

$$Mg = 10 + 5 \Rightarrow M \times 10 = 15 \Rightarrow M = 1.5 \text{ kg}.$$

Question 4: B

When the switch is closed, the two $8\,\Omega$ resistors in parallel have a total resistance given by:

$$\frac{1}{R} = \frac{1}{8} + \frac{1}{8} \Rightarrow \frac{1}{R} = \frac{1}{4} \Rightarrow R = 4\,\Omega$$

The total resistance of the circuit is $R_T = 1 + 3 + 4 = 8\,\Omega$ and so the current in the circuit is:

$$I = \frac{V}{R_T} = \frac{16}{8} = 2 \text{ A}.$$

The voltmeter reading will therefore be:

$$V = 2 \times 3 = 6 \text{ V}.$$

The only possible option is therefore B but calculating the reading when the switch is open is a good way to check. When the switch is open, the circuit will have a total resistance of $1 + 3 + 8 = 12\,\Omega$ and so the current is:

$$I = \frac{16}{12} = \frac{4}{3} \text{ A}$$

The voltmeter reading will therefore be:

$$V = \frac{4}{3} \times 3 = 4 \text{ V}.$$

Question 5: E

The question states that the distance between adjacent nodes is 4 cm, which means that the wavelength, λ, is 8 cm. Since the standing wave is formed by the superposition of two waves of amplitude 1.5 cm, the amplitude of the standing wave is 3 cm. Finally, the frequency of the standing wave can be found:

$$v = f\lambda \Rightarrow f = \frac{3.2}{8 \times 10^{-2}}$$

$$f = 0.4 \times 10^2 \Rightarrow f = 40 \text{ Hz}.$$

In one time period, a particle at an antinode moves a distance of 4 amplitudes: it travels one amplitude to move from the antinode to the origin, and another amplitude to reach the opposite antinode. This is then repeated to reach the original position, giving 4 amplitudes or $4 \times 3 = 12$ cm. In 1 minute, a particle at an antinode therefore moves a total distance given by

Distance per oscillation $\times$ oscillations per second $\times$ time $= 12$ cm $\times 40$ Hz $\times 60$ seconds

$= 12 \times 6 \times 4 \times 10^{-2} \times 10^2 = 72 \times 4 = 288$ m.

Question 6: A

When the ray strikes at an angle θ to the boundary, the angle of incidence (between the ray and the normal) is $90 - \theta$. When the light is at the threshold of being totally internally reflected, the angle of incidence must be equal to the critical angle. The formula for the critical angle is:

$$\frac{v_2}{v_1} = \frac{1}{\sin \theta_c}$$

Using the value of $\theta = 65°$, the angle of incidence is at the critical angle, equal to $90° - 65°$:

$$\frac{v_{air}}{v_{material}} = \frac{1}{\sin(90-65)} \Rightarrow \frac{v_{air}}{v_{material}} = \frac{1}{\cos 65} \Rightarrow v_{material} = v_{air} \cos 65$$

The wavelength and speed are related through the frequency of the wave:

$$v = f\lambda \Rightarrow \lambda_{material} = \frac{v_{material}}{f} = \frac{v_{air} \times \cos 65}{f}.$$

Question 7: E

The average pressure exerted on the ground due to the pyramid is given by:

$$Pressure = \frac{weight}{base\ area}$$

$$p = \frac{volume \times density \times g}{base\ area} \Rightarrow p = \frac{volume}{base\ area} \times density \times g$$

Using the equation given in the question,

$$\frac{Volume}{Base\ area} = \frac{1}{3} \times Height$$

$$\Rightarrow p = \frac{1}{3} \times height \times density \times g = \frac{1}{3} \times 140 \times 2100 \times 10$$

$$= \frac{1}{3} \times 14 \times 21 \times 10^4 = 7 \times 14 \times 10^4 = 98 \times 10^4 = 980 \text{ kPa.}$$

It is important to note that this pressure is in addition to the existing atmospheric pressure, so the total average pressure on the ground is: $980 + 100 = 1080$ kPa.

Question 8: C

The hydrostatic pressure at a distance of x below the surface is $\rho g x$. There is also a constant atmospheric pressure of P, so the total pressure, p, at a distance of x below the surface is equal to $P + \rho g x$.

The question states that the pressure exerted by the gas, p, is directly proportional to its density. This can be written as

$$p \propto \rho \ \Rightarrow p \propto \frac{m}{V}$$

And the mass of the gas in the bubble remains constant:

$$p \propto \frac{1}{V} \ \Rightarrow p \propto \frac{3}{4\pi r^3}$$

Since $\frac{3}{4\pi}$ is a constant,

$$\Rightarrow p \propto \frac{1}{r^3}$$

The proportional relationship can be written using a constant k and the two expressions for p can be equated:

$$p = \frac{k}{r^3} \ \Rightarrow P + \rho g x = \frac{k}{r^3}$$

At a depth h, the radius of the bubble is R:

$$P + \rho g h = \frac{k}{R^3} \ \Rightarrow k = R^3 (P + \rho g h)$$

This expression for k can now be substituted in to find the required expression:

$$\Rightarrow P + \rho g x = \frac{R^3 (P + \rho g h)}{r^3}$$

$$r^3 = \frac{R^3 (P + \rho g h)}{P + \rho g x} \ \Rightarrow r = R \left(\frac{P + \rho g h}{P + \rho g x} \right)^{\frac{1}{3}}.$$

Question 9: G

The easiest way to approach this question is by considering the conservation of energy.

Work Done by Friction $+ \Delta KE + \Delta GPE = 0$.

The work done by friction is given by: $Work = F_{avg} \times distance = 4 \times F_{avg}$.

The block comes to rest, so its change in kinetic energy is:

$$\Delta KE = 0 - \frac{1}{2} \times 2 \times 8^2 \Rightarrow \Delta KE = -64 \text{ J}.$$

The change in gravitational potential energy is given by $-mg\Delta h$, where Δh is the vertical change in height:

$$\sin 30 = \frac{\Delta h}{4} \Rightarrow \Delta h = 4 \times \frac{1}{2} = 2 \text{ m}$$

$$\Delta GPE = -2 \times 10 \times 2 \Rightarrow \Delta GPE = -40 \text{ J}.$$

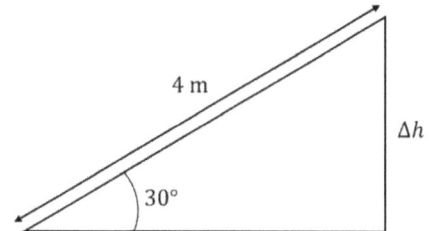

Substituting these values back into the conservation of energy equation:

$$4 \times F_{avg} - 64 - 40 = 0$$

$$\Rightarrow F_{avg} = 16 + 10 = 26 \text{ N}.$$

Question 10: E

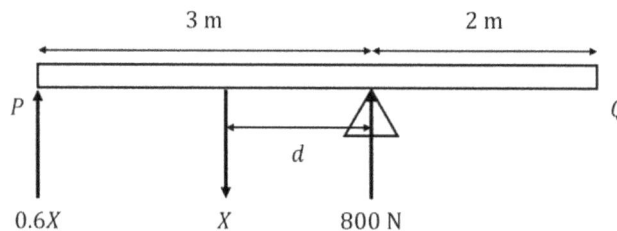

To find the weight of the beam, consider vertical equilibrium:

$$0.6X + 800 = X \Rightarrow 0.4X = 800 \Rightarrow X = 2000 \text{ N}$$

To find the distance from the pivot, consider equilibrium of moments around the pivot:

$$0.6 \times 2000 \times 3 = 2000 \times d \Rightarrow d = 1.8 \text{ m}.$$

Question 11: C

Use the equation $s(t) = ut + \frac{1}{2}at^2$ for the distance travelled to set up simultaneous equations, substituting in $t = 3$ for $s(3)$ and $t = 2$ for $s(2)$:

$$\Rightarrow s(3) - s(2) = 3u + \frac{1}{2}a\left(3^2\right) - \left(2u + \frac{1}{2}a\left(2^2\right)\right) = u + \frac{1}{2}a(9-4) = u + \frac{5a}{2} = 12.2 \ldots (1)$$

$$s(4) - s(3) = 4u + \frac{1}{2}a(4^2) - \left(3u + \frac{1}{2}a(3^2)\right) = u + \frac{1}{2}a(16-9) = u + \frac{7a}{2} = 14.4 \ldots (2)$$

Eq. $(2) - (1)$:

$$\frac{7-5}{2}a = 14.4 - 12.2 \Rightarrow a = 2.2 \text{ ms}^{-2}.$$

Substitute this back into Eq. (1):

$$u + \frac{5 \times 2.2}{2} = 12.2 \Rightarrow u + 5.5 = 12.2 \Rightarrow u = 6.7 \text{ ms}^{-1}.$$

Question 12: A

$$Averg\ power = \frac{strain\ energy}{time\ taken}.$$

In order to calculate the time taken, the final tension force needs to be known. This can be calculated if the extension, x, is known, which in turn can be calculated by considering the strain energy:

$$\frac{1}{2}kx^2 = 0.25 \Rightarrow \frac{1}{2} \times 50 \times x^2 = 0.25$$

$$x^2 = 0.01 \Rightarrow x = 0.1 \text{ m.}$$

Now the force at the final extension can be calculated using Hooke's law:

$$F = kx \Rightarrow F = 50 \times 0.1 = 5 \text{ N.}$$

This means that the tension force is active for a total of $\frac{5}{0.2} = 25$ seconds:

$$Average\ power = \frac{0.25}{25} = 0.01 \text{ W.}$$

Question 13: B

Let the resistance of each resistor be R and the voltage of each cell be V. When the switch is open, the current in the circuit is given by Ohm's law:

$$V_{open} = I_{open} R_{open} \Rightarrow I_{open} = \frac{2V}{3R}$$

So the power dissipated in resistor X is:

$$P = I_{open}^2 R \Rightarrow P = \left(\frac{2V}{3R}\right)^2 \times R$$

$$P = \frac{4V^2}{9R^2} \times R \Rightarrow P = \frac{4V^2}{9R} \Rightarrow P = \frac{4}{9}\left(\frac{V^2}{R}\right).$$

This can be rearranged to give:

$$\Rightarrow \frac{V^2}{R} = \frac{9P}{4}.$$

When the switch is closed, the current in the circuit is:

$$I_{closed} = \frac{V}{2R}.$$

Therefore, the power in the circuit is:

$$P_{closed} = \left(\frac{V}{2R}\right)^2 \times R \Rightarrow P_{closed} = \frac{V^2}{4R^2} = \frac{1}{4}\left(\frac{V^2}{R}\right).$$

Substitute in the expression for $\frac{V^2}{R}$ from above:

$$\Rightarrow P_{closed} = \frac{1}{4}\left(\frac{9P}{4}\right) = \frac{9P}{16}.$$

Question 14: C

Consider just the forces acting on the caravan: drag forces of D_2 and a tension in the bar of T. Newton's second law gives:

$$T - D_2 = Ma \Rightarrow T = D_2 + Ma$$

The Young modulus of the bar can be written as:

$$E = \frac{stress}{strain} = \frac{T}{A} \bigg/ \frac{x}{l} = \frac{Tl}{Ax} \Rightarrow T = \frac{AEx}{l}.$$

Equate the two expressions for tension:

$$D_2 + Ma = \frac{AEx}{l} \Rightarrow x = \frac{(D_2 + Ma)l}{AE}.$$

Question 15: E

Let the battery voltage by V_{in}. Using the potential divider equation, the voltage at the node between P and Q is equal to:

$$V_{PQ} = \frac{V_{in}R_Q}{R_Q + R_P}$$

Similarly, the voltage at the node between R and S is:

$$V_{RS} = \frac{V_{in}R_S}{R_S + R_R}$$

The reading on the voltmeter gives the difference between these voltages:

$$V = V_{PQ} - V_{RS} \Rightarrow V = V_n \left(\frac{R_Q}{R_Q+R_P} - \frac{R_S}{R_S+R_R} \right) \dots (1)$$

$$V = V_{in} \left(\frac{2x}{2x+x} - \frac{x}{x+2x} \right) \Rightarrow V = V_{in} \left(\frac{2x-x}{3x} \right) \Rightarrow V = \frac{V_{in}}{3}$$

Option 1:

Connecting T with a resistance x in parallel with P gives a new effective resistance of P as:

$$\frac{1}{R_P} = \frac{1}{x} + \frac{1}{x} = \frac{2}{x} \Rightarrow R_P = \frac{x}{2}$$

Substituting this new value into Eq. (1) gives:

$$V = V_{in} \left(\frac{2x}{\frac{x}{2}+2x} - \frac{x}{x+2x} \right) \Rightarrow V = V_{in} \left(\frac{4}{5} - \frac{1}{3} \right) \Rightarrow V = \frac{7V_{in}}{15} > \frac{V_{in}}{3}$$

Option 2:

Connecting T in series with Q now gives an effective resistance of Q as:

$$R_Q = x + 2x = 3x$$

Substituting this new value into Eq. (1) gives:

$$V = V_{in} \left(\frac{3x}{x+3x} - \frac{x}{x+2x} \right) \Rightarrow V = V_{in} \left(\frac{3}{4} - \frac{1}{3} \right) \Rightarrow V = \frac{5V_{in}}{12} > \frac{V_{in}}{3}$$

Option 3:

Connecting a resistance x in parallel with R gives a new effective resistance of R as:

$$\frac{1}{R_R} = \frac{1}{2x} + \frac{1}{x} = \frac{3}{2x} \Rightarrow R_R = \frac{2x}{3}$$

Substituting this new value into Eq. (1) gives:

$$V = V_{in} \left(\frac{2x}{x+2x} - \frac{x}{x+\frac{2x}{3}} \right) \Rightarrow V = V_{in} \left(\frac{2}{3} - \frac{3}{5} \right) \Rightarrow V = \frac{1V_{in}}{15} < \frac{V_{in}}{3}$$

Therefore, only options 1 and 2 increase the voltage.

Question 16: F

It is clear from the graph that acceleration is not constant, due to air resistance.

Option 1 is therefore true – as, at any other point, air resistance will have an impact on acceleration.

Option 2 is not true, as this would only be the case if air resistance was neglected and the only force was the weight of the ball.

Option 3 is true, as energy is lost due to resistive forces at each stage.

Question 17: A

Let the stone's initial vertical speed be u. It travels upwards for 0.6 s, at which time it must be at its maximum point, thus $v = 0$ and $a = -10$:

$$v = u + at \Rightarrow 0 = u - 10 \times 0.6 \Rightarrow u = 6 \text{ ms}^{-1}$$

Now substitute h in at time $t = 1$ s, using the SUVAT equation $s = ut + \frac{1}{2}at^2$:

$$h = 6(1) + \frac{1}{2}(-10)(1^2) \Rightarrow h = 6 - 5 = 1 \text{ m}.$$

Question 18: B

First, note that the distance between the hinge and the winch is equal to the length of the ramp, and so an isosceles triangle is formed:

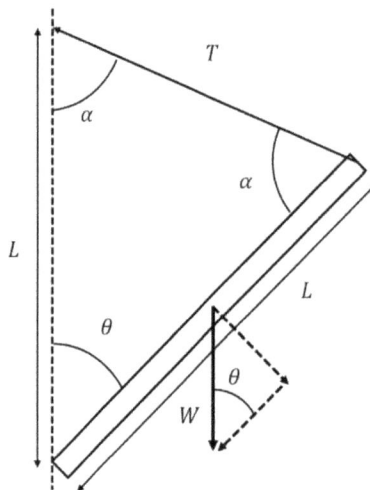

The ramp is lowered at constant speed, so there must be equilibrium of moments about the hinge. The component of the ramp's weight acting perpendicular to the ramp, W_p, is given by:

$$\sin \theta = \frac{W_p}{W} \Rightarrow W_p = W \sin \theta$$

Similarly for the component of the cable tension acting perpendicular to the ramp,

$$\sin \alpha = \frac{T_P}{T} \Rightarrow T_P = T \sin \alpha.$$

Moment is given by: $Moment = Force \times Perpendicular\ distance\ to\ pivot$. So, when the moments are balanced,

$$W \sin \theta \times \frac{L}{2} = T \sin \alpha \times L \Rightarrow T \sin \alpha = \frac{W}{2} \sin \theta.$$

Since internal angles in a triangle sum to 180°,

$$2\alpha + \theta = 180 \Rightarrow \alpha = \frac{180 - \theta}{2} \Rightarrow \alpha = 90 - \frac{\theta}{2}$$

Substitute this back into the equilibrium of moments:

$$T \sin \left(90 - \frac{\theta}{2} \right) = \frac{W}{2} \sin \theta \Rightarrow T \cos \frac{\theta}{2} = \frac{W}{2} \sin \theta$$

Let $\phi = \frac{\theta}{2}$:

$$T \cos \phi = \frac{W}{2} \sin 2\phi$$

Use the double-angle identity for $\sin 2\phi$:

$$T \cos \phi = \frac{W}{2} 2 \sin \phi \cos \phi \Rightarrow T \cos \phi = W \sin \phi \cos \phi$$

$$T = W \sin \phi \Rightarrow T = W \sin \frac{\theta}{2}$$

T therefore increases as θ increases, up to a value of $\frac{\theta}{2} = 90°$. Since the maximum value of θ reached is 90°,

$$T_{max} = W \sin \frac{90}{2} \Rightarrow T_{max} = W \sin 45 \Rightarrow T_{max} = \frac{W}{\sqrt{2}}.$$

Question 19: B

Kinetic energy is given by $KE = \frac{1}{2} m v^2$, so the moving particle has initial velocity given by:

$$\frac{1}{2} m u^2 = E \Rightarrow u = \sqrt{\frac{2E}{m}}$$

Now consider conservation of momentum:

$$m \sqrt{\frac{2E}{m}} = (m + M)v \Rightarrow v = \frac{\sqrt{2Em}}{m + M}$$

The final kinetic energy of the two particles is:

$$KE = \frac{1}{2} (m + M) \frac{2Em}{(m + M)^2} = \frac{Em}{m + M}$$

Kinetic energy transferred to other forms of energy is therefore:

$$E - \frac{Em}{(m + M)} = E \left(\frac{m + M}{m + M} - \frac{m}{m + M} \right) = \frac{EM}{m + M}.$$

Question 20: C

The equation for a critical angle is: $\frac{n_1}{n_2} = \frac{1}{\sin\theta_c}$.

For medium X:
$$\frac{n_X}{n_{air}} = \frac{1}{\sin 45} \Rightarrow n_X = \frac{n_{air}}{\left(\frac{\sqrt{2}}{2}\right)} = n_{air}\sqrt{2}$$

For medium Y:
$$\frac{n_Y}{n_{air}} = \frac{1}{\sin 60} \Rightarrow n_Y = \frac{n_{air}}{\left(\frac{\sqrt{3}}{2}\right)} = n_{air}\frac{2\sqrt{3}}{3}$$

Light can only totally internally reflect when travelling from a more optically dense medium to a less optically dense medium and since $n_X > n_Y$, the direction of incidence must be from X to Y.

When travelling from X to Y, the critical angle is given by:

$$\frac{n_X}{n_Y} = \frac{1}{\sin\theta_c} \Rightarrow \sin\theta_c = \frac{n_Y}{n_X} \Rightarrow \sin\theta_c = \frac{2\sqrt{3}}{3\sqrt{2}} \Rightarrow \sin\theta_c = \frac{\sqrt{3}\sqrt{2}}{3}$$

This expression can also be written as $\frac{\sqrt{2}}{\sqrt{3}}$, and so it is clear than $\sin\theta_c > \frac{\sqrt{2}}{2}$ and therefore $\theta_c > 45°$. To check if θ_c is larger or smaller than $60°$, it is necessary to use an approximation:

$$\frac{\sqrt{3}\sqrt{2}}{3} \approx \sqrt{3} \times \frac{1.4}{3} < \sqrt{3} \times \frac{1}{2}.$$

Since $\sin 60 = \frac{\sqrt{3}}{2}$, $\sin\theta_c < \sin 60$ and therefore: $45° < \theta_c < 60°$.

END OF PAPER

2020

Section 1A

Question 1: C

The increase in thermal energy of the copper tip is

$$\Delta E = mc\Delta T = 2 \times 10^{-3} \times 400 \times 200 = 160 \text{ J}$$

The energy transferred to the tip in the time is

$$E = P \times t = 30 \times 50 = 1500 \text{ J}$$

Hence, the energy transferred to the surroundings is $1500 - 160 = 1340$ J.

Question 2: D

Let the cost of an adult be a, and the cost of a child be c. The information given in the question then forms two simultaneous equations:

$$2a + 3c = 20 \quad (1)$$

$$4a + 4c = 34 \quad (2)$$

$$2 \times (1) - (2):$$

$$2c = 6 \rightarrow c = 3$$

Substituting this back into (1) gives $a = 5.5$. The cost of 6 adults and 2 children is therefore

$$6 \times 5.5 + 2 \times 3 = 33 + 6 = £39$$

Question 3: A

Beta-decay does not affect the mass number but changes the atomic number by $+1$, whilst alpha decay changes the mass number by -4 and the atomic number by -2. For the mass number to have decreased from 238 to 206, there must have been a total $\frac{238-206}{4} = \frac{32}{4} = 8$ alpha emissions. This would change the atomic number to $92 - 2 \times 8 = 92 - 16 = 76$. There must therefore have been $82 - 76 = 6$ beta particles emitted.

Question 4: A

On any given spin, the probability of a 1 is $\frac{3}{8}$, the probability of a 2 is $\frac{1}{8}$ and the probability of a 3 is $\frac{4}{8} = \frac{1}{2}$. For the sum of the two numbers to be 5, the spinner must land on a 2 and a 3. This could occur in either order, so the individual probabilities are multiplied by 2. The probability is therefore $\frac{1}{8} \times \frac{1}{2} \times 2 = \frac{1}{8}$.

Question 5: D

Since power is given by $P = IV$, the efficiency is given by $\frac{160 \times 4000}{400 \times 2000} = \frac{160}{200} = 80\%$. The energy wasted per minute is then

$$E = (1 - 0.8) \times IVt = 0.2 \times 400 \times 10^3 \times 2000 \times 60 = 8 \times 12 \times 10^8 = 96 \times 10^8 = 9.6 \times 10^9 \text{ J}$$

Question 6: B

Rearrange the equation of each line into the form $y = mx + c$:

(1): $y = -\dfrac{1}{3}x + \dfrac{1}{2}$

(2): $y = \dfrac{1}{3}x - \dfrac{4}{9}$

(3): $y = 3x + \dfrac{3}{2}$

(4): $y = -\dfrac{2}{3}x + \dfrac{3}{2}$

Two lines are perpendicular when their gradients multiply to give -1. Therefore, we can see that lines I and 3 are perpendicular.

Question 7: F

The force on the rod is given by $F = BIL$, where the length, L, is the length of the rod through which current travels (12 cm). The acceleration of the rod is then given by $a = \dfrac{F}{m} = \dfrac{BIL}{W/g} = \dfrac{BILg}{W}$. Using the values given in the question gives:

$$a = \frac{0.5 \times 2.4 \times 0.12 \times 10}{0.4} = \frac{1.2 \times 1.2}{0.4} = 3.6 \text{ ms}^{-2}$$

Using the left-hand rule, as the magnetic field is directed into the page and the current in the rod flows downwards, the direction of the acceleration is to the right.

Question 8: F

Use the substitution $y = \dfrac{x}{4} + 3$ to transform the equation into a standard quadratic:

$2y^2 - y - 36 = 0$

$(2y - 9)(y + 4) = 0$

$y = \dfrac{9}{2}, y = -4$

Since $x = 4(y - 3) = 4y - 12$,

$x = 4 \times \dfrac{9}{2} - 12, x = 4 \times -4 - 12$

$x = 6, x = -28$

The sum of the solutions is therefore $6 - 28 = -22$.

Question 9: C

First, consider the conservation of momentum to find the velocity of the combined trolleys. Taking rightwards as positive,

$$8 \times 4 - 2 \times 1 = (8 + 2) \times v$$

$$v = \frac{32 - 2}{10} = 3 \text{ ms}^{-1}$$

The loss in kinetic energy is therefore

$$KE_{loss} = KE_{initial} - KE_{final}$$

$$KE_{loss} = (\frac{1}{2} \times 8 \times 4^2 + \frac{1}{2} \times 2 \times 1^2) - (\frac{1}{2} \times 10 \times 3^2) = 64 + 1 - 45 = 20 \text{ J}$$

Question 10: A

Expand each bracket and then complete the square on the simplified expression:

$$(2x + 3)^2 - (x - 3)^2 = 4x^2 + 12x + 9 - (x^2 - 6x + 9) = 3x^2 + 18x$$

$$= 3(x^2 + 6x) = 3[(x + 3)^2 - 9] = 3(x + 3)^2 - 27$$

Question 11: D

The car accelerates for the first 5 seconds, after which its speed remains constant. This speed can be calculated by considering the graph between 5 and 10 seconds:

$$v = \frac{40 - 10}{10 - 5} = \frac{30}{5} = 6 \text{ ms}^{-1}$$

During the accelerating period, the average acceleration is given by

$$a = \frac{\Delta v}{\Delta t} = \frac{6}{5} = 1.2 \text{ ms}^{-2}$$

Since the car's mass is 800 kg, the average resultant force is

$$F = 800 \times 1.2 = 960 \text{ N}$$

Question 12: B

The number of pairs of boots sold is given by $N = \frac{k}{T^3}$, where T is the temperature and k is a constant. Since 250 pairs are sold when the temperature is 8°C, $k = NT^3 = 250 \times 8^3 = 128000$. When the temperature is x on the following day, 8 times as many pairs of boots are sold:

$$8 \times 250 = \frac{128000}{x^3}$$

$$x^3 = 8^2 = 64$$

$$x = 4°C$$

Question 13: C

Since $v = f\lambda$, $v = \frac{\lambda}{2}$. From $t = 0$ to $t = 0.80$, the wave has travelled a distance of $0.8v = 0.4\lambda$. Therefore, the remaining distance from P to Q must be equal to 0.6λ:

$$0.6\lambda = 6$$

$$= 10 \text{ m}$$

Hence, the speed of the wave is $v = \frac{10}{2} = 5 \text{ ms}^{-1}$.

Question 14: E

Let the pre-sale price of the bicycle be x. The sale price is therefore $\frac{3}{4}x$. The customer's calculated pre-sale price is $\frac{3}{4}x \times \frac{5}{4} = \frac{15}{16}x$. Since this is incorrect by £15,

$$x - \frac{15}{16}x = 15$$

$$\frac{x}{16} = 15$$

$$x = 15 \times 16 = £240$$

Question 15: D

The work done by the parachutist against drag forces is given by

$$W = KE_{initial} - KE_{final} + GPE_{initial} - GPE_{final}$$

$$W = \frac{1}{2} \times 80 \times (40^2 - 5^2) + 80 \times 10 \times 2000 - 0$$

$$W = 40(1600 - 25 + 40000) = 200(320 - 5 + 8000) = 8315 \times 200 = 1663000 \text{ J}$$

Question 16: F

Using Pythagoras' theorem, $QS = \sqrt{y^2 - x^2}$. Using trigonometry on the right triangle,

$$\sin 61 = \frac{QS}{z}$$

$$z = \frac{QS}{\sin 61} = \frac{\sqrt{y^2 - x^2}}{\sin 61}$$

Question 17: G

The maximum extension of the string is $0.30 - 0.10 = 0.20$ m. Using conservation of energy,

$$KE_{gain} + GPE_{gain} = EPE_{loss}$$

$$\frac{1}{2}mv^2 + mg\Delta h = \frac{1}{2}kx^2$$

$$\frac{1}{2} \times 0.05 \times v^2 + 0.05 \times 10 \times 0.2 = \frac{1}{2} \times 20 \times 0.2^2$$

$$0.025v^2 + 0.1 = 0.4$$

$$v^2 = \frac{0.3}{0.025} = 12$$

$$v = \sqrt{12}\text{ ms}^{-1}$$

Question 18: D

The smallest square occurs when these two points are diagonally opposite vertices, whilst the largest square occurs when these two points form the vertices at the ends of one side of the square. The distance between the two points is

$$\sqrt{(5-1)^2 + (3-1)^2} = \sqrt{16+4} = \sqrt{20} = 2\sqrt{5}$$

This is the side length of the largest square, and so its perimeter is $8\sqrt{5}$. This is also equal to the diagonal of the smallest square, so the side length, l, is given by

$$l^2 + l^2 = \left(2\sqrt{5}\right)^2$$

$$l^2 = 10$$

$$l = \sqrt{10}$$

The perimeter of the smallest square is, therefore $4\sqrt{10}$. Hence, the difference between the two perimeters is

$$8\sqrt{5} - 4\sqrt{10} = 4\sqrt{5}\left(2 - \sqrt{2}\right)$$

Question 19: A

Consider Newton's second law on the rocket:

$$F = \frac{d}{dt}(mv)$$

Since both m and v are variables, in this case, apply the product rule:

$$F = m\frac{dv}{dt} + v\frac{dm}{dt}$$

$$a = \frac{dv}{dt} = \frac{F - v\frac{dm}{dt}}{m}$$

Both F and $\frac{dm}{dt}$ are constant, whilst mass decreases at a constant rate. However, as acceleration also increases with velocity ($\frac{dm}{dt}$ is negative and so $\left(-v\frac{dm}{dt}\right)$ is positive), the acceleration increases at an increasing rate.

Question 20: B

Use the quadratic formula to obtain the two solutions:

$$x = \frac{-b \pm \sqrt{b^2 - 4ac}}{2a} = \frac{p \pm \sqrt{(-p)^2 - 4(2)(-4)}}{2 \times 2} = \frac{p \pm \sqrt{p^2 + 32}}{4} = \frac{p}{4} \pm \frac{\sqrt{p^2 + 32}}{4}$$

The difference between the two solutions is therefore

$$\frac{\sqrt{p^2 + 32}}{4} - \left(-\frac{\sqrt{p^2 + 32}}{4}\right) = \frac{\sqrt{p^2 + 32}}{2} = 6$$

$$\sqrt{p^2 + 32} = 12$$

$$p^2 = 144 - 32 = 112$$

$$p = \sqrt{112} = \sqrt{16} \times \sqrt{7} = 4\sqrt{7}$$

Section 1B

Question 21: B

By considering conservation of energy, the work done against friction is given by

$$W = KE_{initial} - KE_{final} + GPE_{initial} - GPE_{final}$$

$$W = \frac{1}{2}mv_1^2 - \frac{1}{2}mv_2^2 + mgh = mgh - \frac{1}{2}m(v_2^2 - v_1^2)$$

Question 22: B

Let $f(x) = x^4 + ax^3 + bx^2 - 12x + 4$. Since $(x - 1)$ and $(x - 2)$ are both factors, $f(1) = 0$ and $f(2) = 0$:

$$f(1) = 1^4 + a(1^3) + b(1^2) - 12 + 4 = 0$$

$$a + b = 7 \ (1)$$

$$f(2) = 2^4 + a(2^3) + b(2^2) - 12(2) + 4 = 0$$

$$16 + 8a + 4b - 24 + 4 = 0$$

$$8a + 4b = 4$$

$$2a + b = 1 \ (2)$$

$$(2) - (1):$$

$$a = -6$$

$$b = 13$$

Question 23: G

The pressure exerted on both pistons must be equal, so

$$p = \frac{F_X}{A_X} = \frac{F_Y}{A_Y}$$

$$\frac{36}{\pi \times 0.02^2} = \frac{F_Y}{\pi \times 0.06^2}$$

$$F_Y = 36 \times 3^2 = 324 \text{ N}$$

The work done on piston X must also be equal to the work done on piston Y:

$$F_X \times d_X = F_Y \times d_Y$$

$$36 \times 5.4 = 36 \times 3^2 \times d_Y$$

$$d_Y = \frac{5.4}{9} = 0.6 \text{ cm}$$

Question 24: E

A sketch of the curve and line is shown below:

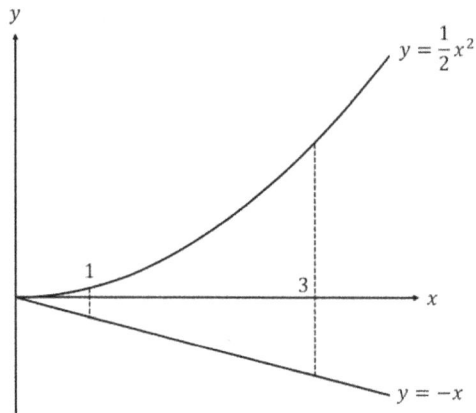

The area between the curve $y = \frac{1}{2}x^2$ and the line $y = -x$ is equal to the area under the curve plus the area above the line. The area is therefore given by

$$A = \int_1^3 \left(\frac{1}{2}x^2 - (-x)\right) dx = \int_1^3 \left(\frac{1}{2}x^2 + x\right) dx$$

$$A = \left[\frac{x^3}{6} + \frac{x^2}{2}\right]_1^3 = \left(\frac{3^3}{6} + \frac{3^2}{2}\right) - \left(\frac{1}{6} + \frac{1}{2}\right)$$

$$A = \frac{9}{2} + \frac{9}{2} - \frac{1}{6} - \frac{1}{2} = \frac{51 - 1}{6} = \frac{50}{6} = \frac{25}{3}$$

Question 25: H

Using the information given, the speed of the wave in water is $300 \times 5 = 1500 \text{ ms}^{-1}$. From the graph, the period of the wave is equal to 4 ms, so the frequency is 250 Hz. The frequency remains constant when moving from air to water, so the wavelength is

$$\lambda = \frac{v}{f} = \frac{1500}{250} = 6 \text{ m}$$

Question 26: C

Let the equation of the line be $y = mx + c$. When it is reflected in the line $y = x$, the equation of the new line can be written as $x = my + c$. Rearranging this equation gives $y = \frac{1}{m}(x - c)$. The gradient of the reflected line is therefore $\frac{1}{m}$.

Question 27: D

The force on the axle is

$$F = BIL = 0.05 \times 0.6 \times 50 \times 0.04 = 30 \times 20 \times 10^{-4} = 0.06 \text{ N}$$

Since this force acts at a distance of 15 cm from the axle, the moment produced is

$$0.06 \times 15 = 0.90 \text{ Ncm}$$

Question 28: A

The sum of the first n terms of an arithmetic progression is given by

$$S_n = \frac{n}{2}(2a + (n-1)d)$$

$$S_{20} = 10(2a + 19d) = 50$$

$$2a + 19d = 5 \ (1)$$

$$S_{40} - S_{20} = 20(2a + 39d) - 50 = -50$$

$$20(2a + 39d) = 0$$

$$2a + 39d = 0 \ (2)$$

$$(2) - (1):$$

$$20d = -5$$

$$d = -\frac{1}{4}$$

$$2a = 5 + \frac{19}{4} = \frac{39}{4}$$

$$S_{100} = 50(2a + 99d) = 50\left(\frac{39}{4} - \frac{99}{4}\right) = 50\left(-\frac{60}{4}\right) = 50 \times -15 = -750$$

Question 29: B

First, calculate the acceleration of the box up the slope:

$$v^2 - u^2 = 2as$$

$$a = \frac{v^2 - u^2}{2s} = \frac{7^2 - 3^2}{2 \times 5} = \frac{40}{10} = 4 \text{ ms}^{-2}$$

Now apply Newton's second law parallel to the slope, where W_P is the component of the weight acting down the slope:

$$F - W_P = ma$$

$$W_P = F - ma = 30 - 3 \times 4 = 18 \text{ N}$$

Question 30: D

A sketch of the lines is shown to the right:

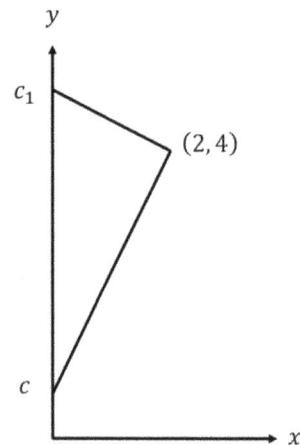

Substituting the point $(2, 4)$ into the equation of L gives

$$4 = m(2) + c$$

$$c = 4 - 2m$$

The perpendicular line must have a gradient of $-\frac{1}{m}$, and so the equation of the line is

$$y = -\frac{1}{m}x + c_1$$

Again, substituting the point $(2, 4)$ gives

$$4 = -\frac{1}{m}(2) + c_1$$

$$c_1 = 4 + \frac{2}{m}$$

As shown on the diagram, the area of the triangle is

$$A_t = \frac{1}{2} \times (c_1 - c) \times 2 = 5$$

$$4 + \frac{2}{m} - (4 - 2m) = 5$$

$$\frac{2}{m} + 2m = 5$$

$$2m^2 - 5m + 2 = 0$$

$$(2m - 1)(m - 2) = 0$$

$$m = \frac{1}{2}, m = 2$$

Hence, the larger value is $m = 2$.

Question 31: F

The total resistance of the wires is

$$R = \frac{\rho L}{A} = \frac{2.5 \times 10^{-7} \times 8 \times 10^3 \times 2}{(1 \times 10^{-2})^2} = \frac{40 \times 10^{-4}}{10^{-4}} = 40\ \Omega$$

The current in the wire is $I = \frac{P}{V} = \frac{120 \times 10^3}{24 \times 10^3} = 5$ A. Hence, the power wasted as heat is

$$P = I^2 R = 5^2 \times 40 = 1000\ \text{W}$$

Question 32: E

In a geometric progression, there is a common ratio, r, between each term. Hence, from the third to the fifth term:

$$r^2 = \frac{2}{4} = \frac{1}{2}$$

$$r = \pm\frac{1}{\sqrt{2}}$$

The first term of each progression is therefore

$$a = \frac{4}{r^2} = 4 \times 2 = 8$$

The sum to infinity of a geometric progression is

$$S_\infty = \frac{a}{1-r}$$

Hence, the modulus of the difference between the sums to infinity is

$$\left|S_P - S_Q\right| = \frac{8}{1-\frac{1}{\sqrt{2}}} - \frac{8}{1+\frac{1}{\sqrt{2}}} = \frac{8\sqrt{2}}{\sqrt{2}-1} - \frac{8\sqrt{2}}{\sqrt{2}+1}$$

$$\left|S_P - S_Q\right| = 8\sqrt{2}\left(\frac{\sqrt{2}+1-(\sqrt{2}-1)}{(\sqrt{2}-1)(\sqrt{2}+1)}\right) = 8\sqrt{2}\left(\frac{2}{1}\right) = 16\sqrt{2}$$

Question 33: C

Consider the maximum displacement of the ball whilst in contact with the racket when its velocity is zero. Defining the initial direction of the ball's motion as positive, the ball has an initial velocity 24 ms^{-1}, final velocity 0 ms^{-1} and acceleration -6000 ms^{-2}:

$$v^2 - u^2 = 2as$$

$$0 - 24^2 = 2 \times (-6000) \times s$$

$$s = \frac{24^2}{12000} = \frac{24 \times 2}{1000} = 0.048 \text{ m} = 4.8 \text{ cm}$$

Since this is half the total distance the ball travels whilst in contact with the racket, the total distance is 9.6 cm.

Question 34: A

At a turning point, the differential of the curve is zero:

$$\frac{dy}{dx} = 3x^2 + 6\sqrt{5}px + 3p = 0$$

Since there are two distinct points, the discriminant must be greater than zero:

$$b^2 - 4ac > 0$$

$$\left(6\sqrt{5}p\right)^2 - 4(3)(3p) > 0$$

$$180p^2 - 36p > 36p$$

$$5p^2 - p > 0$$

$$p(5p - 1) > 0$$

$$p < 0, p > 0.2$$

Question 35: C

In the first configuration, an added load of 0.8 kg causes an extension of 4 cm:

$$F = kx$$

$$0.8 \times 10 = k \times 0.04$$

$$k = 200 \text{ Nm}^{-1}$$

When the two springs are connected in parallel, their effective spring constant is doubled, so $k_2 = 400 \text{ Nm}^{-1}$. Now, the extension, x_2, can be found:

$$2 \times 10 = 400x_2$$

$$x_2 = \frac{1}{20} = 0.05 \text{ m}$$

The strain energy is therefore

$$E = \frac{1}{2}k_2x_2^2 = \frac{1}{2} \times 400 \times 0.05^2$$

$$E = 2 \times 10^2 \times 25 \times 10^{-4} = 50 \times 10^{-2} = 0.5 \text{ J}$$

Question 36: E

First, make the substitution $\sin^2 x = 1 - \cos^2 x$:

$$14\cos^3 x + 10(1 - \cos^2 x)\cos x = 13\cos x$$

Now, rearrange and factorise the equation, making sure not to divide through by $\cos x$ as this would lose solutions.

$$14\cos^3 x + 10\cos x - 10\cos^3 x - 13\cos x = 0$$

$$4\cos^3 x - 3\cos x = 0$$

$$\cos x(4\cos^2 x - 3) = 0$$

$$\cos x = 0 \ (1) \text{ or } 4\cos^2 x - 3 = 0 \ (2)$$

$\cos x = 0$ has two solutions in every period, so in the range $-2\pi \leq x \leq 2\pi$ there are four solutions to (1).

$$(2) \cos^2 x = \frac{3}{4}$$

$$\cos x = \pm\frac{\sqrt{3}}{2}$$

For each value of $\cos x$, there are two solutions in each period. As there are two possible values and two periods, there are 8 solutions to (2). Hence, there are a total of 12 solutions overall.

Question 37: B

The area under a force-time graph is equal to the change in momentum of the object during the period. Hence,

$$\Delta p = \frac{1}{2} \times 0.2 \times 20 = 2$$

Since initial velocity is zero,

$$\Delta p = mv = 2$$

$$v = \frac{2}{2.5} = 0.8 \text{ ms}^{-1}$$

The kinetic energy is therefore

$$KE = \frac{1}{2}mv^2 = \frac{1}{2} \times 2.5 \times \left(\frac{2}{2.5}\right)^2 = \frac{2}{2.5} = 0.80 \text{ J}$$

Question 38: D

$$\log_{10} x^2 = 2\log_{10} x$$

Now, substitute $y = \log_{10} x$:

$$(2y)^2 + y = 3$$

$$4y^2 + y - 3 = 0$$

$$(4y - 3)(y + 1)$$

$$y = \frac{3}{4}, y = -1$$

$$\log_{10} x = \frac{3}{4}, \log_{10} x = -1$$

$$x = 10^{\frac{3}{4}}, x = 10^{-1}$$

Hence, the product of the roots is $10^{\frac{3}{4}} \times 10^{-1} = 10^{-\frac{1}{4}}$.

Question 39: D

There is an error in the diagram for this question. The terminals of the batteries should be the other way around, as in the diagram below.

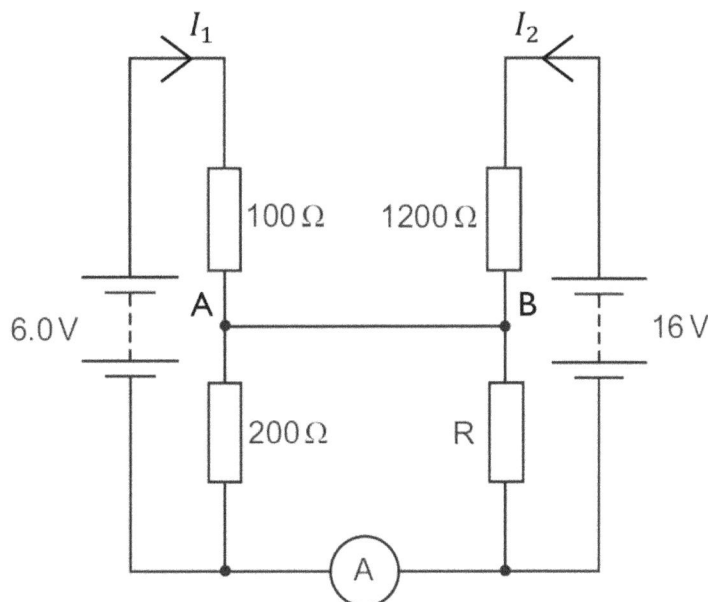

If the current through the ammeter is zero, there must also be zero current flowing between nodes A and B. As shown in the diagram, let the current flowing through the left circuit be I_1 and the current in the right circuit be I_2.

$$I_1 = \frac{6}{100 + 200} = \frac{6}{300} = 0.02 \text{ A}$$

$$I_2 = \frac{16}{R + 1200}$$

As there is no resistance between A and B, the voltage at each node must be equal.

$$V_A = 0.02 \times 200 = 4 \text{ V}$$

$$V_B = I_2 R = \frac{16R}{R + 1200} = 4 \text{ V}$$

$$16R = 4R + 4800$$

$$12R = 4800$$

$$R = 400 \ \Omega$$

Question 40: C

The gradient function is given by differentiating the equation of the curve:

$$\frac{dy}{dx} = -4 + 6x^{\frac{1}{2}} - 2x$$

The x-coordinate at the maximum gradient is then found by differentiating again and setting to zero:

$$\frac{d^2y}{dx^2} = 3x^{-\frac{1}{2}} - 2 = 0$$

$$\frac{3}{\sqrt{x}} = 2$$

$$\sqrt{x} = \frac{3}{2}$$

$$x = \frac{9}{4}$$

Substituting back into $\frac{dy}{dx}$ gives

$$\frac{dy}{dx} = -4 + 6\sqrt{\frac{9}{4}} - 2\left(\frac{9}{4}\right) = -4 + 6\left(\frac{3}{2}\right) - \frac{9}{2} = \frac{1}{2}$$

END OF SECTION

Section 2

Question 1: E

Let spring P and extensions p and q respectively. Since they both have the same tension, their extension will be inversely proportional to their spring constant. Hence, $p = 3q$. Since $p + q = 6$,

$$3q + q = 6$$

$$q = \frac{6}{4} = \frac{3}{2} \rightarrow p = \frac{9}{2} = 4.5 \text{ cm}$$

Question 2: D

The resistance of each strand is

$$R = \frac{\rho L}{A} = \frac{4.8 \times 10^{-7} \times 15}{\pi \times (2 \times 10^{-4})^2} = \frac{720}{4\pi} = \frac{180}{\pi}$$

Since the strands are in parallel, the total resistance of the cable is

$$\frac{1}{R_T} = 12 \times \frac{\pi}{180}$$

$$R_T = \frac{15}{\pi}$$

Question 3: C

Let the refractive index of the block be n. Since the refractive index of air is 1 and the critical angle is given by $\sin \theta_c = \frac{n_2}{n_1}$,

$$\sin 60 = \frac{1}{n} = \frac{\sqrt{3}}{2}$$

$$n = \frac{2}{\sqrt{3}}$$

The velocity of light in the block is $v = \frac{c}{n} = \frac{c\sqrt{3}}{2}$. Since PQ is equal to twice the radius of the semicircle, $PQ = XY = L$. The time taken for the light to travel from P to Q is therefore

$$t = \frac{L}{c\sqrt{3}/_2} = \frac{2L}{c\sqrt{3}}$$

Question 4: E

The mass of the cube is

$$(20 \times 10^{-2})^3 \times 2000 = 8 \times 10^3 \times 10^{-6} \times 2 \times 10^3 = 16 \text{ kg}$$

The tension in the spring is therefore 160 N. Since the spring's strain energy is given by $E = \frac{1}{2}Fx$,

$$3.2 = \frac{1}{2}(160)x$$

$$x = \frac{3.2}{80} = \frac{0.4}{10} = 0.04 \text{ m}$$

The spring constant is therefore

$$k = \frac{F}{x} = \frac{160}{0.04} = 4000 \text{ Nm}^{-1}$$

Question 5: G

The initial vertical component of speed is $20\cos 60 = 10 \text{ ms}^{-1}$. Consider the vertical motion, taking upwards as positive:

$$v^2 - u^2 = 2as$$

$$(-4)^2 - 10^2 = 2(-10)h$$

$$16 - 100 = -20h$$

$$h = \frac{100 - 16}{20} = \frac{84}{20} = 4.2 \text{ m}$$

The time taken is given by

$$v = u + at$$

$$-4 = 10 - 10t$$

$$t = \frac{10 + 4}{10} = 1.4 \text{ s}$$

The initial horizontal component of speed is $20\sin 60 = 10\sqrt{3} \text{ ms}^{-1}$. Since there is no horizontal acceleration, this remains constant. The horizontal distance, d, is therefore

$$d = 1.4 \times 10\sqrt{3} = 14\sqrt{3} \text{ m}$$

Question 6: D

The amplitude of the wave is equal to the maximum displacement of particles from their rest position. From the diagrams, we can see that the particles with the greatest displacement are those initially at 2 m and 6 m. These particles are each displaced a magnitude of 0.7 m, and so this is the amplitude of the wave.

Question 7: B

The spaceship's velocity can be calculated by considering its mass and kinetic energy:

$$KE = \frac{1}{2}mu^2$$

$$\frac{1}{2} \times 8 \times 10^4 \times u^2 = 10^{12}$$

$$u^2 = 0.25 \times 10^8$$

$$u = 0.5 \times 10^4 = 5000 \text{ ms}^{-1}$$

The applied impulse is equal to the change in momentum of the spaceship:

$$m(v - u) = -8 \times 10^7$$

$$v - 5000 = \frac{-8 \times 10^7}{8 \times 10^4} = -1000$$

$$v = 4000 \text{ ms}^{-1}$$

Hence, the average rate of loss of kinetic energy is

$$\frac{\Delta KE}{t} = \frac{KE_{initial} - KE_{final}}{t} = \frac{\frac{1}{2} \times 8 \times 10^4 \times (5000^2 - 4000^2)}{2}$$

$$\frac{\Delta KE}{t} = 2 \times 10^4 \times 10^6 \times (5^2 - 4^2) = 2 \times 10^{10} \times 9 = 1.8 \times 10^{11} \text{ W}$$

Question 8: F

First, calculate the mass of the prism. Since the triangular cross-section is an equilateral triangle, each angle is $60°$, and so the area of the triangle is $\frac{1}{2}x^2 \sin 60 = \frac{\sqrt{3}x^2}{4}$. The mass of the prism is therefore

$$\frac{\sqrt{3}x^2}{4} \times 3x \times \rho = \frac{3\sqrt{3}\rho x^3}{4}$$

The prism exerts the minimum pressure when the largest area is in contact with the ground, so

$$p_{min} = \frac{\frac{3\sqrt{3}\rho x^3}{4} \times g}{3x \times x} = \frac{\sqrt{3}\rho g x}{4}$$

Question 9: F

The two situations are shown in the diagrams below, where the mass of the ruler is M. In each case, the weight of the ruler acts through its centre of gravity at the 50 cm mark.

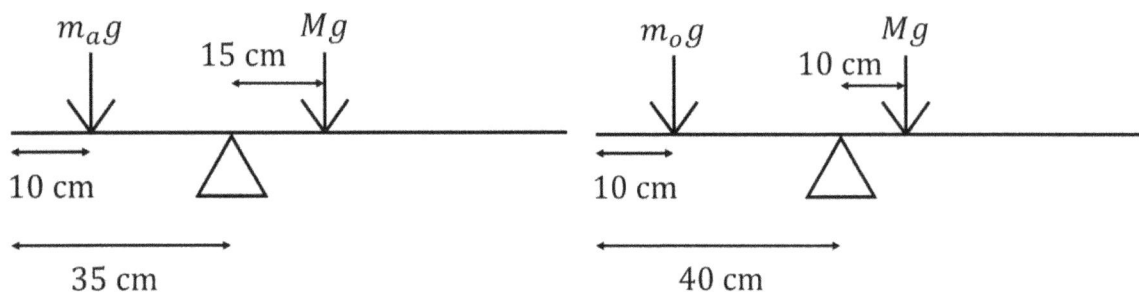

In the first case, $25m_a = 15M$. In the second case, $30m_o = 10M$. Divide the two equations:

$$\frac{25m_a}{30m_o} = \frac{15M}{10M}$$

$$\frac{5m_a}{6m_o} = \frac{3}{2}$$

$$\frac{m_a}{m_o} = \frac{9}{5}$$

Question 10: C

Since $P = Fv$, the driving force on level ground is $F_1 = \frac{900}{12} = 75$ N. When climbing the slope, the added driving force is equal to the component of the weight acting parallel to the slope:

$$F_2 = 850 \sin 30 = 850 \times 0.5 = 425 \text{ N}$$

Hence the total driving force is $425 + 75 = 500$ N.

Question 11: D

The four possible arrangements are shown below:

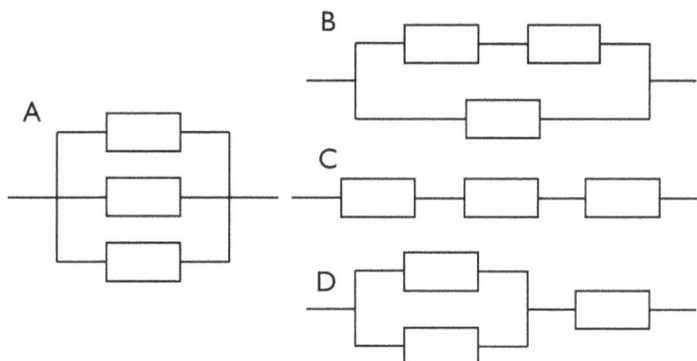

Let each resistor have resistance R:

$$R_A = \frac{R}{3}$$

$$R_B = \frac{R(2R)}{R + 2R} = \frac{2R}{3}$$

$$R_C = 3R$$

$$R_D = \frac{R}{2} + R = \frac{3R}{2}$$

Since $\frac{18}{8} = \frac{9}{4}$, the two arrangements specified in the question must be B and D, as $R_D = \frac{9}{4} R_B$.

$$R_B = \frac{2R}{3} = 8$$

$$R = 12\ \Omega$$

The resistances of the other two arrangements are therefore $R_A = 4\ \Omega$ and $R_C = 36\ \Omega$.

Question 12: D

Let the initial resistance of the LDR be R_1:

$$100 = 5 \times 10^{-3}(4 \times 10^3 + R_1)$$

$$R_1 = 20 \times 10^3 - 4 \times 10^3 = 16 \text{ k}\Omega$$

The voltage across the fixed resistor is $V_1 = 5 \times 10^{-3} \times 4 \times 10^3 = 20$ V. When the intensity of light decreases, the resistance of the LDR increases. This means that the fixed resistor will have a decreased share of the voltage, so $V_2 = 20 \times 0.5 = 10$ V., The current in the circuit is, therefore;

$$I = \frac{10}{4 \times 10^3} = 0.25 \times 10^{-2} = 2.5 \text{ mA}$$

Hence, the new resistance of the LDR is given by

$$100 = 2.5 \times 10^{-3}(4 \times 10^3 + R_2)$$

$$R_2 = 40 \times 10^3 - 4 \times 10^3 = 36 \text{ k}\Omega$$

The change in resistance is, therefore, $36 - 16 = 20$ kΩ.

Question 13: F

The diagram of the new configuration is shown below:

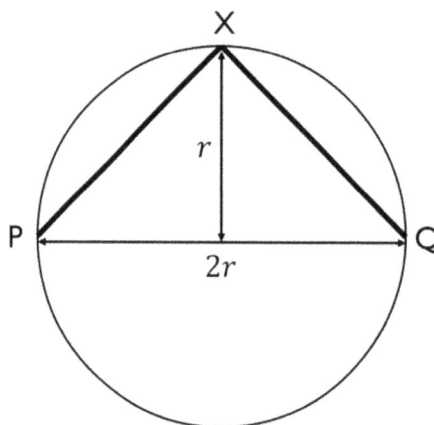

The length PX is found using Pythagoras' theorem:

$$PX^2 = r^2 + r^2$$

$$PX = \sqrt{2}r$$

Hence the new length of the string is $2\sqrt{2}r$ and so the extension is $x = 2\sqrt{2}r - 2r = 2r(\sqrt{2} - 1)$. The energy stored is therefore

$$E = \frac{1}{2}kx^2 = \frac{1}{2}k\left[2r(\sqrt{2} - 1)\right]^2 = \frac{1}{2}k(4r^2)(2 - 2\sqrt{2} + 1) = 2(3 - 2\sqrt{2})kr^2$$

Question 14: A

Since force is the rate of change of momentum, momentum is the integral of force with respect to time.

$$p = \int_0^T (X + Y\sqrt{t})dt = \int_0^T \left(X + Yt^{\frac{1}{2}}\right)dt$$

$$p = \left[Xt + \frac{2}{3}Yt^{\frac{3}{2}}\right]_0^T = XT + \frac{2}{3}YT\sqrt{T} = T\left(X + \frac{2}{3}Y\sqrt{T}\right)$$

Question 15: E

Calculate the trolley's acceleration by differentiating x with respect to time twice:

$$x = 8 + 4t + 2t^2$$

$$\frac{dx}{dt} = 4 + 4t$$

$$a = \frac{d^2x}{dt^2} = 4$$

Since there is a constant acceleration, there must be a constant force of $F = 3 \times 4 = 12$ N acting on the trolley. Work done is given by $W = F \times d$, so the total distance is required. When $t = 0$, $x = 8$ and when $t = 5$, $x = 8 + 20 + 50 = 78$. Hence the distance travelled is 70 m.

$$W = F \times d = 12 \times 70 = 840 \text{ J}$$

Question 16: E

Wave speed is related to refractive index by $v = \frac{c}{n}$, therefore $n_Q : n_P = 2 : \sqrt{5}$ and $n_R : n_Q = 3 : \sqrt{6}$. Let $n_P = n$:

$$n_Q = \frac{2n}{\sqrt{5}}, n_R = \frac{3n_Q}{\sqrt{6}} = \frac{6n}{\sqrt{30}}$$

Applying Snell's law,

$$n_P \sin\theta = n_Q \sin\theta_Q = n_R \sin 45$$

$$\sin\theta = \frac{n \quad \sin 45}{n_P} = \frac{6}{\sqrt{30}} \times \frac{\sqrt{2}}{2} = \frac{3}{\sqrt{15}} = \frac{\sqrt{15}}{5}$$

Question 17: D

The actual path that the swimmer must aim to take to overcome the current is shown in the diagram below:

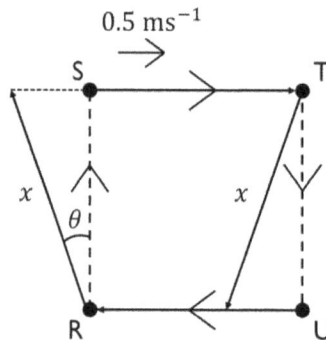

$$\sin \theta = \frac{0.5}{1} = \frac{1}{2} \rightarrow \theta = 30°$$

Hence, the length marked x is given by

$$\cos 30 = \frac{30}{x}$$

$$x = \frac{30}{\sqrt{3}/_2} = \frac{60}{\sqrt{3}} = 20\sqrt{3} \text{ m}$$

Consider each side of the square individually:

$$t_{ST} = \frac{30}{1 + 0.5} = \frac{30}{1.5} = 20 \text{ s}$$

$$t_{UR} = \frac{30}{1 - 0.5} = \frac{30}{0.5} = 60 \text{ s}$$

$$t_{RS} = t_{TU} = \frac{20\sqrt{3}}{1} = 20\sqrt{3} \text{ s}$$

The total time is, therefore, $80 + 40\sqrt{3}$ s.

Question 18: B

At a stress of 8×10^8 Pa,

$$\epsilon = \frac{\sigma}{E} = \frac{8 \times 10^8}{2 \times 10^{11}} = 4 \times 10^{-3} = \frac{x}{L}$$

$$x = 4 \times 10^{-3} \times 4 = 16 \times 10^{-3} \text{ m}$$

The tension at the maximum stress is

$$T = 8 \times 10^8 \times 2 \times 10^{-4} = 16 \times 10^4 \text{ N}$$

Hence, the work done is

$$W = \frac{1}{2}Tx = \frac{1}{2} \times 16 \times 10^4 \times 16 \times 10^{-3} = 128 \times 10^1 \text{ J} = 1.28 \text{ kJ}$$

Question 19: C

Velocity is given by the gradient of the displacement-time graph. Initially, as the displacement increases, there is a constant, positive velocity. When the displacement stops increasing, there is zero velocity. As the displacement then decreases, there is a constant, negative velocity. Since the negative gradient is steeper than the positive gradient, the negative velocity has a greater magnitude than the positive velocity. The correct option is, therefore, C.

Question 20: D

Let the resistance of the variable resistor be R, and the voltage across the resistor be V.

$E = V + Ir$

$V = E - Ir = E - kI^3$

The power dissipated in the variable resistor is

$P = IV = EI - kI^4$

For maximum power, differentiate and set to zero:

$\dfrac{dP}{dI} = E - 4kI^3 = 0$

$I^3 = \dfrac{E}{4k}$

$I = \left(\dfrac{E}{4k}\right)^{\frac{1}{3}}$

Since the voltage across the resistor is given by $V = IR$,

$E = IR + kI^3$

$E = \left(\dfrac{E}{4k}\right)^{\frac{1}{3}} R + \dfrac{E}{4}$

$R = \dfrac{3E}{4} \times \left(\dfrac{4k}{E}\right)^{\frac{1}{3}} = 3\left(4^{-\frac{2}{3}}\right)E^{\frac{2}{3}}k^{\frac{1}{3}} = 3\left(\dfrac{kE^2}{4^2}\right)^{\frac{1}{3}} = 3\left(\dfrac{kE^2}{16}\right)^{\frac{1}{3}}$

END OF PAPER

2021

Section 1

Question 1: E

$$= \frac{5xy^2 \times (5x^2y)^{-3} \times 5x^2y}{25x^3y^2} = \frac{1}{5x^3}$$
$$\frac{25x^3y^2}{125x^6y} = \frac{1}{5x^3}$$

Question 2: E

$p_1v_1 = p_2v_2$

$p_2 = 1.2p_1$

$\therefore v_2 = \dfrac{p_1v_1}{1.2p1} = \dfrac{v_1}{1 \cdot 2}$

$p_1 = \dfrac{m}{v_1} \quad p_2 = \dfrac{m}{v_2} = \dfrac{m}{\frac{v_1}{1 \cdot 2}} = 1.2\dfrac{m}{v_1}$

$\therefore P_2 = 1.2P_1$

Question 3: D

$\dfrac{p}{2} + \dfrac{3}{q} = \dfrac{4}{r}$

$\dfrac{3}{q} = \dfrac{4}{r} - \dfrac{p}{2}$

$\Rightarrow \dfrac{3}{q} = \dfrac{8 - pr}{2r}$

$\therefore q = \dfrac{6r}{8 - pr}$

Question 4: F

$$\frac{V_p}{n_p} = \frac{V_s}{n_s}$$

$$P_p = V_p I_p$$

$$\Rightarrow \quad 3000 = 12.5 \times V_p$$

$$\therefore V_p = 240 \text{ V}$$

$$\frac{240}{100} = \frac{V_s}{25} \quad \therefore V_s = 60 \text{ V}$$

$$P_s = 40 \times 60 = 2.4 \text{kw}$$

$$\therefore \text{Efficiency} = \frac{2.4}{3} \times 100\%$$

$$= 80\%$$

Question 5: C

Total surface area of $P = 2\pi \left(\frac{x}{2}\right)^2 + \pi xy$

Total surface area of $Q = 2\pi \left(\frac{y}{2}\right)^2 + \pi xy$

Since $x > y$,

Surface area of P – Surface Area of Q

$$= 2\pi \left(\frac{x}{2}\right)^2 + \pi xy - 2\pi \left(\frac{y}{2}\right)^2 - \pi xy$$

$$= \frac{\pi}{2}(x^2 - y^2)$$

Question 6: C

$$E \cdot P \cdot E \quad = \frac{1}{2} F \, x$$

$$= \frac{1}{2} \times mg \times x$$

$$= \frac{1}{2} \times 0.5 \times 10 \times (0.1 - 0.08)$$

$$= 0.05 \text{ J}$$

Question 7: A

Original price of item $P = P$
After reduction $= 0.9\,p$
After increase $= 1.1 \times 0.9p = 0.99p$.
Original price of $Q = q$
After reducton increase $= 1.1q$
After reduction $= 1.1 \times 0.9q = 0.99q$
$\therefore$ Both lower than original price

Question 8: D

Use $P = \dfrac{V^2}{R}$ (same dic voltage, so V is constat)
Initially 20 lamps assume idertial resistance.

$$\therefore P_1 = \frac{v^2}{20R} = P$$

After one lamp fails,

$$P_2 = \frac{V^2}{19R}$$
$$\therefore P_2 = \frac{20}{19} \cdot \frac{v^2}{20R} = \frac{20}{19}p$$

Question 9: D

Length of $SQ = \sqrt{8^2 - 4^2} = 4\sqrt{3}$ cm
$\therefore$ length of $RQ = \dfrac{3}{4} \cdot SQ = 3\sqrt{3}$ cm.
$\therefore$ length of $PQ = \sqrt{8^2 - (3\sqrt{3})^2} = \sqrt{37}$ cm/

Question 10: E

Newton's II Law, Force is the rate of change of momentum.

$$\therefore F = \frac{(12 - 0) \times 10^6}{4} = 3 \times 10^6 \text{ N}$$

Recall, this force is the resultant force.

$$\therefore \text{ Force from motor } = 3 \times 10^6 + 1.8 \times 10^7$$
$$= 2.1 \times 10^7 \text{ N}$$

Question 11: F

$$y = x^2 - 4x + 5$$
$$y = 2x + c$$
$$x^2 - 4x + 5 = 2x + c$$
$$\Rightarrow x^2 - 6x + 5 = c$$

$$x = q \qquad\qquad x = p, \qquad\qquad p^2 - 6p + 5 = c - 00$$
$$q^2 - 6q + 5 = c - Q$$
$$q^2 - p^2 - 6q + 6p = 0$$
$$\Rightarrow (q + p)(q - p) - 6(q - p) = 0$$

$$\Rightarrow 8(q + p) - 48 = 0$$
$$\therefore q + p = 6 \text{ (3)}$$
$$\frac{q - p = 8}{2q = 14} \quad \therefore q = 7$$
$$\therefore p = -1$$
$$\therefore c = 7^2 - 6(7) + 5 = 12$$

Question 12: H

Speed of the wave $= 8 + 3 = 11 \text{ ms}^{-1}$

$$v = f\lambda$$
$$11 = \frac{1}{T} \cdot \lambda$$
$$\therefore \lambda = 11 \times 8 = 88 \text{ m}$$

Question 13: F

$$\left(\frac{\sqrt{3}/2 - 1}{\frac{1}{2}}\right)^3 = 15\sqrt{3} - 26$$

Question 14: B

Voltage across resistor = $8 \times 0.25 = 2V$

∴ Voltage across lamp = $4V$

Only two options with correct voltage value but B is correct since resistance increases as filament lamp heats up.

Question 15: B

Assume total number of green sweets initially $= x$.

∴ total number of red sweets initially $= 9 - x$

$$\therefore \quad p(\text{red, red}) = \frac{5}{12}$$
$$\frac{9-x}{9} \times \frac{8-x}{8} = \frac{5}{12}$$
$$(9-x)(8-x) = 30$$
$$\Rightarrow \quad 72 - 9x - 8x + x^2 = 30$$
$$\Rightarrow \quad x^2 - 17x + 42 = 0$$
$$\Rightarrow \quad x^2 - 3x - 14x + 42 = 0$$
$$\Rightarrow \quad x(x-3) - 14(x-3) = 0$$
$$\therefore x = 3, 14$$
$$\therefore x = 3$$
$$\therefore \text{p(green, green)} = \frac{3}{9} \times \frac{2}{8} = \frac{1}{12}$$

$\frac{1}{2}R$	$\frac{1}{4}R$	$\frac{1}{8}R$	$\frac{1}{16}R$

Question 16: C

$$\times \xrightarrow{T} \frac{1}{2} \times \xrightarrow{2T} \frac{1}{4} \times \xrightarrow{3T} \frac{1}{8} \times \xrightarrow{4T} \frac{1}{16} \times$$

$$\boxed{\frac{1}{2}S} \qquad \boxed{\frac{1}{4}S}$$

$$Y \xrightarrow{2T} \frac{1}{2}Y \qquad \xrightarrow{4T} \frac{1}{4}Y$$

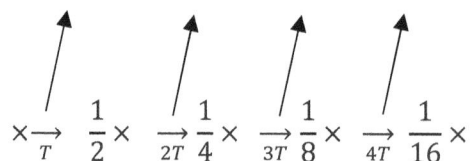

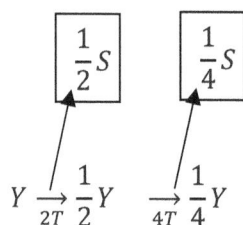

$\therefore$ After 4T years total $R = \frac{1}{2}R + \frac{1}{4}R + \frac{1}{8}R + \frac{1}{16}R + \underset{c_{\text{initial}}}{\frac{R}{}} = \frac{31}{16}R$

$\therefore$ After $4T$ year totel $S = \frac{1}{2}S + \frac{1}{4}S + 1S$

$$= \frac{7}{4}S$$

$$\therefore \frac{\text{number of atoms of } R}{\text{number of atoms of } S} = \frac{31/16R}{7/4\ S}$$

$$= \frac{31}{28}$$

Question 17: A

$$\sqrt{4^2 + 6^2 + (2x)^2} = \sqrt{77}$$
$$\therefore 4x^2 = 25$$
$$\therefore x = \sqrt{\frac{25}{4}} = \frac{5}{2} \text{ cm}$$

Question 18: D

Energy for ice to melt $= 300 \times 20$

$$= 6000 \text{ J}$$

Energy for water to freeze $= 180 \times 4 \times 25$

$$= 18000 \text{ J}$$

$\therefore$ To find the final temperature of water $\rightarrow 18000 - 6000 = mc\Delta T$

$$\therefore 200 \times 4 \times \Delta T = 12000$$
$$\therefore \Delta T = 15$$
$$\therefore T_f = 15°\text{C}$$

Question 19: H

m miles total journey
$1\text{km} = x$ miles
$\therefore$ I mile $= \dfrac{1}{x}$ km
$\therefore m$ miles $= \dfrac{m}{x}$ km
$f km/l$ of fuel $\quad \therefore$ f km using $1l$

$\therefore$ Ikm using $\dfrac{1}{F} l$

$$\therefore \text{Cost} = \frac{mp}{f_x} \text{ pence } \rightarrow \frac{mp}{100 f_x}$$

Question 20: E

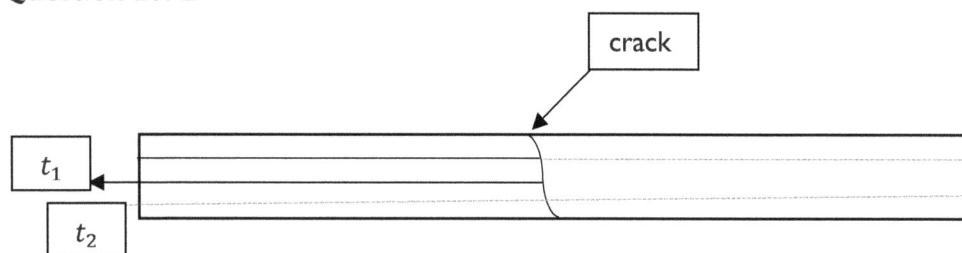

$vt_1 = 2x$, $vt_2 = 2L$ (v is the speed of ultrasound in the material)

$$\therefore x = \frac{vt_1}{2} \quad v = \frac{2L}{t_2}$$

$$\therefore x = \frac{xLt_1}{xt_2} = \frac{Lt_1}{t_2}$$

distance behween crack and far end $= L - Lt_1$

$$= L\frac{(t_2 - t_1)}{t_2} \Bigg)$$

Question 21: C

$$y = \left(2\sqrt{x} - \frac{1}{2\sqrt{x}}\right)^2$$

$$\frac{dy}{dx} = 2\left(2\sqrt{x} - \frac{1}{2x^{1/2}}\right) \cdot \left(x^{\frac{1}{2}-1} - \frac{1}{2}x - \frac{1}{2}x^{-\frac{3}{2}}\right)$$

$$x = \frac{1}{2},$$

$$\frac{dy}{dx} = 3.$$

Question 22: G

$$\rightarrow F \quad = ma$$

$$\rightarrow 12\sin 60° \quad = \frac{3}{4}a$$

$$\therefore a \quad = 8\sqrt{3} \text{ ms}^{-2}$$

Question 23: G

$$S_n = \frac{n}{2}\{2a + C_{n-1})d\}$$

$$S_{n+1} = \frac{n+1}{m2}\{2a + nd\}$$

$$S_{n-1} = \frac{n-1}{2}\{2a + (n-2)d\}$$

$$S_{n+1} - S_{n-1}$$

$$= \frac{n+1}{2}\{2a + nd\} - \frac{n-1}{2}\{2a + (n-2)d\}$$

$$= \frac{n}{2}\{2a + nd\} + \frac{1}{2}(2a + nd) - \frac{n}{2}(2/a + nd - 2d)$$

$$+ \frac{1}{2}(2a + nd - 2d)$$

$$= \frac{1}{2}(2a + nd + 2nd) + \frac{1}{2}(2a + nd - 2d)$$

$$= \frac{1}{2}[4a + 4nd - 2d]$$

$$= 2a + 2nd - d = 2a + (2n - 1)d$$

Question 24: B

$$2.5 \times \text{wavelength} = 33 \text{ cm}$$

$$\therefore \lambda = \frac{33}{2.5 \times 100} = 0.132 \text{ m}$$

$$\therefore v = f\lambda \Rightarrow f = \frac{330}{0.132} = 2500\text{H}_2$$

Wave is travelling left to right, therefore cir particles moving right to left.

Question 25: D

$$(x^2 + 4x + 3)^2 = 1$$

$$x^2 + 4x + 3 - 1 = 0 \text{ or } x^2 + 4x + 3 + 1 = 0$$
$$x^2 + 4x + 2 = 0, \; x^2 + 4x + 4 = 0$$
$$b^2 - 4ac > 0 \Rightarrow (x + 2)^2 = 0$$
$$\therefore 2 \text{ real distinct solutions}, \; \therefore x = -2$$

∴ 3 rexl distinct solutions in total

Question 26: D

Current through the heater: $P = I^2 R$

$$P = I^2 R$$
$$\therefore \quad = I^2 \times 4$$
$$\therefore I = \sqrt{\frac{9}{4}} = 1.5 \text{ A.}$$

Voltage across heater:

$$V = IR = 1.5 \times 4 = 6 \text{ V}$$

∴ voltage across 8Ω resictor = 24 V

∴ current through 8Ω resistor = $\frac{24}{8} = 3A$.

∴ 3 A is getting divided equally beturen 4Ω heater and resistor $R(1.5 \text{ A})$

∴ R Las resistence 4Ω.

Question 27: H

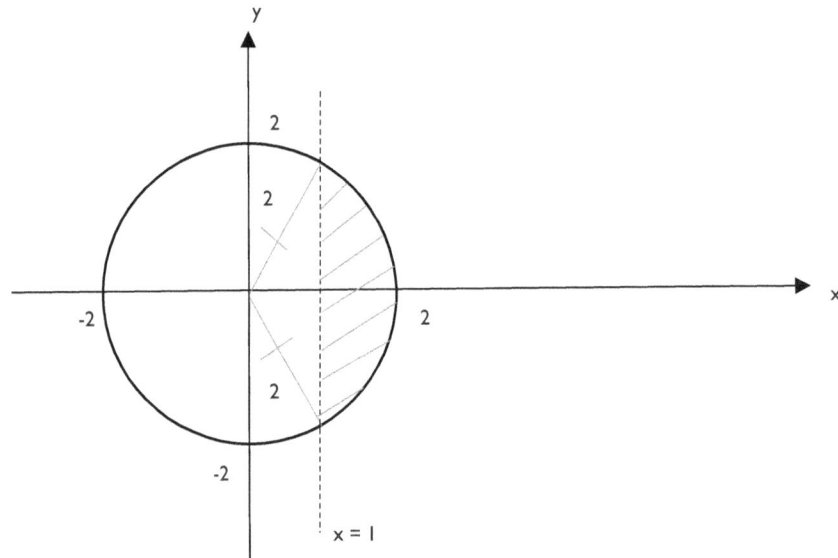

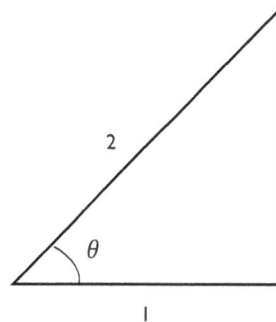

Cos $\theta = 1/2$

Therefore, $\theta = \cos^{-1}(1/2)$

$\theta = 60°$

$\therefore$ Area of smaller segment $= \frac{1}{2} \times 2^2 \times \frac{2}{3}\pi$

$$-\frac{1}{2} \times 2^2 + \sin\left(\frac{2}{3}\pi\right)$$
$$= \frac{4}{3}\pi - \sqrt{3}\ H$$

Question 28: D

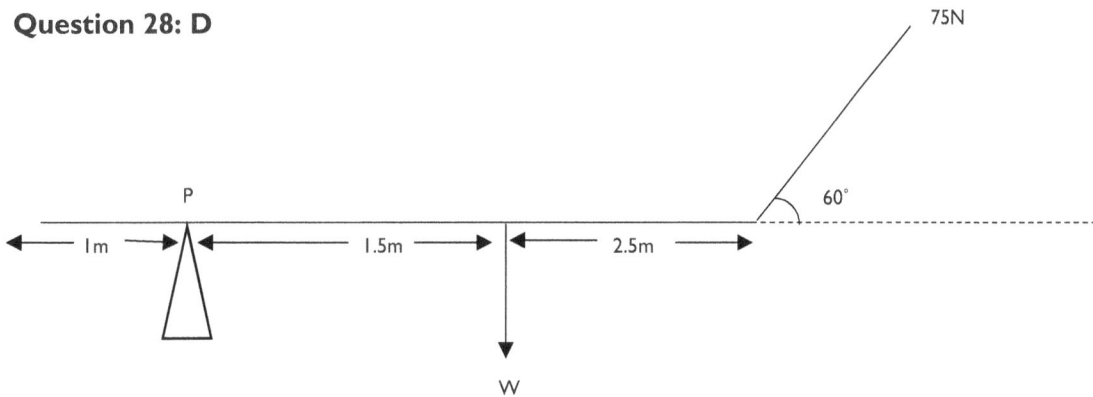

Moments about $P(V+)$:

$$W \times 1.5 - 75\sin\ 60° \times 4 = 0$$
$$\therefore W = 100\sqrt{3}\ \text{N}$$

Question 29: E

$\dfrac{\log_{10} 27 \times 64 \times 216}{3}$

$= \dfrac{\log_{10} 3^3 \times 64 \times 3^3(8)}{3}$

$= 2\log_{10} 3 + \log_{10} 4 + \log_{10} 2$

$= \log_{10} 9 \times 4 \times 2$

$= \log_{10} 72$

Question 30: B

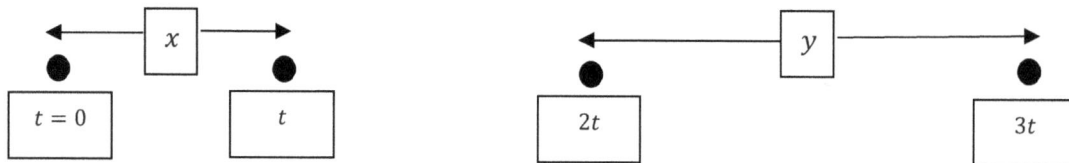

Using $s = ut + \frac{1}{2}at^2$

$$x = u_0 t + \frac{1}{2}at^2 = \quad (1)$$

Finding. u_2;

$$u_2 \quad = u_0 + at \therefore u_2 = u_0 + 2at$$
$$= u_0 + a(2t) \therefore$$

Final interval:

$$y = \quad u_2 t + \frac{1}{2}at^2$$
$$\Rightarrow y = \quad (u_0 + 2at)t^2 + \frac{1}{2}at^2$$
$$\text{From (1) } \frac{1}{2}at^2 = x - u_0 t$$
$$y = \quad (u_0 + 2at)t + x - u_0 t$$
$$\therefore \ y = x + 2at^2$$
$$\therefore a = \frac{y - x}{2t^2}$$

Question 31: A

Can use trial and error

Question 32: C

$$E = \frac{\sigma}{\varepsilon}$$

$$\sigma = \frac{F}{A} = \frac{7.2 \times 10}{1.8 \times 10^{-6}}$$

$$= 40 \text{MPa}$$

$$\varepsilon = \frac{0.5 \times 10^{-3}}{2.5} = 2 \times 10^{-4}$$

$$E = \frac{\sigma}{\varepsilon} = 2 \times 10^{11} P_a$$

$$E = \frac{1}{2} \text{Fx} = \frac{1}{2} \times 72 \times 0.5 \times 10^{-3}$$

Enersy stored $= 0.018$ J

Question 33: B

$$a, ar, ar^2, ar^3, \ldots . S_\infty = \frac{a}{1-r} = \frac{8}{5}$$

$$ar, ar^3, ar^5, \ldots . \delta\infty = \frac{ar}{1-r^2} = \frac{3}{5}$$

$$\frac{a}{1-r} = \frac{8}{5} = 0$$

$$\frac{ar}{1-r^2} = \frac{3}{5} = 2$$

$$\frac{\cancel{a}}{\cancel{1-r}} \times \frac{(1+r)\cancel{(1+r)}}{\cancel{a}r} = \frac{8}{5} \times \frac{5}{3}$$

$$\Rightarrow \frac{1+r}{r} = \frac{8}{3}$$

$$\therefore 1+r = \frac{8}{3}r$$

$$\therefore \frac{5}{3}r = 1 \quad \therefore r = \frac{3}{5}$$

$$\therefore a = \frac{16}{25}$$

$$\therefore a + r = \frac{31}{25}$$

Question 34: D

Conservation of momentum.
Momentum before = momentum after

$$(30 + 10) \times 4 = 30 \times 0 + 10 \times v$$
$$\therefore V = \frac{160}{10} = 16 \text{ ms}^{-1}$$

Question 35: C

$$y = (x - 4)(x^2 - 2x - 8)$$
$$y = -x^2 + 8x + 16$$

Recognise that the first curve is a cubic curve. The second curve is a quadratic curve. Therefore, must intersect at two points only.

Question 36: G

Volume is conserved:

$$A_1 L_1 = A_2 L_2$$
$$A_1 L_1 = \frac{1}{4} A_1 L_2 \therefore L_2 = A_2 4 L_1$$
$$R = \frac{\rho L_1}{A_1}, \; A_2 = \frac{1}{4} A_1$$
$$\therefore \; R_2 = \frac{\rho L_2}{A_2} = \frac{\rho \cdot 4 L_1}{\frac{1}{4} A_1}$$
$$\therefore \; R_2 = 16 \frac{\rho L_1}{A_1} = 16R$$

Question 37: H

$$\frac{3}{\sqrt{3}\sqrt{9}+\sqrt{3}\sqrt{7}}+\frac{3}{\sqrt{3}\sqrt{8}+\sqrt{3}\sqrt{6}}+\frac{3}{\sqrt{3}\sqrt{7}+\sqrt{3}\sqrt{5}}$$
$$+\cdots+\frac{3}{\sqrt{3}\sqrt{3}+\sqrt{3}}$$

$$=\frac{3}{\sqrt{3}}\left(\frac{1}{\sqrt{9}+\sqrt{7}}+\frac{1}{\sqrt{8}+\sqrt{6}}+\frac{1}{\sqrt{7}+\sqrt{5}}+\frac{1}{\sqrt{6}+\sqrt{4}}\right.$$
$$\left.+\frac{1}{\sqrt{5}+\sqrt{3}}+\frac{1}{\sqrt{4}+\sqrt{2}}+\frac{1}{\sqrt{3}+\sqrt{1}}\right)$$

$$=\frac{3\times\sqrt{3}}{3}\left(\frac{\sqrt{9}-\sqrt{7}}{2}+\frac{\sqrt{8}-\sqrt{6}}{2}+\frac{\sqrt{7}-\sqrt{5}}{2}\right.$$

$$=\quad+\frac{\sqrt{6}-\sqrt{4}}{2}+\frac{\sqrt{4}-\sqrt{3}}{2}+\frac{\sqrt{4}-\sqrt{2}}{2}+\frac{\sqrt{3}-1}{2})$$

$$=\sqrt{3}\left(\frac{3}{2}+\frac{2\sqrt{2}}{2}-\frac{\sqrt{2}}{2}-\frac{1}{2}\right)$$

$$-\sqrt{3}\left(1-\frac{\sqrt{2}}{2}\right)$$

Question 38: D

$$a=4.0-0.36t$$
$$v=\int a\,dt$$
$$v=4t-\frac{0.36}{2}t^2+c$$
$$t=0, v=0 \therefore c=0$$
$$s=\int_0^{10}(4t-0.18t^2)dt$$
$$=2[t^2]_0^{10}-\frac{0.18}{3}[t^3]_0^{10}$$
$$=140\text{ m}$$

Question 39: B

$$
\begin{aligned}
\text{Area} \quad &= \sqrt{x}(15 - x) \\
\therefore A \quad &= 15x^{1/2} - x^{3/2} \\
\frac{dA}{dx} \quad &= \frac{15}{2}x^{-1/2} - \frac{3}{2}x^{1/2} \\
\frac{dA}{dx} \quad &= 0 \\
\frac{15}{x\sqrt{x}} \quad &= \frac{3}{x}\sqrt{x} \\
\therefore (\sqrt{x})^2 \quad &= \frac{15}{3} \\
\therefore x \quad &= 5 \\
\therefore A_{\max} \quad &= 10\sqrt{5}
\end{aligned}
$$

Question 40: E

Conservation of momentum:

$$
\begin{aligned}
v_2 &\leftrightarrow \frac{4m}{0} mv_1 + 4mv_2 \\
\therefore v_2 &= -\frac{1}{4}v_1 \\
E_{k_1} &= \frac{1}{2}mv_1^2 \\
E_{k_2} &= \frac{1}{2} \cdot 4m \cdot \left(\frac{1}{4}v_1\right)^2 = \frac{1}{4}E_{k_1} \\
W &= E_{k_1} + E_{k_2} \\
\Rightarrow W &= E_{k_1} + \frac{1}{4}E_{k_1} \\
\Rightarrow E_{k_1} &= \frac{4}{5}W \\
\therefore E_{k_1} &- \frac{4}{4}E_{k_2} \\
&= \frac{4}{5}W - \frac{1}{4}E_{k_1} \\
&= \frac{4}{5}W - \frac{1}{4}W \\
&= \frac{3}{5}W
\end{aligned}
$$

Section 2

Question 1: D

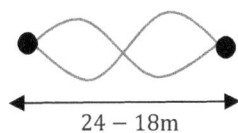

∴ node to node = 1λ

$$\therefore \quad \therefore \lambda = 6m$$
$$v = f\lambda$$
$$336 = f \times 6$$
$$\therefore f = \frac{336}{6} = 56 \text{ Hz}_z$$

Question 2: C

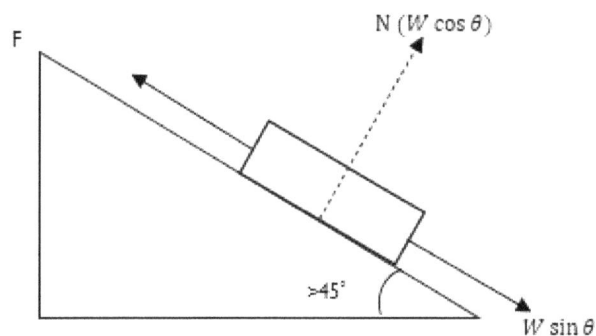

$W\sin\theta \leq F$

And since $W = \sqrt{(W_{\cos\theta}\theta)^2 + (W_{\sin}\theta)^2}$

$\therefore W > F$

If $F \geq W\sin\theta$, then $F > N$

$\therefore N\,N < F < W$

Question 3: G

Resistance of wire $= \frac{Pl}{A}$

$$= \frac{10^{-7} \times 20}{0.1 \times (10^{-3})^2}$$
$$= 20\Omega$$
$$V = IR = 200 \times 10^{-3}(20 + 50) = 14 \text{ V}$$

Question 4: D

$$Mg\sin \theta - mg = (M + m)a$$
$$\sin \theta = \frac{4}{5}$$
$$Mg \cdot \frac{4}{5} - mg = (m + M)a \quad \prod_{\text{Newton's I Law}}$$

$\therefore M$ must be greeter than $\frac{5}{4}$ m in order for M to accelerate.

$$\therefore M > \frac{5}{4}m \text{ D}$$

Question 5: F

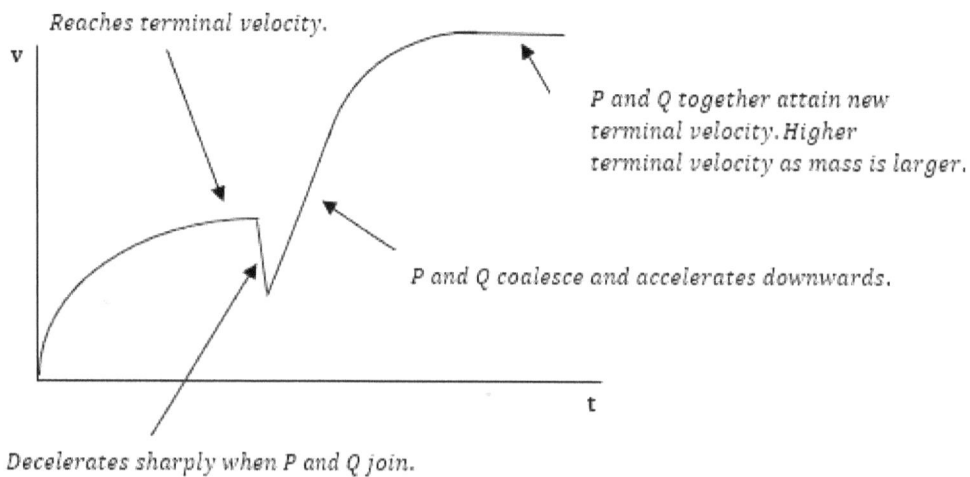

Reaches terminal velocity.

P and Q together attain new terminal velocity. Higher terminal velocity as mass is larger.

P and Q coalesce and accelerates downwards.

Decelerates sharply when P and Q join.

Question 6: A

$R = kv^2$

$P = R_{v_{max}} v_{max} = cP^n$

$P = kv_{max}{}^3$

$P = k(cP_a^n)^3 \therefore P = kc^3 p^{3n}$

Powher $e^{n/3}$ To balance powers,

$n = \dfrac{1}{3}$

Question 7: D

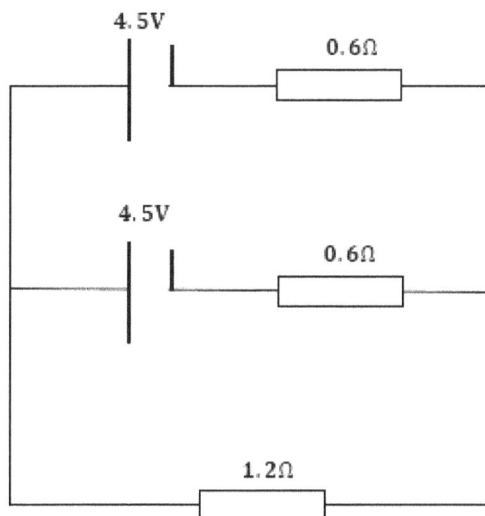

Consider one voltage source $\therefore$ 4.5 V

0.6Ω internal resistance in parallel $\therefore$ 0.3Ω.

$$\text{Current } = \frac{4.5}{1.2 + 0.3} = 3 \text{ A.}$$

$$\therefore P = I^2 R = 3^2 \times 1.2 = 10.8 \text{ W}$$

Question 8: B

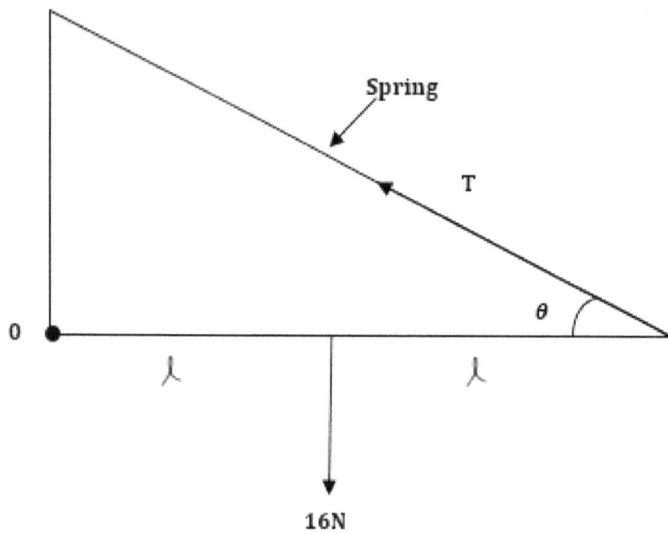

Moments about 0 :

$$16 \times \text{ } = T \times \frac{4}{5} \times 2 \text{ }$$
$$\therefore T = 10 \text{ N}. \; h = 20 \text{Nm}^{-1}$$
$$\therefore x = \frac{10}{h} = 0.5 \text{ m}.$$
$$\therefore E \cdot P \cdot E = \frac{1}{2}T_x = \frac{1}{2} \times 10 \times 0.5 = 2.5 \text{ J}$$

Question 9: D

$$x = 4t^3 \qquad v = \frac{dx}{dt} \qquad a = \frac{dv}{dt}$$
$$v = 12t^2$$
$$a = 24t$$
$$\therefore \quad \text{at } t = 5, a = 120 \text{ ms}^{-2}$$
$$\therefore F = \text{ rate of chenge of monentum}$$
$$\therefore F = ma = 2 \times 120 = 240 \text{ kg ms}^{-2}$$

Question 10: C

The wavefronts are more curved than the ones in the gel, so the x cannot be in the metal. Since they are more rounded than the wavefronts in the gel, their epicentre must lie in the gel below U.

Question 11: E

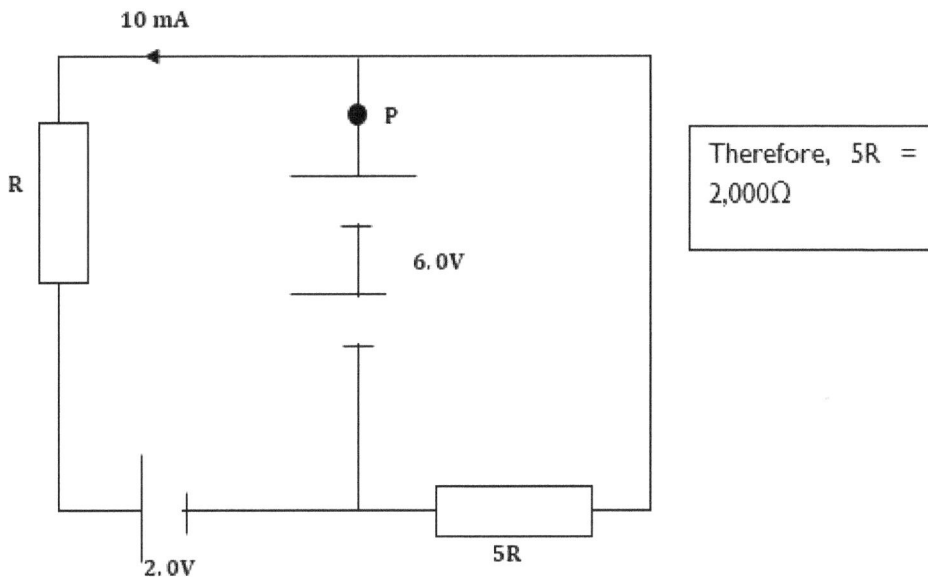

Therefore, $5R = 2,000\Omega$

Through resistor R,

$$6 - 2 = 10 \times 10^{-3} \times R$$
$$\therefore R = 400\Omega$$

Current through $5R$:

$$I = \frac{6}{5 \times 400} = 3 \text{ mA}$$

$\therefore$ Cerrar at $P = 10_{mA} + 3mA$

$$= 13mA$$

Question 12: E

Since they have different unstretched lengths, but equal masses, this means they must have different thicknesses since they are the same metal.   Therefore, upon applying same tensile force they must stretch by different amounts.

Question 13: C

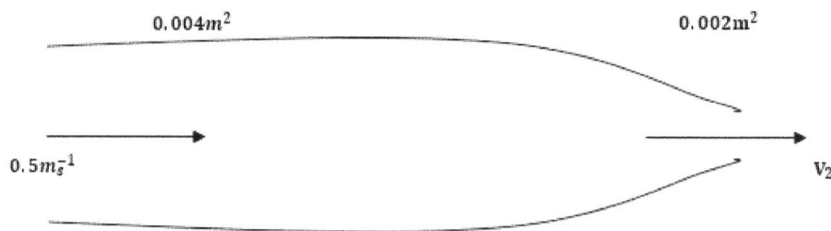

$$\underset{5\ \text{ms}^{-1}}{\overset{0.004\ \text{m}^2}{\longrightarrow}} \longrightarrow \underset{v_2}{0.002\ \text{m}^2}$$

$$A_1 v_1 = A_2 v_2$$

$$\therefore v_2 = \frac{0.004 \times 0.5}{0.002} = 1\ \text{ms}^{-1}/$$

$$P = F \cdot v_{\text{avg}}$$

$$+P = 0.75\ \text{W}$$

$$= \frac{ma \cdot v_{\text{ung}}}{t \cdot (v_2 - v_1)} \cdot \frac{(1 + 0.5)}{2}$$

$$= \frac{1000 \times 0.5 \times 0.004 \times (1 - 0.5)}{1} \times 0.75$$

Question 14: E

$$Q \cos 60° = W \qquad\qquad P = Q\sin 60°$$

$$\therefore W = \frac{Q}{2} \qquad P = 2w\frac{\sqrt{3}}{2} \qquad\qquad \therefore P = \frac{}{Q} = \sqrt{3}W$$

$$\therefore Q = 2W$$

$$E = \frac{\sigma}{\varepsilon}$$

$$\therefore \varepsilon_2 = \frac{\sigma}{E} = \frac{2W}{4A_p} = \frac{1}{2}\frac{W}{EA_p}$$

$$\varepsilon_p = \frac{\dfrac{\sqrt{3}W}{A_p}}{E} = \sqrt{3}\frac{W}{EA_p}$$

$$\varepsilon_p : \varepsilon_Q = \frac{\sqrt{3}}{\dfrac{1}{2}} = 2\sqrt{3}$$

Question 15: H

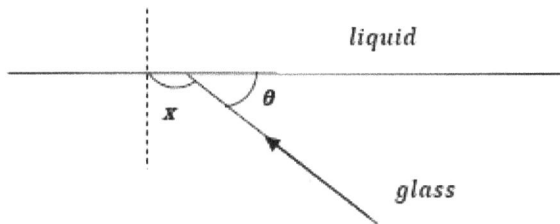

Therefore $\theta_c < \theta < 90°$

$$\cos^{-1}\left(\frac{4}{5}\right) < \theta < 90°$$

Angle between normal and ray must be > critical angle

$$n_1\sin \theta_1 = n_2\sin \theta_2$$

$$\Rightarrow \left(\frac{3 \times 10^8}{2 \times 10^8}\right)\sin x = 1.2\sin(90)$$

$$\Rightarrow 1.5\sin(90 - \theta_c) = 1.2$$

$$\cos \theta_c = \frac{4}{5} \qquad\qquad \therefore \theta_c = \cos^{-1}\left(\frac{4}{5}\right)$$

Question 16: E

From $t = [0,10]$

$$v = \int a\,dt = 2t + c.$$

$\therefore v = 2t$

$s = t^2 + k \ (s = 0, t = 0)$

$\therefore s = t^2$

From $t = [10,20]$

$$v = \int -1.5\,dt = -1.5t + 20$$

$$s = \int_{10}^{20} (-1.5t + 20)t$$

$$= -0.75t^2 + 20t + 100$$

$$= -0.75[t^2]_{10}^{20} + 20[t]_{10}^{20} + 100$$

$$= 225 \text{ m}$$

Question 17: G

$$\rho = \frac{m}{V}$$

$$= \frac{8.7}{2.5}$$

$$= 3.48 \text{ g cm}^{-3}$$

Question 18: C

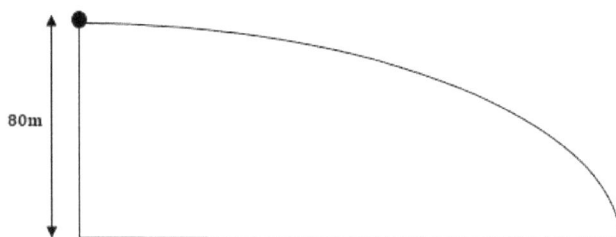

Vertical velocity:

$$s = ut + \frac{1}{2}at^2$$

$$-80 = 0 + \frac{1}{2}(-10)t^2$$

$$\therefore t = 4s.$$

$$v = u + at = 0 + 10 \times 4 = 40 \text{ ms}^{-1}$$

Horizontal velocity:

$$E_k = \frac{1}{2}mv^2$$

$$125 = \frac{1}{2} \times 0.1v^2$$

$$V = 50 \text{ ms}^{-1}$$

$$v = \sqrt{v_n^2 + v_v^2}$$

$$v_n = \sqrt{50^2 - 40^2} = 30_{ms^{-1}}$$

$$\therefore \text{ distance } = 30 \times 4 = 120 \text{ m}$$

Question 19: A

$$\text{Energy transferred } = \int P \, dt$$

$$= V \int I(t) \, dt$$

$$= Vk \int t^2 \, dt$$

$$= \frac{Vk}{3}[t^3]_6^{t_f}$$

$$= \frac{Vk}{3}t_F{}^3$$

$$I_F = kt_F{}^2$$

$$\therefore \left(\frac{I_F}{k}\right)^{\frac{1}{2}} = t_F$$

$$\therefore E = \frac{Vk}{3}\left(\frac{I_F}{k}\right)^{\frac{3}{2}}$$

Question 20: C

$$
\begin{aligned}
F \quad &= P \cdot A \\
&= PghdA \\
&= P_g h(x)hdx \\
&= \frac{Pg}{2}\int x^2 dx \\
&= \frac{Pg}{6}[x^3]_0^{0.6} = 360 \text{ N}
\end{aligned}
$$

END OF PAPER

ENGAA PRACTICE PAPERS

Already seen them all?

So, you've run out of past papers? Well hopefully that is where this book comes in. It contains two unique mock papers - each compiled by ENGAA Expert tutors at *UniAdmissions* and available nowhere else.

Having successfully gained a place at Cambridge for engineering, our tutors are intimately familiar with the ENGAA and its associated admission procedures. So, the novel questions presented to you here are of the correct style and difficulty to continue your revision and stretch you to meet the demands of the ENGAA.

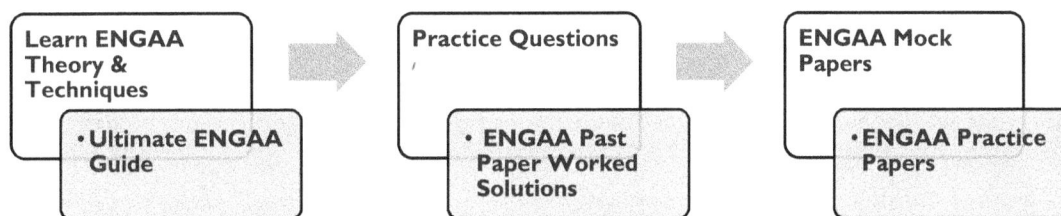

Learn ENGAA Theory & Techniques		Practice Questions		ENGAA Mock Papers
• Ultimate ENGAA Guide	→	• ENGAA Past Paper Worked Solutions	→	• ENGAA Practice Papers

Start Early

It is much easier to prepare if you practice little and often. Start your preparation well in advance; ideally ten weeks but at the latest within a month. This way you will have plenty of time to complete as many papers as you wish to feel comfortable and won't have to panic and cram just before the test, which is a much less effective and more stressful way to learn. In general, an early start will give you the opportunity to identify the complex issues and work at your own pace.

Prioritise

Some questions can be long and complex – and given the intense time pressure you need to know your limits. It is essential that you don't get stuck with very difficult questions. If a question looks particularly long or complex, mark it for review and move on. You don't want to be caught five questions short at the end just because you took more than 3 minutes in answering a challenging multi-step question. If a question is taking too long, choose a sensible answer and move on. Remember that each question carries equal weighting and therefore, you should adjust your timing accordingly. With practice and discipline, you can get very good at this and learn to maximise your efficiency.

Positive Marking

There are no penalties for incorrect answers; you will gain one mark for each right answer and will not get one for each wrong or unanswered one. This provides you with the luxury that you can always guess should you not be able to figure out the right answer for a question or run behind on time. Since each question provides you with 4 to 6 possible answers, you have a 16-25% chance of guessing correctly. Therefore, if you aren't sure (and are running short of time), then make an educated guess and move on. Before 'guessing' you should try to eliminate a couple of answers to increase your chances of getting the question correct. For example, if a question has 5 options and you manage to eliminate 2 options - your chances of getting the question increase from 20% to 33%!

Avoid losing easy marks on other questions because of poor exam technique. Similarly, if you have failed to finish the exam, take the last 10 seconds to guess the remaining questions to at least give yourself a chance of getting them right.

Practice

This is the best way of familiarising yourself with the style of questions and the timing for this section. Although the exam will essentially only test GCSE & AS level knowledge, you are unlikely to be familiar with the style of questions in all sections when you first encounter them. Therefore, you want to be comfortable at using this before you sit the test.

Practising questions will put you at ease and make you more comfortable with the exam. The more comfortable you are, the less you will panic on the test day and the more likely you are to score highly. Initially, work through the questions at your own pace, and spend time carefully reading the questions and looking at any additional data. When it becomes closer to the test, **make sure you practice the questions under exam conditions**.

Repeat Questions

When checking through answers, pay particular attention to questions you have gotten wrong. If there is a worked answer, look through that carefully until you feel confident that you understand the reasoning, and then repeat the question without help to check that you can do it. If only the answer is given, have another look at the question and try to work out why that answer is correct. This is the best way to learn from your mistakes, and means you are less likely to make similar mistakes when it comes to the test. The same applies for questions you were unsure of and made an educated guess for, even if you got it right. When working through this book, **make sure you highlight any questions you are unsure of**; this means you know to spend more time looking over them once marked.

Use the Options

Some questions may try to overload you with information. When presented with large tables and data, it's essential you look at the answer options so you can focus your mind. This can allow you to reach the correct answer a lot more quickly. Consider the example below:

The table below shows the results of a study investigating antibiotic resistance in staphylococcus populations. A single staphylococcus bacterium is chosen at random from a similar population. Resistance to any one antibiotic is independent of resistance to others.

Antibiotic	Number of Bacteria tested	Number of Resistant Bacteria
Benzyl-penicillin	1011	98
Chloramphenicol	109	1200
Metronidazole	108	256
Erythromycin	105	2

Calculate the probability that the bacterium selected will be resistant to all four drugs.

A. 1 in 10^6
B. 1 in 10^{12}
C. 1 in 10^{20}
D. 1 in 10^{25}
E. 1 in 10^{30}
F. 1 in 10^{35}

Looking at the options first makes it obvious that there is **no need to calculate exact values** - only in powers of 10. This makes your life a lot easier. If you hadn't noticed this, you might have spent well over 90 seconds trying to calculate the exact value when it wasn't being asked for.

******

In other cases, you may actually be able to use the options to arrive at the solution quicker than if you had tried to solve the question as you normally would. Consider the example below:

A region is defined by the two inequalities: $x - y^2 > 1$ and $xy > 1$. Which of the following points is in the defined region?

A. (10,3)
B. (10,2)
C. (-10,3)
D. (-10,2)
E. (-10,-3)

Whilst it's possible to solve this question both algebraically and graphically by manipulating the identities, by far **the quickest way is to actually use the options**. Note that options C, D and E violate the second inequality, narrowing down to answer to either A or B. For A: $10 - 3^2 = 1$ and, thus, this point is on the boundary of the defined region and not actually in the region. Therefore, the answer is B (as $10 - 4 = 6 > 1$.)

In general, it pays dividends to look at the options briefly and see if they can be help you arrive at the question more quickly. Get into this habit early – it may feel unnatural at first, but it's guaranteed to save you time in the long run.

Manage your Time

It is highly likely that you will be juggling your revision alongside your normal school studies. Whilst it is tempting to put your A-levels on the back burner, falling behind in your school subjects is not a good idea, don't forget that to meet the conditions of your offer, should you get one, you will need at least one A*. So, time management is key!

Make sure you set aside a dedicated 90 minutes (and much more closer to the exam) to commit to your revision each day. The key here is not to sacrifice too many of your extracurricular activities, everybody needs some down time, but instead to be efficient. Take a look at our list of top tips for increasing revision efficiency below:

1. Create a comfortable workstation.
2. Declutter and stay tidy.
3. Treat yourself to some nice stationery.
4. See if music works for you → if not, find somewhere peaceful and quiet to work.
5. Turn off your mobile or at least put it into silent mode.
6. Silence social media alerts.
7. Keep the TV off and out of sight.
8. Stay organised with to do lists and revision timetables – more importantly, stick to them!
9. Keep to your set study times and don't bite off more than you can chew!
10. Study while you're commuting.
11. Adopt a positive mental attitude.
12. Get into a routine.
13. Consider forming a study group to focus on the harder exam concepts.
14. Plan rest and reward days into your timetable – these are excellent incentive for you to stay on track with your study plans!

Keep Fit & Eat Well

'A car won't work if you fill it with the wrong fuel' - your body is exactly the same. You cannot hope to perform unless you remain fit and well. The best way to do this is not underestimate the importance of healthy eating. Beige, starchy foods will make you sluggish; instead start the day with a hearty breakfast like porridge. Aim for the recommended 'five a day' intake of fruit/veg and stock up on the oily fish or blueberries – the so called "super foods".

When hitting the books, it's essential to keep your brain hydrated. If you get dehydrated, you'll find yourself lethargic and possibly developing a headache, neither of which will do any favours for your revision. Invest in a good water bottle that you know the total volume of and keep sipping throughout the day. Don't forget that the amount of water you should be aiming to drink varies depending on your mass, so calculate your own personal recommended intake as follows: 30 ml per kg per day.

It is well known that exercise boosts your wellbeing and instils a sense of discipline. All of which will reflect well in your revision. It's well worth devoting half an hour a day to some exercise; get your heart rate up, break a sweat, and get those endorphins flowing.

Sleep

It's no secret that when revising you need to keep well rested. Don't be tempted to stay up late revising as sleep actually plays an important part in consolidating long term memory. Instead aim for a minimum of 7 hours of sleep each night, in a dark room without any glow from electronic appliances. Install flux (https://justgetflux.com) on your laptop to prevent your computer from disrupting your circadian rhythm. Aim to go to bed the same time each night and no hitting snooze on the alarm clock in the morning!

REVISION TIMETABLE

Still struggling to get organised? Then try filling in the example revision timetable below, remember to factor in enough time for short breaks, and stick to it! Remember to schedule in several breaks throughout the day and actually use them to do something you enjoy e.g. TV, reading, YouTube etc.

	8AM	10AM	12PM	2PM	4PM	6PM	8PM
MONDAY							
TUESDAY							
WEDNESDAY							
THURSDAY							
FRIDAY							
SATURDAY							
SUNDAY							
EXAMPLE DAY	School				Statistics	Pure maths	Mech-anics

Top tip! Ensure that you take a watch that can show you the time in seconds into the exam. This will allow you have a much more accurate idea of the time you're spending on a question. In general, if you've spent >120 seconds on a section 1 question, move on regardless of how close you think you are to solving it.

GETTING THE MOST OUT OF MOCK PAPERS

Mock exams can prove invaluable if tackled correctly. Not only do they encourage you to start revision earlier, they also allow you to **practice and perfect your revision technique**. They are often the best way of improving your knowledge base or reinforcing what you have learnt. Probably the best reason for attempting mock papers is to familiarise yourself with the exam conditions of the ENGAA as they are particularly tough.

Start Revision Earlier

Thirty five percent of students agree that they procrastinate to a degree that is detrimental to their exam performance. This is partly explained by the fact that they often seem a long way in the future. In the scientific literature this is well recognised; Dr Piers Steel, an expert on the field of motivation, states that *'the further away an event is, the less impact it has on your decisions'*.

Mock exams are therefore a way of giving you a target to work towards and motivate you in the run up to the real thing – every time you do one, treat it as the real deal! If you do well, then it's a reassuring sign; if you do poorly, then it will motivate you to work harder (and earlier!).

Practice and perfect revision techniques

In case you haven't realised already, revision is a skill which can take some time to learn. For example, the most common revision techniques, including **highlighting and/or re-reading are quite ineffective**, ways of committing things to memory. Unless you are thinking critically about something you are much less likely to remember it or indeed understand it.

Mock exams, therefore, allow you to test your revision strategies as you go along. Try spacing out your revision sessions so you have time to forget what you have learnt in-between. This may sound counterintuitive but the second time you remember it for longer. Try teaching another student what you have learnt, this forces you to structure the information in a logical way that may aid memory. Always try to question what you have learnt and appraise its validity. Not only does this aid memory but it is also a useful skill for Oxbridge interviews and beyond.

Improve your knowledge

The act of applying what you have learnt reinforces that piece of knowledge. A question may ask you to think about a relatively basic concept in a novel way (not cited in textbooks), and so deepen your understanding. Exams rarely test word for word what is in the syllabus, so, when running through mock papers, try to understand how the basic facts are applied and tested in the exam. As you go through the mocks or past papers, take note of your performance and see if you consistently under-perform in specific areas, thus highlighting areas for future study.

Get familiar with exam conditions

Pressure can cause all sorts of trouble for even the most brilliant students. The ENGAA is a particularly time-pressured exam with high stakes – your future (without exaggeration) does depend on your result to a great extent. The real key to the ENGAA is overcoming this pressure and remaining calm to allow you to think efficiently.

Mock exams are therefore an excellent opportunity to devise and perfect your own exam techniques to beat the pressure and meet the demands of the exam. **Don't treat mock exams like practice questions – it's imperative you do them under time conditions.**

THINGS TO HAVE DONE BEFORE USING THIS BOOK

Do the groundwork

- Read in detail: the background, methods, and aims of the ENGAA as well logistical considerations such as how to take the ENGAA in practice. A good place to start is an ENGAA textbook like The Ultimate ENGAA Guide (flick to the back to get a free copy!) which covers all the necessary groundwork.

- It is generally a good idea to start re-capping all your GCSE maths and physics.

- Practice substituting formulas together to reach a more useful expressions involving known variables e.g. $P = IV$ and $V = IR$ can be combined to give $P = V2/R$ and $P = I2R$. Remember that calculators are not permitted in the exam, so get comfortable doing more complex long addition, multiplication, division, and subtraction.

- Get comfortable rapidly converting between percentages, decimals, and fractions.

- These are all things which are easiest to do alongside your revision for exams before the summer break. Not only gaining a head start on your ENGAA revision but also complimenting your year 12 studies well.

- Discuss scientific problems with others - propose experiments and state what you think the result would be. Be ready to defend your argument. This will rapidly build your scientific understanding for section 2 but also prepare you well for an Oxbridge interview.

- Read through the ENGAA syllabus before you start tackling whole papers. This is absolutely essential. It contains several stated formulae, constants, and facts that you are expected to apply - or may just be an answer in their own right. Familiarising yourself with the syllabus is also a quick way of teaching yourself the additional information other exam boards may learn which you do not. Sifting through the whole ENGAA syllabus is a time-consuming process so we have done it for you. Be sure to flick through the syllabus checklist later on, which also doubles up as a great revision aid for the night before!

> **Remember!** It's better that you make all the mistakes you possibly can now in mock papers and then learn from them so as not to repeat them in the real exam.

Ease in gently

With the groundwork laid, there's still no point in adopting exam conditions straight away. Instead invest in a beginner's guide to the ENGAA, which will not only describe in detail the background and theory of the exam, but take you through, section by section, what is expected. The Ultimate ENGAA Guide is the most popular ENGAA textbook – you can get a free copy by flicking to the back of this book.

When you are ready to move on to past papers, take your time and puzzle your way through all the questions. Really try to understand solutions. A past paper question won't be repeated in your real exam, so don't rote learn methods or facts. Instead, focus on applying prior knowledge to formulate your own approach.

If you're really struggling and have to take a sneak peek at the answers, then practice thinking of alternative solutions. It is unlikely that your answer will be more elegant or succinct than the model answer, but it is still a good task for encouraging creativity with your thinking. Get used to thinking outside the box!

Accelerate and Intensify

Start adopting exam conditions after you've done two past papers. Don't forget that **it's the time pressure that makes the ENGAA hard** – if you had as long as you wanted to sit the exam, you would probably get 100%. If you're struggling to find comprehensive answers to past papers, then ENGAA *Past Papers Worked Solutions* contains detailed, explained answers to every ENGAA past paper question and essay (flick to the back to get a free copy).

Doing all the past papers is a good target for your revision. In any case, choose a paper and proceed with strict exam conditions. Take a short break and then mark your answers before reviewing your progress. For revision purposes, as you go along, keep track of those questions that you guess – these are equally as important to review as those you get wrong.

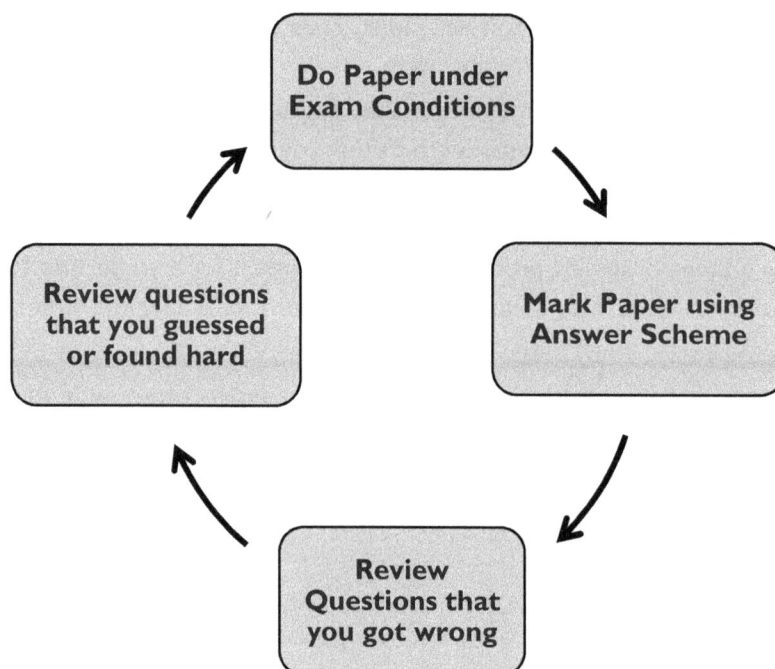

Once you've exhausted all the past papers, move on to tackling the unique mock papers in this book. In general, you should aim to complete one to two mock papers every night in the ten days preceding your exam.

Section 1: An Overview

What will you be tested on?	No. of Questions	Duration
The ability to apply scientific knowledge including knowledge up to A Level Physics/Maths	54 MCQs	80 Minutes

Section 1 of the exam involves short MCQ questions relating to Mathematics and Physics that are designed to see if you can quickly apply the principles that you have learnt in school in a time-pressured exam. You will have, on average, 90 seconds per question so it is vital to work very quickly - some questions later tend to be harder so you should be doing the initial questions in under a minute. It cannot be emphasised enough that the limiting factor in this test is time, not your ability. Practice is therefore crucial to learn the technique, skills and tricks to answer section 1 questions quickly. There are two parts to this section:

- Part A - Mathematics and Physics (28 questions)
- Part B - Advanced Mathematics and Advanced Physics (26 questions)

Each part has its own syllabus which can be found on the Cambridge Undergraduate Admissions website. We will provide a quick summary of the syllabus to give you an overview.

Part A

- **Number** – you should be confident performing a wide range of numerical calculations without the use of a calculator. As this is a MCQ exam, producing order of magnitude estimates will be very useful.
- **Algebra** – you should be competent at basic algebra taught up to AS maths. You will already be at this standard, but the key is to practice lots of questions, so you manipulate your algebra at the required speed.
- **Geometry** – you should know the basic geometry taught at GCSE level. As this is a MCQ exam, you will not need to provide geometric proofs.
- **Measure** – this is linked to "number" but we recommend that you become fully confident when dealing with scale factors. A question can often be simplified by working with this approach.
- **Statistics** – a very basic knowledge of GCSE statistics is all that is necessary. It is however important to know how to combine different parts of statistics together and not get bogged down in long calculations.
- **Probability** – a basic GCSE level of knowledge of probability is required but you will need to work through these questions quickly. We recommend that you practice drawing out 'tree diagrams' quickly to solve these problems.
- **Electricity** – only AS Level physics is needed. However, expect questions to be significantly more challenging requiring you to combine Kirchoff's voltage and current laws. These questions may contain significantly more algebra than A level exams so it is important that you can perform the necessary calculations quickly.
- **Motion and Energy** (i.e. M1) – this will be at the standard of an AS Level Physics exam but remember that the questions are short. There are often little tricks that can only be learned by practice. If stuck, try drawing out a diagram and resolving forces, or even write down some conservation equations.
- **Thermal Physics** – we are not taught much (if any) quantitative thermal physics at school. This means that you should expect conceptual questions on conduction, convection and radiation.

- **Waves** – this is a complicated topic that is touched on in GCSE and AS physics. It is important to know they key words and what they represent: wavelength, frequency, period, speed and amplitude. Expect some basic calculations or some tricky conceptual questions.
- **Electromagnetic Spectrum** - very similar to GCSE physics. It is just a case of memorising the different parts of the EM spectrum and their different dangers/applications.
- **Radioactivity** - another conceptual topic that requires you know about the different kinds of radioactive decay. We recommend that you also learn to balance the basic decay equations for alpha, beta, gamma, decay as well as nuclear fusion/fission.

Part B

The syllabus for part B includes all of the knowledge from part A as well as some additional topics. Most of these are taken from the AS Level syllabus and these will be important for Part B, section 2 of the ENGAA as well as the Cambridge engineering interview.

- **Algebra and Functions**- This is very similar to part A but requires the solution of more complex algebraic equations (including inequalities). The inequalities can be challenging to do under time pressure, so we recommend quickly drawing out the x-y plane and identifying the region of interest. The factor/remainder theorem from A2 maths also appear in the syllabus so you may be tested on this.
- **Sequences and Series**- This covers the content of AS level maths - two important things to note are sums of series and the binomial expansion which can often provide a shortcut to MCQ questions.
- **Coordinate Geometry** in the (x, y) plane- This is also content covered in AS Level maths and the challenge will be completing questions under time pressure. Practise converting equations to a standard form and then sketching them on the x-y plane- this will often help you spot the solution.
- **Trigonometry** - Basic trigonometry covering material tested in AS level maths. You are expected to know two basic trig formulae as well as the values of sine, cosine and tangent for the angles 0°, 30°, 45°, 60°, 90°. As you will not have a calculator, it is crucial to memorise these values. This will also be very useful for interviews.
- **Exponential and Logarithms**- be confident at using the log formulae that you learnt in AS level maths. Using the formula will often simplify a question and with practice you will be able to determine whether to solve an equation in exponential form or logarithmic form.
- **Differentiation**- when sitting your ENGAA exam, you will likely have covered advanced topics in differentiation including the product, chain and quotient rule. However, the exam itself only tests very basic differentiation taught in AS maths so try not to overcomplicate these questions. It is also important to appreciate the use of differentiation in physics questions.
- **Integration**- this once again contains the basic integration taught as part of AS level maths. It will be important to practice definite integration without a calculator as it is very easy to make a simple mistake. We also recommend understanding the application of AUC to solving physics questions.
- **Graphs of Functions**- this is a very important topic as it provides a lot of tricks to solve maths and physics questions. We recommend you know the C3 transformations of graphs inside out and how to draw out sketches of common functions.

- **Forces and Equilibrium-** this is very similar to the content of Part A. It requires you to be confident as resolving forces in 2D. With practice, you will be able to decide quickly which axes to resolve along. We recommend covering this topic in an M1 textbook and doing practice questions until you can resolve diagrams without thinking.

- **Kinematics-** learn the SUVAT equations! These will come up again and again and there are only 4 (maybe 5) to memorise. For any question in kinematics, we recommend writing out the known quantities in SUVAT, figuring out the unknowns and then deciding on which equation to use.

- **Newton's Laws-** this links very closely to the forces and equilibrium but may include motion. With questions considering more than one body (e.g. a car towing a trailer), make sure you know which forces are acting on which body - a diagram can often save the day!

- **Momentum-** using conservation of momentum to solve basic 1D problems will be necessary for the exam - writing down the conservation equations appreciating that momentum is a vector quantity will be sufficient.

- **Energy-** practice writing out the kinetic and potential energies of a system in a given coordinate system. You are often told which coordinates to use. Once you have a conservation equation for energy and one for momentum, solving these simultaneous equations will lead you to the correct answer.

Although we have separated section one into part A and B just like it will be done in the exam, it often makes sense to revise for both parts at the same time as there is significant overlap.

Section 2: An Overview

What will you be tested on?	No. of Questions	Duration
The ability to apply scientific knowledge including knowledge up to A Level Physics/Maths	Approximately 20 structured MCQs	40 minutes

Section two will include some structured MCQ questions that are slightly more challenging, and you may be required to show some working. You will have the added benefit of a calculator, but you will still need to solve equations quickly and efficiently.

The specification lists out six additional areas that are tested on Section 2 of the ENGAA exam. These are, in principle, overlapping with much of the content of Section 1 but we recommend that you focus on the quantitative and algebraic aspects of these topics:

- Vectors and scalars - this is a section in the syllabus, but it is really covered by all of the other topics. You should be able to resolve a vector into its components and perform basic operations on vectors. It is worth knowing that the dot product of two perpendicular vectors is 0 and the formula $P = \mathbf{F} \cdot \mathbf{v}$.

- Mechanics - mechanics is heavily tested on the ENGAA because you have the knowledge to analyse problems in a quantitative manner. This is once again very similar to the material in Section 1 but there are 4 additional areas listed on the syllabus:
  - **Kinematics**- learn how to interpret velocity-time and displacement-time graphs.
  - **Dynamics**- the key additional topic is projectile motion. Remember, that you can use SUVAT independently in both dimensions.
  - **Energy**- no real additional topics, although the principle of conservation of energy is explicitly mentioned here. You should be confident writing down the total energy of the system and using this as a conserved quantity.
  - **Momentum**- this is the same as Section 1, but questions are likely to be algebraic in nature.

- **Mechanical Properties** of matter- this contains the basic content that will be important in many branches of mechanical engineering. The content is mainly covered in AS Physics and may be tested with a quantitative approach. The area under the curve of a force extension graph has a physical meaning showing a link between integration and physics. Make sure that you are comfortable using integration in this context.

- Electric **Circuits**- you need to be able to completely solve any resistor circuit they put in front of you. This means that you need to be able to figure out the voltage, current and power going through any resistor in a circuit. This will often be a case of solving simultaneous equations and this needs to be done quickly and accurately given the time pressure. It is worth looking at some common circuits (e.g. Wheatstone bridges and Potential Dividers) as these do appear frequently.

- **Waves**- there are some additional concepts taught in either AS/A2 physics. Waves are a difficult topic to understand conceptually and you might not have to time to do so before the exam - do not worry too much as most students will be in the same position. Try to understand the following key words from a conceptual perspective: polarisation, diffraction, path difference, phase, coherence and interference.

- **Quantum and Nuclear Physics**- some additional concepts surrounding wave-particle duality. Once again, there are not many formulae in this section, so it is all about the understanding. A few practice questions on these topics will help significantly.

As you can see, there is a significant overlap between Sections 1 and 2, but the questions in Section 2 are slightly longer and more involved. This means that you can expect a larger focus on some of the quantitative aspects of physics and some slightly more unusual questions.

REVISION CHECKLIST

The material for the overviews of Sections 1 and 2 have mainly been taken from the 2017 syllabus - this may change in the future. We recommend you consult the most up-to-date syllabus to see if there are any differences.

Maths

Syllabus Point	What to Know
1. Number	Understand and use BIDMAS. Define: factor, multiple, common factor, highest common factor, least common multiple, prime number, prime factor decomposition, square, positive and negative square root, cube and cube root. Use index laws to simplify multiplication and division of powers. Interpret, order and calculate with numbers written in standard index form. Convert between fractions, decimals and percentages. Understand and use direct and indirect proportion; apply the unitary method. Use surds and π in exact calculations, simplify expressions that contain surds. Calculate upper and lower bounds to contextual problems. Round to a given number of decimal places or significant figures.
2. Algebra	Simplify rational expressions by cancelling or factorising. Set up quadratic equations and solve them by factorising. Set up and use equations to solve problems involving direct and indirect proportion. Use linear expressions to describe the nth term of a sequence. Use Cartesian coordinates in all four quadrants. Equation of a straight line, $y=mx+c$; parallel lines have the same gradient. Graphically solve simultaneous equations. Recognise and interpret graphs of simple cubic functions, the reciprocal function, trigonometric functions and the exponential function; $y=kx$ for integer values of x and simple positive values of k. Draw transformations of $y = f(x)$ [$(y=af(x)$, $y=f(ax)$, $y=f(x)+a$, $y=f(x-a)$ only].
3. Geometry	Recall and use properties of angles at a point, on a straight line, perpendicular lines and opposite angles at a vertex, and the sums of the interior and exterior angles of polygons. Understand congruence and similarity; use Pythagoras' theorem in 2-D and 3-D. Use the trigonometric ratios, between 0° and 180°, to solve problems in 2-D and 3-D. Understand and construct geometrical proofs, including using circle theorems: a. the angle subtended at the circumference in a semicircle is a right angle. b. the tangent at any point on a circle is perpendicular to the radius at that point. Describe and transform 2-D shapes using single or combined rotations, reflections, translations, or enlargements, including the use of vector notation.

4. Measures	Calculate perimeters and areas of shapes made from triangles, rectangles, and other shapes; find circumferences and areas of circles, including arcs and sectors. Calculate the volumes and surface areas of prisms, pyramids, spheres, cylinders, cones and solids made from cubes and cuboids (formulae given for the sphere and cone). Use vectors, including the sum of two vectors, algebraically and graphically. Discuss the inaccuracies of measurements; understand and use three-figure bearings.
5. Statistics	Identify possible sources of bias in experimental methodology. Discrete vs. continuous data; design and use two-way tables. Interpret cumulative frequency tables and graphs, box plots and histograms. Define mean, median, mode, modal class, range, and inter-quartile range. Interpret scatter diagrams and recognise correlation, drawing and using lines of best fit. Compare sets of data by using statistical measures.
6. Probability	List all the outcomes for single and combined events. Identify different mutually exclusive outcomes and know that the sum of the probabilities of all these outcomes is 1. Construct and use Venn diagrams. Know when to add or multiply two probabilities and understand conditional probability. Understand the use of tree diagrams to represent outcomes of combined events. Compare experimental and theoretical probabilities. Understand that if an experiment is repeated, the outcome may be different.

Physics

Syllabus Point	What to Know
1. Electricity	Electrostatics: charging of insulators by friction, gain of electrons induces negative charge, uses in paint spraying and dust extraction. Conductors vs. insulators. Current = charge/time. Resistance = voltage/current; how to connect ammeters and voltmeters. V–I graphs for a fixed resistor and a filament lamp. Series vs. parallel circuits. Resistors in series (but not parallel). Voltage = energy/charge. Basic circuit symbols and diagrams. Power = current x voltage. Energy = power x time. $\left(\frac{V_p}{V_s} = \frac{n_p}{n_s}\right)$ thus when 100% efficient VpIp = VsIs. Method of electromagnetic induction, applied to a generator.
2. Motion and Energy	Speed = distance/time; difference between speed and velocity. Acceleration = change in velocity/time. Distance-time vs. velocity-time graphs (including calculation and interpretation of gradients and average speed). Newtons laws: - First law: momentum = mass x velocity, conservation of momentum. - Second law: force = mass x acceleration, force = rate of change of momentum; resultant force, W = mg, gravitational field strength, free fall acceleration, terminal velocity. - Third law: every action has an equal and opposite reaction. Work = force x distance = transfer of energy. Potential energy = mgh and Kinetic energy = $1/2mv^2$. Crumple zones and road safety. Power = Energy transferred / Time. Conservation of energy, forms of energy, useful and wasted energy, percentage efficiency.

3. Thermal physics	Conduction: factors affecting rate of conduction. Convection: temperature and density of fluids. Radiation: infrared, absorption and re-emission. Particle models of solids, liquids, gases, and state changes. Density = mass/volume. Experimental methods of determining densities.	
4. Waves	Transfer of energy without net movement of matter, transverse (electromagnetic) vs. longitudinal (sound). Define amplitude, wavelength, frequency (1Hz = 1 wave/second), and period. Wave speed = frequency x wavelength. Reflection and refraction (including ray diagrams), and Doppler effect. Application of ultrasound.	
5. Electromagnetic Spectrum	Properties of electromagnetic waves (speed of light, transverse). Distinguished by wavelength, longest to shortest: radio, microwaves, infrared, visible light, ultraviolet, x-ray, gamma. Applications and dangers.	
6. Radioactivity	Atomic structure, charges and mass of subatomic particles; ionisation. Radioactive decay: alpha vs. beta vs. gamma emission, decay equations, define activity of a sample. Ionising radiation: penetrating ability, ionising ability, presence of background radiation (including origin), applications and dangers. Define half-life and interpret from graphs. Nuclear fission: absorption of thermal neutrons, uranium-235 (decay equation), chain reaction. Nuclear fusion: hydrogen to form helium, requires significant temperature and pressure, significance as a possible energy sauce.	

How to use Practice Papers

If you have done everything this book has described so far then you should be well equipped to meet the demands of the ENGAA, and therefore **the mock papers in the rest of this book should ONLY be completed under exam conditions**.

This means:

- Absolute silence – no TV or music.
- Absolute focus – no distractions such as eating your dinner.
- Strict time constraints – no pausing half-way through.
- No checking the answers as you go.
- Give yourself a maximum of three minutes between sections – keep the pressure up.
- Complete the entire paper before marking.
- Mark harshly.

In practice, this means setting aside two hours in an evening to find a quiet spot, without interruptions, to tackle the paper. Completing one mock paper every evening in the week running up to the exam would be an ideal target.

- Tackle the paper as you would in the exam.
- Return to mark your answers but mark harshly if there's any ambiguity.
- Highlight any areas of concern.
- If warranted, read up on the areas you felt you underperformed on to reinforce your knowledge.
- If you inadvertently learnt anything new by muddling through a question, go and tell somebody about it to reinforce what you've discovered.

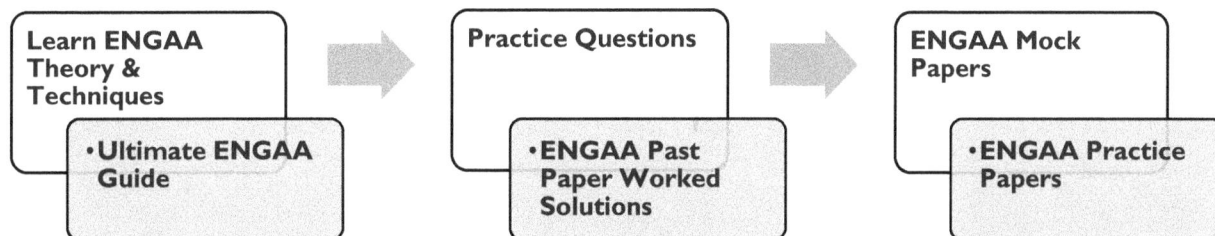

Learn ENGAA Theory & Techniques	Practice Questions	ENGAA Mock Papers
•Ultimate ENGAA Guide	•ENGAA Past Paper Worked Solutions	•ENGAA Practice Papers

Finally relax... the ENGAA is an exhausting exam, concentrating so hard continually for two hours will take its toll. So, being able to relax and switch off is essential to keep yourself sharp for exam day! Make sure you reward yourself after you finish marking your exam.

Top Tip! You can get free copies of *The Ultimate ENGAA Guide* and *ENGAA Past Paper Worked Solutions* books by flicking to the back of this book.

Mock Paper A

Section 1A

Q1. An astronaut on a rogue planet is doing an experiment to measure the acceleration due to gravity *g*. He drops a small ball a distance of 15 metres, vertically downwards. He is able to accurately measure the velocity of the ball when it hits the surface and records a value of 20 ms^{-1}. What is his first estimate of *g*?

A. 0.6 ms^{-2}
B. 10.0 ms^{-2}
C. 13.3 ms^{-2}
D. 26.6 ms^{-2}

Q2. A motor on a crane is used to lift a section of steel in a building project. The steel has a mass of 5000 kg and is lifted 3m in a time of 2s. The motor has a constant current of 50A running through it and the entire system has an efficiency of 75%. What voltage is applied to the motor?

A. 1.3kV
B. 1.5kV
C. 2kV
D. 8kV

Q3. A student owns a rare vintage comic worth £25. The comic is appreciating in value at a rate of 10% each year. How much will the comic be worth in 2 years?

A. £30
B. £30.25
C. £30.50
D. £30.75
E. £40

Q4. The acceleration g due to gravity on any planet in our universe is given by:

$$g = \frac{GM}{R^2}$$

where G is the universal gravitational constant, M is the mass of the planet and R is the radius of the planet.

In a different universe, the universal gravitational constant G' is twice that of our universe. Find the ratio $\frac{g_{planet}}{g_{earth}}$ for a planet in the other universe that has half the radius and twice the density of earth.

The volume of a sphere of radius R is given by: $V = \frac{4}{3}\pi R^3$.

A. $\frac{g_{planet}}{g_{earth}} = 2$

B. $\frac{g_{planet}}{g_{earth}} = 1$

C. $\frac{g_{planet}}{g_{earth}} = \frac{1}{2}$

D. $\frac{g_{planet}}{g_{earth}} = 4$

E. $\frac{g_{planet}}{g_{earth}} = \frac{1}{4}$

Q5. A lamp rated at 12V 60W is connected to the secondary coil of a step-down transformer and is at full brightness. The primary coil is connected to a supply of 200V. The transformer is 75% efficient.

What is the current in the primary coil?

A. 0.2A

B. 0.4A

C. 3.5A

D. 5A

Q6. Rearrange the equation $y = (5x - 1)^2 - 7$ to make x the subject.

A. $x = \frac{1}{5} \pm \sqrt{y + 7}$

B. $x = \frac{1}{5} \pm \sqrt{y - 7}$

C. $x = \frac{1}{5}\left(1 + \sqrt{y - 7}\right)$

D. $x = \frac{1}{5}\left(1 \pm \sqrt{y + 7}\right)$

Q7. The displacement-time graph shown represents a wave of wavelength 2cm.

What is the speed of the wave?

A. 0.33 cms^{-1}
B. 0.67 cms^{-1}
C. 0.75 cms^{-1}
D. 1.33 cms^{-1}
E. 1.50 cms^{-1}
F. 3.00 cms^{-1}

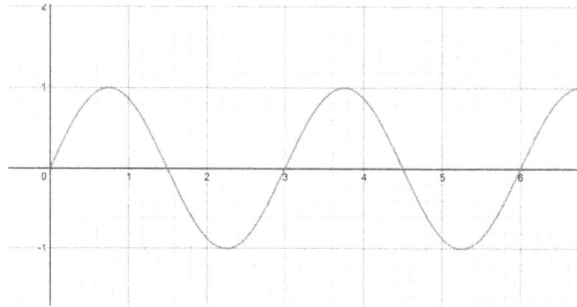

Q8. Which of the following quantities has the same unit as the rate of change of momentum?

A. Work
B. Energy
C. Acceleration
D. Weight

Q9. Solve the inequality $3x^2 \leq 45 - 6x$.

A. $x \geq -5$ and $x \geq 3$
B. $x \geq -5$ and $x \leq 3$
C. $x \leq -5$ and $x \geq 3$
D. $x \leq 5$ and $x \geq -3$

Q10. A car of mass m is travelling along a straight and narrow road at speed u. A cat walks into the road a distance d in front of the car and stops in the middle of the road. What constant force F must be applied to the car so that it does not hit the cat? The road is too narrow for the car to drive past the cat without hitting it.

A. $F = \frac{-mu^2}{2d}$
B. $F = \frac{-u^2}{2d}$
C. $F = \frac{-mu^2}{2}$
D. $F = \frac{-mu^2}{d}$
E. The collision is unavoidable.

Q11. A radioactive nucleus emits a beta particle, then an alpha particle and finally another beta particle. The final nuclide is:

A. An isotope of the original element.
B. The same element with a different proton number.
C. A new element of higher proton number.
D. A new element of lower nucleon number.

Q12. The ratio of A:B is $(x + 1) : 2$ and the ratio of B:C is $x : 4$. Given that the ratio of A:C = 3:4, what are the possible values of x?

A. $x = 1, 2$
B. $x = -1, -2$
C. $x = -2, 3$
D. $x = 2, -3$

Q13. The variables x and y represent positive quantities, where x is inversely proportional to y^2. When $x = 12, y = 2\sqrt{3}$.

What is the value of y when $x = 9$?

A. 2
B. 3.5
C. 4
D. $3\sqrt{2}$
E. $2\sqrt{6}$
F. $4\sqrt{3}$

Q14. A kettle is used to boil an unknown mass of water initially at 25°C. The circuit within the kettle contains a heater rated at 14 kW. Given that it takes two minutes for the water to boil and the specific heat capacity c of water is $4200 \, \text{Jkg}^{-1}\text{K}^{-1}$, calculate the mass of water. Assume there is no heat loss to the surroundings.

A. 0.5kg
B. 3kg
C. 5kg
D. 10kg

Q15. In a cathode ray tube, 7.5×10^{15} electrons strike the screen in 40 s. Given the charge of an electron is 1.6×10^{-19}C, what current does this represent?

A. 1.3×10^{-16}A
B. 5.3×10^{-15}A
C. 3.0×10^{-5}A
D. 1.2×10^{-3}A

Q16. Which of the following is a simplification of $(\sqrt{5} - \sqrt{3})^2$?

A. $2 - 2\sqrt{15}$
B. $8 + 2\sqrt{15}$
C. 8
D. $8 - 2\sqrt{15}$

Q17. A bag contains 2 red, 3 green and 2 blue balls. Two balls are drawn at random without replacement. What is the probability that none of the balls drawn are blue?

A. $\frac{10}{21}$

B. $\frac{11}{21}$

C. $\frac{2}{7}$

D. $\frac{5}{7}$

Q18. A pyramid is formed from four equilateral triangles of unit length, on a square base. Calculate its height h.

A. $\sqrt{2}$

B. $\sqrt{3}$

C. $\frac{1}{\sqrt{2}}$

D. $\frac{1}{\sqrt{3}}$

E. None of the above.

Q19. Two identical lamps are connected to a battery in different arrangements. In the first arrangement, C_1, they are connected in series. In the second arrangement, C_2, they are connected in parallel. How does the brightness of the lamps in each arrangement compare?

A. Brighter in C_1 (series)

B. Brighter in C_2 (parallel)

C. Same brightness in both arrangements.

D. There is not enough information.

Q20. A mass m hangs from a force meter attached to the ceiling of an elevator. When the elevator is stationary at the top of the building, the force meter reads mg. What is the reading on the force meter while the elevator descends towards the ground floor at constant speed?

A. Less than mg.

B. More than mg.

C. mg.

D. Zero.

E. This cannot be answered without knowing the speed of the elevator.

Q21. Which of the following options is equivalent to $\left(2\sqrt{3} - 3\sqrt{2}\right)^2$?

A. $30 - 6\sqrt{6}$

B. $4\sqrt{3} - 12\sqrt{6} + 9\sqrt{2}$

C. 30

D. $30 - 12\sqrt{6}$

E. $4\sqrt{3} - 24 + 9\sqrt{2}$

Q22. The air resistance, F, acting on a skydiver is given by the equation $F = kv^2$, where k is a constant and v is their speed in ms^{-1}. If a skydiver has a mass of 75 kg and reaches terminal velocity at a speed of 50 ms^{-1}, what is the value of the constant, k? The gravitational field strength, g, is 10 Nkg^{-1}.

A. 1.5
B. 3.0
C. 0.3
D. 0.03
E. 0.15

Q23. A shop sells three types of pens: blue, red and black. The ratio of blue to red pens is 4:9, whilst the ratio of red to black pens is 3:10. What is the ratio of blue to black pens?

A. 2:15
B. 2:5
C. 3:9
D. 5:2
E. 2:9

Q24. In the diagram below, what is the overall resistance between A and B?

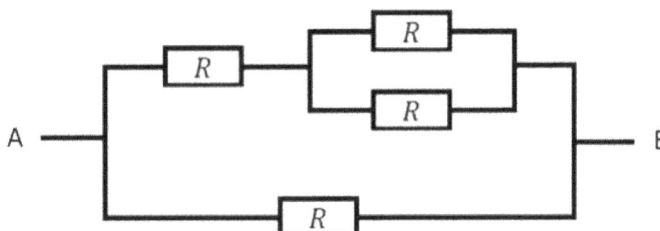

A. $\frac{3R}{5}$
B. 2R
C. $\frac{5R}{2}$
D. $\frac{3R}{4}$
E. $\frac{5R}{8}$

Q25. The mean of five numbers is 24. If one number is excluded, the mean is reduced by 4. What number is excluded?

A. 8
B. 16
C. 28
D. 32
E. 40

Q26. A car of mass m is travelling along a straight road at a constant speed, u, when it suddenly brakes, applying a constant braking force, F. What magnitude of braking force is required for the car to completely stop in a distance $2L$?

A. $\frac{2L}{mu^2}$

B. $\frac{mu^2}{4L}$

C. $\frac{mu}{4L}$

D. $\frac{mu^2}{L}$

E. $\frac{2L}{mu^2}$

Q27. A is directly proportional to the square root of B and B is inversely proportional to C. When A is 6, B is 4 and when B is 8, C is 2. What is A in terms of C?

A. $A = \frac{12}{\sqrt{C}}$

B. $A = 12\sqrt{C}$

C. $A = 6C$

D. $A = \frac{8C}{3}$

E. $A = \frac{6}{\sqrt{C}}$

Q28. There are some rocks that contain lead as a product of radioactive decay of polonium. In these rocks, a fixed quantity of polonium decays to a stable isotope of lead. Which of the graphs below best represents the number of lead atoms (N) as time progresses?

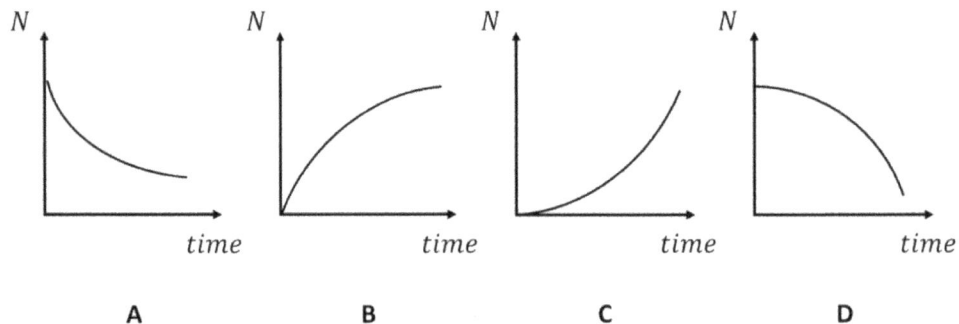

A. A

B. B

C. C

D. D

END OF SECTION

Section 1B

1. There is a large, stationary ship which is docked at a port and is partially submerged below the water level. The ship remains afloat. Which of the statements below must be true about the ship?

1. The upwards force on the ship is equal to the weight of the ship.
2. The volume of water displaced by the ship is the same as the volume of the ship.
3. The density of the ship is the same as the density of the water.
4. The weight of the water displaced by the ship is the upwards force on the ship.

A. 1 and 3
B. 2 and 3
C. 1 and 4
D. 1, 2 and 3
E. 2, 3 and 4

2. A collision occurs between a rubber ball and a wall. The ball is travelling at speed v before the collision. If half of its kinetic energy is converted into heat energy, what is the temperature increase of the ball? You may assume the rubber has a specific heat capacity of c and mass m.

A. $\dfrac{mv}{2c}$

B. $\dfrac{v^2}{2c}$

C. $\dfrac{v^2}{4c}$

D. $\dfrac{mv^2}{2c}$

E. $\dfrac{mv^2}{4c}$

3. In the diagram below, there are two masses hanging from a pulley. Their respective masses are M_A and M_B respectively. The mass of the pulley can be neglected. The wire may be treated as light and inextensible.

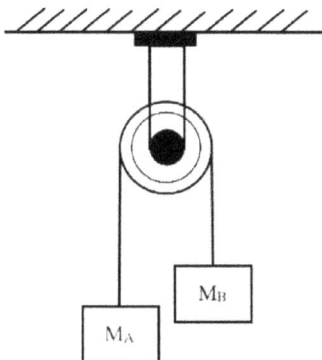

Determine the acceleration of the system of masses in terms of M_A and M_B, where you may assume that $M_A > M_B$.

A. $a = \frac{(2M_A - M_B)}{M_A M_B}$ ms^{-2}

B. $a = \frac{(M_A - M_B)}{(M_A + M_B)}$ ms^{-2}

C. $a = \frac{(M_A + M_B)^2}{(M_A - M_B)}$ ms^{-2}.

D. $a = \frac{(M_A M_B)}{(M_A + M_B)}$ ms^{-2}.

E. $a = \frac{M_A^2}{(M_A + M_B)}$ ms^{-2}.

F. $a = M_A^2 - M_B^2$ ms^{-2}.

4. As shown in the diagram below, there are two equal masses attached to two identical springs, with the same spring constant.

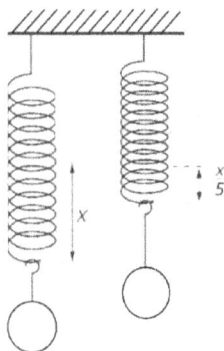

If the elastic potential energy in the first spring is given by E, what is the energy stored in the second spring, given that it is extended by a fifth of the original amount?

A. $\frac{E^2}{125}$

B. $\frac{E}{\sqrt{5}}$

C. $\frac{E}{25}$

D. $25E^2$

E. $5E^2$

5. The diagram below shows a circuit in which the primary coil of a transformer is connected to a 10V source. The secondary coil is connected to a 5Ω filament bulb and 15Ω resistor. The turns ratio is 3:1. What is the current through the filament lamp?

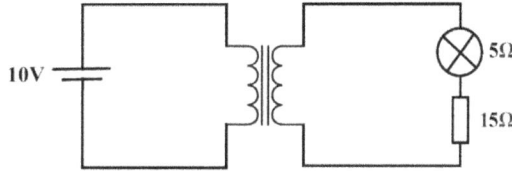

A. 5 A

B. 0.7 A

C. 0.2 A

D. 10 A

E. 6 A

F. 15 A

10V — 5Ω — 15Ω

6. A pendulum consists of a mass m attached to the end of a string of length l. The period of the pendulum is given by the following equation:

$$P = 2\pi\sqrt{\frac{L}{g}}$$

If P is the period of the pendulum of length L, what is the ratio of P and the period of a pendulum of twice the length L?

A. $\sqrt{2}$

B. $2\sqrt{2}$

C. $\frac{\sqrt{2}}{2}$

D. $\frac{\sqrt{2}}{4}$

E. 2

7. A charge, q_B, with mass m, travels at a velocity v. There is another charge, q_A, on its path of travel which is stationary. As they have similar charge, there is an electrostatic repulsion between them. Determine the minimum distance of approach between the two charges.

A. $\dfrac{q_A q_B}{2\pi\epsilon_0 m v^2}$

B. $\dfrac{q_A^2}{2\epsilon_0 m v^2}$

C. $\dfrac{2q_B}{\pi\epsilon_0 m v^2}$

D. $\dfrac{q_A m v^2}{2\pi}$

E. $\dfrac{q_A q_B^2}{2\pi m v^2}$

F. $\dfrac{\epsilon_0 q_A q_B}{2\pi\epsilon_0 m v^2}$

8. Which of the following statements is not an assumption made in the ideal gas model?

A. The size of the molecules is negligible compared to intermolecular distances.
B. The collisions between molecules are perfectly elastic.
C. The volume occupied by the gas is a constant quantity.
D. The intermolecular forces are negligible in an in ideal gas.
E. The collisions between the molecules and the walls of its container are perfectly elastic.

9. The displacement of a vibrating string is given by the following function.

$$s(x,t) = \frac{1}{5} \cos\frac{2\pi}{\lambda}(x - vt)$$

Determine the maximum acceleration of the particles.

A. $\frac{4\pi^2 v^2}{25\lambda^2}$

B. $\frac{4v^2}{5\lambda^2}$

C. $\frac{4v^2}{\lambda^2}$

D. $\frac{4\pi^2 v^2}{5\lambda^2}$

E. $\frac{\pi^2}{25\lambda^2}$

10. There is a sample of a radioactive substance which has a half-life of 45 seconds. Determine the percentage of the original sample which remains after 110 seconds.

A. 9%
B. 11%
C. 19%
D. 26%
E. 35%
F. 45%

11. A ray of light is incident upon a surface, as shown in the diagram below. The angle of incidence has been labelled, and the normal is indicated by the dotted line. The refractive index of each material, n_1 and n_2, is shown in the diagram.

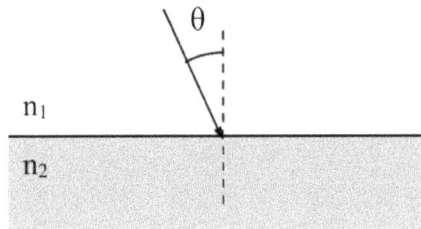

Which of the following statements is true?

A. The light ray will only reflect until the critical angle is reached.
B. If the incident angle is less than the critical value, the ray will undergo total internal reflection.
C. The light ray will only refract until the critical angle is reached.
D. The light ray will reflect and refract until the critical angle is reached.
E. If the incident angle is equal to the critical value, the ray will undergo total internal reflection.

12. A gardener decides to install solar panels onto the roof of his greenhouse to power the lights. Each solar panel has an area of 0.25 m^2 each. If the solar panels have an efficiency of 20%, how many solar panels are required to light five bulbs, each with a power output of 60 W?

A. The intensity of sunlight at the surface of the Earth is approximately 1360 Wm^{-2}, where intensity is defined as the power pet unit area.3
B. 4
C. 5
D. 6
E. 7
F. 8

13. The time of decay for different radioactive substances varies greatly. A property of a radioactive substance which allows scientists to quantify this rate of decay is known as the 'half-life' of that substance.

Which statement below correctly defines half-life?

A. It is the time taken for half of the radioactive nuclei in a sample to emit alpha radiation.
B. It is the time taken for half of the radioactive nuclei in a sample to emit gamma radiation.
C. It is the time taken for half of the radioactive nuclei in a sample to undergo radioactive decay.
D. It is the time taken for the mass of the radioactive sample to halve.
E. It is the time taken for a radioactive sample to completely undergo radioactive decay.

14. The temperature against time graph is shown below for a gas which is sealed within a container. On the graph, the region where the temperature is constant for a long period of time represents a physical process. Which process does it represent?

A. Evaporation.
B. Vaporisation.
C. Condensation.
D. Boiling.
E. Solidification.
F. Melting.

15. There is an object which is accelerating, due to a constant force, in the positive x-direction and its motion is purely one-dimensional.

Which of the following statements are true regarding its motion?

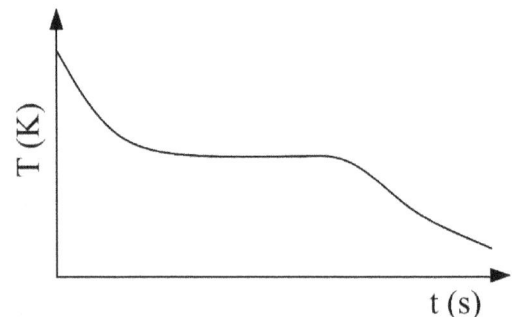

1. The velocity-time graph forms a curved graph.
2. The acceleration-time graph forms a linear graph.
3. It may be travelling in the negative x-direction.

A. Only 1
B. Only 2
C. Only 3
D. Both 1 and 2
E. Both 1 and 3
F. Both 2 and 3

16. The displacement, from a fixed origin O, of an object is given below:

$$s(t) = \frac{2}{3}t^3 - \frac{7}{2}t^2 + 3t + 17$$

Determine the earliest time at which the object comes to rest.

A. 0.10 seconds
B. 0.25 seconds
C. 0.50 seconds
D. 0.75 seconds
E. 1.25 seconds
F. 2.50 seconds

17. A ball is thrown straight up into the air. By setting upwards to be the positive direction, which of the following statements is true regarding its motion? You may assume air resistance is negligible.

A. The ball's velocity is negative and the acceleration is negative.
B. The ball's velocity is positive and the acceleration is positive.
C. The ball's velocity is positive and the acceleration is zero.
D. The ball's velocity is positive and the acceleration is negative.
E. The ball's velocity is negative and the acceleration is zero.

18. For a light wave, the intensity is proportional to the square of its amplitude. In the vicinity of a star, an intensity, for the light emitted by the star, of I is measured. This corresponds to an amplitude of A of the light wave. What is the corresponding amplitude of the light wave at the position where double this intensity is measured?

A. $\frac{\sqrt{2}}{2}A$
B. $\frac{1}{2}A$
C. $2A$
D. $\frac{3\sqrt{2}}{2}A$
E. $\frac{\sqrt{2}}{4}A$
F. $\sqrt{2}A$

19. Which statement below best explains why the sun can be seen from the Earth but cannot be heard from the Earth?

A. Sound is a longitudinal wave, whereas light is a transverse wave.
B. Light is a progressive wave.
C. The wavelength of light is smaller than the wavelength of sound.
D. The frequency of light is greater than the frequency of sound.
E. Diffraction only occurs in light and not sound.
F. Light travels faster than sound.

20. A boy is kicking a football against a wall. He kicks the ball at an angle $\theta = 30°$ such that the ball is perched exactly on top of the wall, as shown in the diagram below.

If the height of the wall is 2 metres, determine the initial velocity, v_0, the boy kicked the ball with for this to occur. You may assume that air resistance is negligible.

A. $4\sqrt{2}$

B. $\sqrt{13}$

C. $2\sqrt{10}$

D. $2\sqrt{5}$

E. $4\sqrt{10}$

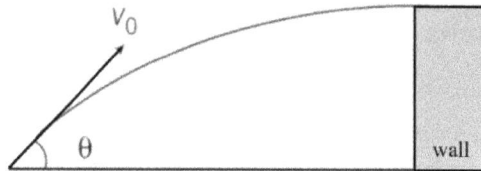

21. A potential difference V is applied across two resistors in parallel, R_1 and R_2. What is the total power dissipated in the circuit?

A. $V^2 \times \left(\frac{1}{R_1} + \frac{1}{R_2}\right)$

B. $V^2(R_1 + R_2)$

C. $\frac{V^2}{R_1 + R_2}$

D. $V^2 \div \left(\frac{1}{R_1} + \frac{1}{R_2}\right)$

E. $\frac{V^2}{(R_1 + R_2)^2}$

22. A playground slide has a steep initial slope which gradually becomes a gentler slope. If a child goes down the slide, what happens to their speed v and the magnitude of their acceleration a?

	v	a
A	Increases	Increases
B	Decreases	Decreases
C	Increases	Decreases
D	Decreases	Increases
E	Increases	Constant

23. A 10 kg block sits 1 m from the end of a 4 m-long plank of wood. How much force needs to be applied at the other end of the plank to lift the block off the ground? The gravitational field strength is 10 Nkg⁻¹.

A. 1 N
B. 2 N
C. 2.5 N
D. 10N
E. 20N
F. 25 N
G. 50 N
H. 100 N

24. Which of the following is a correct unit for potential difference?

A. Cs⁻¹
B. JC⁻¹
C. VA⁻¹
D. JA⁻¹
E. Vs⁻¹

25. Immediately after opening their parachute, a skydiver has a constant deceleration of 5 ms⁻². The mass of just the skydiver is 70 kg. Assume that the strings connecting the parachute to the skydiver are massless and inextensible. What is the total tension in the parachute strings? The gravitational field strength is 10 Nkg⁻¹.

A. 350 N
B. 400 N
C. 700 N
D. 1050 N
E. 1200 N
F. 1550 N

26. A battery is connected in series across a single bulb, before being disconnected and connected across two bulbs in series. All bulbs are identical. Which of the following statements is correct?

A. The current is lower in the circuit with two bulbs.
B. The potential difference across the battery is lower in the circuit with two bulbs.
C. The current is higher in the circuit with two bulbs.
D. The potential difference across the battery is higher in the circuit with two bulbs.
E. The current is the same in both circuits.

END OF SECTION

Section 2

Question 1

An electric field, E, is set up between two parallel plates with separation d. A proton of charge e and mass m is fired from the centre of these two plates, at speed 5 ms^{-1}. A gravitational field with g = 10 Nkg^{-1} acts.

Hint: The electric field provides a force qE on a particle of charge q in the direction of the field.

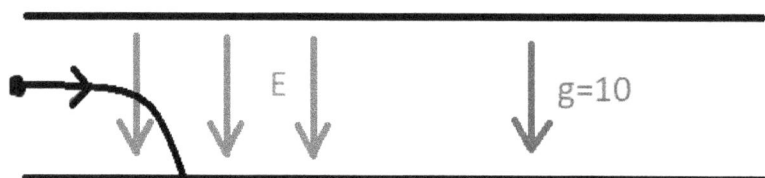

a. Which of the following graphs describes the time at which the electron collides with the plates as a function of the electric field strength E?

A

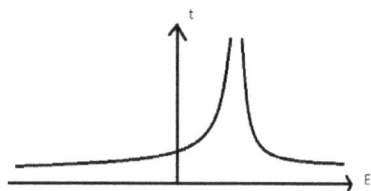

C

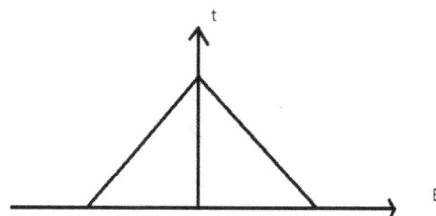

B

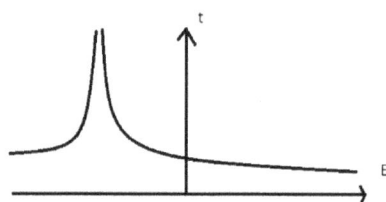

D

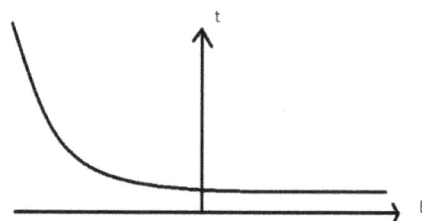

b. Given that $E = -\frac{gm}{2e}$ and d = 5m, find the horizontal distance travelled before a plate is hit.

A. 3m

B. 4m

C. 5m

D. 7m

E. 6m

c. What is an equivalent SI unit for the electric field strength E?

A. $kgms^{-3}A^{-1}$

B. NmC^{-1}

C. $kgms^{-3}C$

D. JsAC

E. J

d. Two radioactive elements need to be identified. One emits electrons, and the other positrons (positively charged electrons) particles. Can these two elements be distinguished using the set-up provided?

A. Yes.

B. No.

e. Many protons are fired through every second with $E = -\frac{4gm}{e}$ for a time T. Consider the $T \to \infty$ limit. A negative particle is placed in the centre of the plates. What will happen to it?

A. It will remain in the centre.

B. It will move to the left.

C. It will move up.

D. It will move down.

Question 2

A truck of mass 5 tonnes rolls down a slope of angle $sin^{-1}(\frac{1}{40})$ to the horizontal at a constant speed. Assume that the resistive force is constant and that $g = 10 \, Nkg^{-1}$.

a. Calculate the magnitude of the resistive forces.

A. 1250N

B. 2000N

C. 800N

D. 1000N

E. 1500N

b. The truck is equipped with new tyres, and now the resistive forces are 500 N. Find the power the engine must produce in order for the truck to ascend the slope at a speed of 12 kmh⁻¹.

A. 5.8 kW

B. 21 kW

C. 18 kW

D. 4.5 kW

E. 3.2 kW

c. The truck runs on diesel. At a certain speed, it requires 10 kW of power to climb the slope. The burning of each tonne of diesel produces x joules of energy. Given that half of the 1 tonne tank is used to climb the slope, in a time of 1000 seconds, calculate x.

A. 3×10^8
B. 2×10^8
C. 2×10^7
D. 30×10^7
E. 3×10^4

d. The truck now accelerates up the slope, with a driving force of 1000N. Supposing it starts at the bottom at rest, and the slope is 50 m in length, how long will it take to reach the top?

A. 15s
B. 32s
C. 9s
D. 27s
E. 10s

Question 3

a. Which of the following rows is correct?

	Highest Frequency	Lowest Energy
A	Radio	Gamma
B	Visible	Radio
C	Gamma	Radio
D	Visible	Gamma
E	Radio	Ultraviolet

b. Which of the following statements regarding EM waves are true?

1. EM waves propagate at the same speed in all materials.
2. EM waves are longitudinal.
3. EM waves cannot diffract.
4. EM waves can neither be created nor destroyed.

A. Only 1
B. All of the above.
C. Only 4
D. 3 and 4
E. None of the above.

c. Which of the following statements regarding EM waves are true?

1. EM waves propagate at the same speed in all materials.
2. EM waves are longitudinal.
3. EM waves cannot diffract.
4. EM waves can neither be created nor destroyed.

A. Only 1
B. All of the above.
C. Only 4
D. 3 and 4
E. None of the above.

d. A distant star is moving towards us. What observations would we draw, compared to the same stationary star?

1. The star appears larger.
2. The star appears redder.
3. The star appears bluer.
4. The star appears brighter.
5. The star appears dimmer.

A. Only 1
B. 1, 3 and 5
C. 3 and 5
D. 2 and 4
E. Only 3

Question 4

A circuit is constructed as below. Each resistor has resistance R. Some points of interest are labelled on the circuit. The resistance of the bulb is R/2.

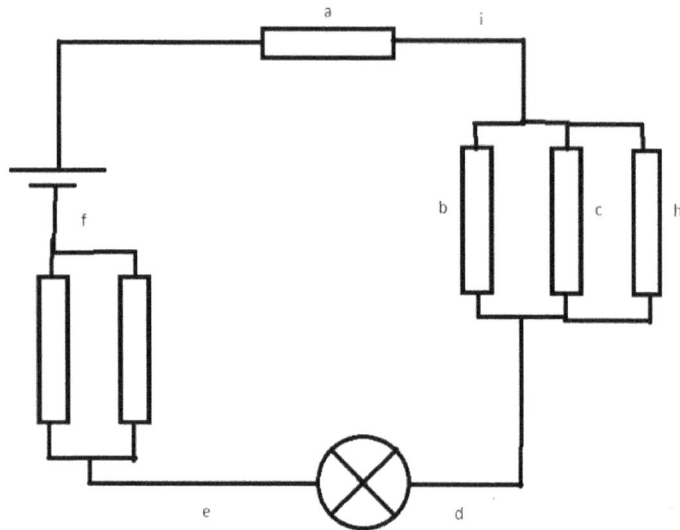

a. Which of the following statements are true?

1. The current through *a* and *b* is the same.
2. The current through *a* and *f* is the same.
3. The voltage across (*d*, *e*) is the same as across (*e*, *f*).
4. The voltage across a is the same as across (*e*, *f*).

A. Only 1
B. 2 and 3
C. 1, 2, 3 and 4
D. Only 2
E. Only 3

b. Given that the current through the point labelled *b* in the circuit is I, what is the voltage across the cell?

A. IR
B. $\frac{3}{IR}$
C. 5IR
D. 7IR
E. 2R

c. The voltage across the cell is V. Suppose the resistor labelled *h* malfunctions and doubles in resistance. Calculate the original and new power outputted by the bulb.

	Old	New
A	$\dfrac{9V^2}{98R}$	$\dfrac{25V^2}{288R}$
B	$\dfrac{25V^2}{288R}$	$\dfrac{9V^2}{98R}$
C	$\dfrac{9V^2}{95R}$	$\dfrac{29V^2}{95R}$
D	$\dfrac{11V^2}{98R}$	$\dfrac{25V^2}{288R}$

d. To fix the faulty resistor, the circuit from i to d is removed and connected to a cell directly, and the current through *h* measured. The faulty resistor is then replaced by an identical resistor, of half the original diameter, and the current remeasured. Calculate the percentage change in current through the resistor *h*.

A. +300%

B. +200%

C. +50%

D. −50%

E. −75%

END OF PAPER

Mock Paper B

Section 1A

Q1. Solve the inequality: $x^2 + 5x \geq -4$.

A. $x \leq -4$ or $x \geq -1$
B. $x \leq 4$ and $x \geq -1$
C. $x \geq -4$ and $x \leq -1$
D. $x \geq -4$ and $x \leq 1$

Q2. A transformer has 1200 turns on the primary coil and 500 turns on the secondary coil. The primary coil draws a current of 0.25A from a 240V AC supply. If the efficiency of the transformer is approximately 83%, what is the current in the secondary coil?

A. 0.1A
B. 0.21A
C. 0.5A
D. 0.6A

Q3. A cone has a height h equal to the diameter of a sphere. If the volume of the two objects are equal, and the radius of the sphere is r, calculate the radius R of the base of the cone. You are given that the volume of a cone is $\frac{\pi R^2 h}{3}$, the volume of a sphere is $\frac{4\pi r^3}{3}$.

A. $R = r\sqrt{2}$
B. $R = 2r$
C. $R = 4r$
D. $R = 8r$

Q4. What is the equation, $f(x)$, of the line that intersects $y = 2x - 2$ at right angles at position $x = 1$?

A. $f(x) = -\frac{1}{2}x$
B. $f(x) = -\frac{1}{2}x + \frac{1}{2}$
C. $f(x) = \frac{1}{2}x - \frac{1}{2}$
D. $f(x) = x$

Q5. Rearrange the equation $y = (7x - \frac{1}{2})^{\frac{1}{2}} - 2$ to make y the subject of x.

A. $x = (y+2)^2 + \frac{1}{14}$

B. $x = \frac{(y+2)^2}{7} + \frac{1}{2}$

C. $x = \frac{(y+2)^2}{7} - \frac{1}{14}$

D. $x = \frac{(y+2)^2}{7} + \frac{1}{14}$

E. $x = \frac{(y-2)^2}{7} + \frac{1}{14}$

Q6. Which of the following is **not** a unit of power?

A. N m s^{-1}

B. kg m^2 s^{-3}

C. J s^{-1}

D. kg m^{-1} s^{-1}

Q7. A circuit consists of a 10V battery connected in series with a 500Ω resistor. Calculate the number of electrons that flow through the resistor in five seconds given that the charge of an electron is (1.6×10^{-19})C.

A. 6.25×10^{17}

B. 6.25×10^{18}

C. 62.5×10^{17}

D. 3.50×10^{17}

Q8. A circuit consists of a battery, fixed resistor and NTC thermistor all connected in series. If the temperature of the NTC thermistor is increased, what happens to the power dissipated by the fixed resistor?

A. It increases.

B. It decreases.

C. It stays the same.

Q9. A ball is released from rest upon a cliff at a height of 35 m above the ground. What speed is it travelling when it hits the ground?

A. 700 ms^{-1}

B. $10\sqrt{7} \text{ ms}^{-1}$

C. $5\sqrt{14} \text{ ms}^{-1}$

D. $100\sqrt{7} \text{ ms}^{-1}$

E. $5\sqrt{7} \text{ ms}^{-1}$

Q10. A stone is projected horizontally by a catapult consisting of two rubber cords. The chords, which obey Hooke's Law, are stretched and released. When each cord is extended by x, the stone is projected with speed v. What is the speed of the stone when each cord is stretched instead by 2x?

A. v

B. $\sqrt{2v}$

C. 2v

D. 4v

Q11. If you look at the hands of a clock at 6.15, what is the angle between the minute and hour hand?

A. 80°

B. 97.5°

C. 95°

D. 22.5°

E. 30°

Q12. Two points on a progressive wave are one-eighth of a wavelength apart. The distance between them is 0.5m and the frequency of oscillation is 10Hz. What is the minimum speed of the wave?

A. $0.2 \, \text{ms}^{-1}$

B. $10 \, \text{ms}^{-1}$

C. $20 \, \text{ms}^{-1}$

D. $40 \, \text{ms}^{-1}$

Q13. An electric motor of input power $P_{in} = 100W$ raises a 10kg mass at a constant speed of 0.5ms^{-1}. What is the efficiency of the motor?

A. 5%

B. 12%

C. 50%

D. 100%

Q14. A recent graduate buys a new car worth £25,000. The value of the car depreciates by 10% each year, by how much has the car reduced in value after 3 years?

A. £6000

B. £6775

C. £7500

D. £8325

E. £9875

F. £18,225

Q15. A small steel ball of mass 100g is falling downwards at terminal velocity through a jar of honey. Which of the following statements are true?

1. The kinetic energy of the ball is increasing.
2. The drag force on the ball is equal to 1 N.
3. There is no net force on the ball.

A. None of the above
B. 1 only
C. 2 only
D. 3 only
E. 1 and 2 only
F. 1 and 3 only
G. 2 and 3 only
H. All of the above

Q16. The ratio of A:B is 4:3 and the ratio of B:C is 5:6, which of the following gives the ratio of A:C in it's simplest form?

A. 10:9
B. 9:10
C. 8:5
D. 5:8

Q17. Calculate the area of the triangle shown below, all measurements are in cm

A. $\left(6 + \sqrt{3}\right) \text{cm}^2$
B. $\left(6 - \sqrt{3}\right) \text{cm}^2$
C. $\left(9 + \sqrt{3}\right) \text{cm}^2$
D. $\left(12 + \sqrt{3}\right) \text{cm}^2$
E. $\left(12 - \sqrt{3}\right) \text{cm}^2$

$3 - \sqrt{3}$

$5 + \sqrt{3}$

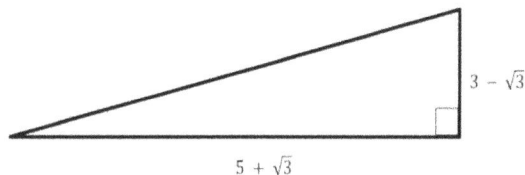

Q18. When a uranium-235 nucleus absorbs a neutron, it undergoes nuclear fission. One particular reaction produces technetium and indium as daughter nuclei as shown below. Which of the following equations is true?

A. $^1_0 n + {}^{235}_{92} U \rightarrow {}^x_{43} Tc + {}^y_z In + w {}^1_0 n$
B. $x + y = 236$
C. $z + w = 92$
D. $x = 236 - (y + w)$
E. $y = 235 - x$

Q19. A car is travelling along a horizontal straight road. A velocity-time graph for part of the car's journey is given below:

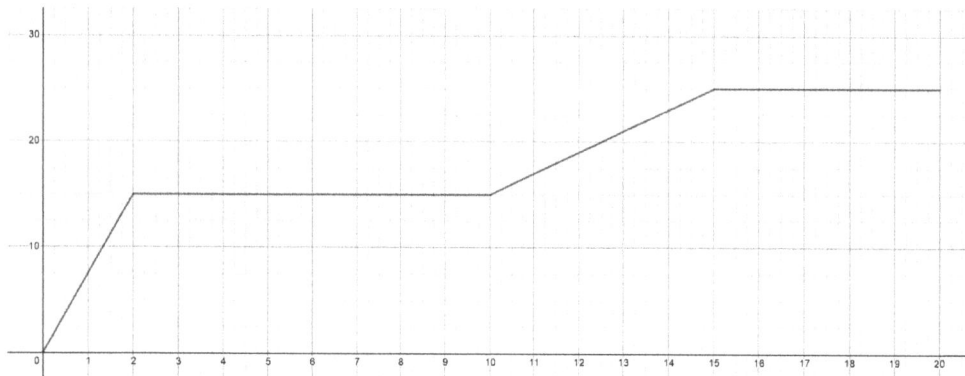

During this part of the journey, what is the distance travelled by the car while it is accelerating?

A. 15 m
B. 100 m
C. 115 m
D. 120 m
E. 125 m
F. 225 m

Q20. A stationary unstable nucleus of mass M emits an alpha particle of mass m with kinetic energy E, what is the speed of recoil of the daughter nucleus?

Before | After

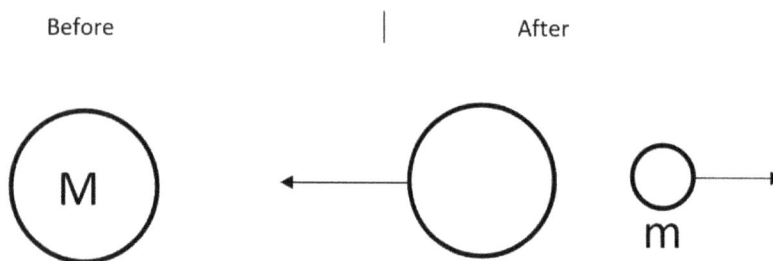

A. $\dfrac{\sqrt{2mE}}{(M-m)}$

B. $\dfrac{\sqrt{2mE}}{M}$

C. $\dfrac{(M-m)}{\sqrt{2mE}}$

D. $\dfrac{2mE}{(M-m)^2}$

Q21. Simplify $\left(\dfrac{8x^{\frac{5}{2}}y^{-\frac{1}{2}}}{2\sqrt{x}y^{\frac{3}{2}}}\right)^{\frac{1}{2}}$.

A. $\dfrac{\sqrt{2}x}{y}$

B. $\dfrac{2x}{y}$

C. $\dfrac{2y}{x}$

D. $\dfrac{2x^2}{y}$

E. $\dfrac{\sqrt{2}x}{y^2}$

Q22. An empty lift, with a mass of 400 kg, accelerates upwards at 2 ms^{-2} as shown in the diagram. What is the tension, T, in the cable? The gravitational field strength, g, is 10 Nkg^{-1}.

A. 800 N

B. 2000 N

C. 2400 N

D. 4000 N

E. 4800 N

F. 6000 N

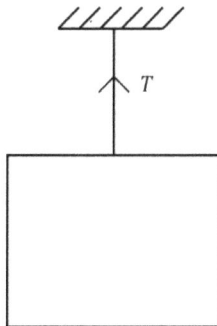

Q23. A school year has 200 pupils, 112 of which are girls. During an afternoon of sports:

1. 20 boys play tennis,

2. 60 pupils go swimming, $\frac{3}{4}$ of which are boys,

3. 34 more girls do athletics than boys.

Assuming all pupils take part in one sport, how many girls play tennis?

A. 10

B. 25

C. 40

D. 63

E. 36

Q24. After an accident, the level of radiation in a nuclear laboratory was tested to be 64 times the maximum permissible level. The radiation was due to a radioactive material with a half-life life of 15 days. After how many days will the laboratory be safe to use again?

A. 6

B. 8

C. 64

D. 75

E. 90

F. 105

Q25. Solve the inequality $3x^2 \geq 2 - 5x$.

A. $x \geq 3, x \leq -2$

B. $-2 \leq x \leq \frac{1}{3}$

C. $x \geq \frac{1}{3}, x \leq -2$

D. $-\frac{1}{2} \leq x \leq 3$

E. $x \geq 2, x \leq -\frac{1}{3}$

F. $x \geq 3, x \leq -\frac{1}{2}$

Q26. A motor with input power 200 W raises a mass of 15 kg at a constant speed of 0.5 ms⁻¹. What is the efficiency of the system? The gravitational field strength, g, is 10 Nkg⁻¹.

A. 5%

B. 10%

C. 15%

D. 25%

E. 37.5%

F. 75%

G. 100%

Q27. A cylindrical pipe has outer radius 1.5 m, inner radius 1 m and length 2 m. If the mass of the pipe is 5000 kg, which of the following expressions is correct for its density, ρ?

A. 2000π

B. 5000π

C. 10000π

D. $\frac{2000}{\pi}$

E. $\frac{5000}{\pi}$

F. $\frac{10000}{\pi}$

Q28. The nucleus of a radioactive isotope **X** is at rest. **X** decays by emitting an alpha particle so that it becomes a new, stable nuclide **Y**. Which of the following statements about the decay is correct?

A. The momentum of **Y** is equal and opposite to the momentum of the α-particle.

B. The momentum of **Y** is equal to the momentum of **X**.

C. The kinetic energy of **Y** is equal to the kinetic energy of the α particle.

D. The total kinetic energy is the same before and after the decay.

E. **Y** is at rest and the energy released by the decay causes the α-particle to move away

END OF SECTION

Section 1B

1. A ball of mass 500 g is thrown horizontally at a wall with a speed of 12 ms⁻¹ and rebounds with a speed of 6 ms⁻¹. If the ball is in contact with the wall for 0.3 s, what is the average force exerted on the ball?

A. 0.1N
B. 5N
C. 8N
D. 10 N
E. 15 N
F. 30 N
G. 50 N
H. 100 N

2. A 12 V solar-powered battery can produce a current of 0.5 A for 30 minutes. The battery is charged using a square solar panel of side length 20 cm. Assuming a constant incident solar power of 1 kWm⁻² and a charging efficiency of 10%, how long does the battery need to be charged for?

A. 5 minutes   B. 15 minutes   C. 30 minutes
D. 45 minutes   E. 60 minutes   F. 90 minutes

3. A ray of light passes through a glass block surrounded by air, taking the path as shown in the diagram below:

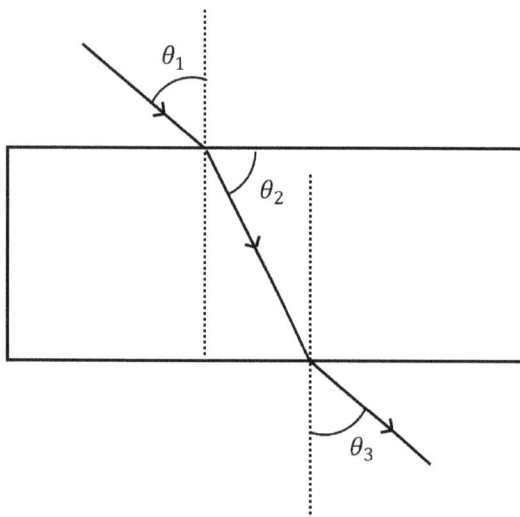

Which of the following is **false**?

A. $\theta_1 = \theta_3$.
B. The glass block is more optically dense than the air.
C. $\theta_2 > \theta_1$.
D. The light has the same frequency in both the air and glass.
E. The light travels more quickly in the glass.

4. A mass of 60 kg is placed on top of a vertical spring, with spring constant $k = 6000$ Nm^{-1}. After it has fully compressed, how much energy is stored in the spring? The gravitational field strength is 10 Nkg^{-1}.

A. 5 J
B. 8 J
C. 10 J
D. 12 J
E. 15 J
F. 20 J
G. 30 J

5. A rock is thrown directly upwards with a speed of 4 ms^{-1} from a height of 1 m. What will the rock's speed be just before it hits the ground on the way back down? The acceleration due to gravity is 10 ms^{-2}.

A. 2 ms^{-1}
B. 3 ms^{-1}
C. 4 ms^{-1}
D. 6 ms^{-1}
E. 8 ms^{-1}
F. 10 ms^{-1}
G. 14 ms^{-1}

6. A wire with resistance R has constant cross-sectional area A and length L. What would be the resistance of a wire made of the same material but with twice the cross-sectional area and half the length?

A. $\frac{R}{4}$
B. $\frac{R}{2}$
C. R
D. 2R
E. 4R

7. A battery is connected across a fixed resistor and a light dependent resistor in series. A voltmeter is connected across the fixed resistor and an ammeter in series with both resistors. When the incident light intensity to the light dependent resistor falls, what happens to the voltage and current measured on the two instruments?

	Voltage	Current
A	Constant	Increases
B	Constant	Decreases
C	Increases	Constant
D	Increases	Increases
E	Increases	Decreases
F	Decreases	Constant
G	Decreases	Increases
H	Decreases	Decreases

8. In which of the following situations will total internal reflection occur? Assume that $n_{air} = 1$, $n_{glass} = 1.5$ and that, under the correct conditions, the critical angle for total internal reflection is 42°.

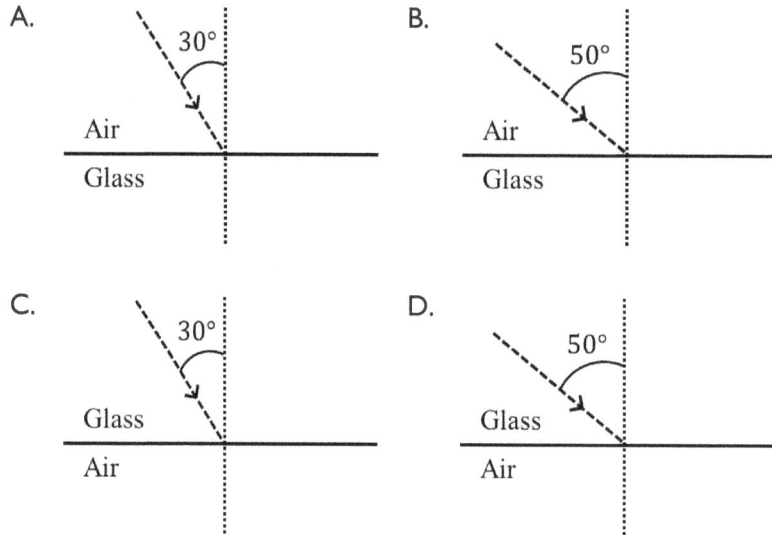

A.

30°

Air

Glass

B.

50°

Air

Glass

C.

30°

Glass

Air

D.

50°

Glass

Air

9. A pulley is connected to the edge of a table, as shown in the diagram below. All strings are massless and inextensible, and the pulley is perfectly smooth.

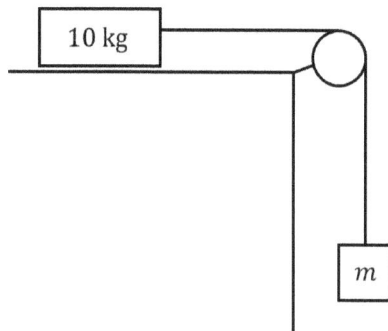

10 kg

m

The coefficient of friction between the table and the block resting on the table is $\mu = 0.8$. What mass, m, of the second block is required to accelerate the 10 kg block at 2 ms^{-2}? The gravitational field strength is 10 Nkg^{-1}.

A. 2 kg
B. 8 kg
C. 10 kg
D. 18 kg
E. 20 kg

10. A ball is released from rest on a slide at point X in the diagram below.

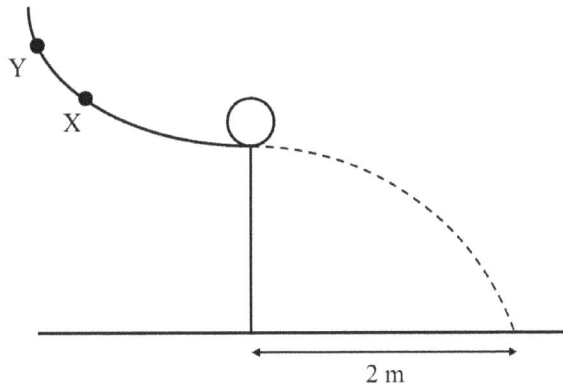

The ball leaves the slide horizontally at the position shown and lands 3 seconds later. The process is then repeated with the ball released from point Y, again leaving the slide horizontally. Which of the following options correctly describes the ball's motion after leaving the slope in the second case?

	Time to land after leaving slope	Horizontal distance to landing point
A	Less than 3 s	Less than 2 m
B	Less than 3 s	2 m
C	Less than 3 s	More than 2 m
D	3 s	Less than 2 m
E	3 s	More than 2 m
F	More than 3 s	Less than 2 m
G	More than 3 s	2 m
H	More than 3 s	More than 2 m

11. A rock is launched vertically upwards by a catapult and reaches a maximum height of 85 m before starting to fall. Neglecting air resistance and assuming that the rock leaves the catapult when it is 5 m above the ground, what speed is the rock launched at? The acceleration due to gravity is 10 ms^{-2}.

A. $4\sqrt{10}$ ms^{-1}

B. $30\sqrt{2}$ ms^{-1}

C. 40 ms^{-1}

D. $15\sqrt{10}$ ms^{-1}

E. 50 ms^{-1}

12. An electric motor pulls a block of mass 8 kg horizontally along a rough surface at a constant speed of 0.5 ms⁻¹. Assuming that the motor has an efficiency of 75% and the coefficient of friction between the block and the surface is 0.6, what input power to the motor is required? The gravitational field strength is 10 Nkg⁻¹.

A. 1.8 W

B. 2.4 W

C. 18 W

D. 24 W

E. 32 W

F. 48 W

13. A person standing on a pier by the sea observes that the period of the waves on the surface of the sea is 1.5 seconds, and that it takes 18 seconds for an individual wave to travel the length of the 60 m pier. What is the wavelength of the waves?

A. 0.2 m

B. 0.45 m

C. 2.22 m

D. 5 m

E. 7.25m

14. All resistors in the circuit have the same resistance of R. What is the total resistance of the following network of resistors?

A. $\frac{2R}{7}$

B. $\frac{6R}{13}$

C. R

D. $\frac{3R}{2}$

E. 2R

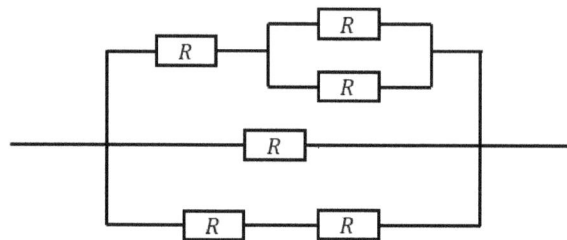

15. A small ball is projected from a point O, which is height h above horizontal ground. The ball has an initial speed of 20 ms^{-1} and is projected at an angle of 30° above the horizontal. If the ball strikes the ground 3 seconds after reaching its maximum height, what was its starting height, h? Assume $g = 10 \text{ ms}^{-2}$ and that there are no resistive forces acting on the ball.

A. 15m
B. 20m
C. 25m
D. 35m
E. 10m
F. 40m

16. Given that $\sec x - \tan x = -5$, find the value of cos x.

A. -0.2
B. 0.2
C. $-\frac{13}{5}$
D. $\frac{-5}{13}$
E. 0.5
F. -0.5

17. Consider the line with equation $y = 2x + k$ where k is a constant, and the curve $y = x^2 + (3k - 4)x + 13$. Given that the line and the curve do not intersect, what are the possible values of k?

A. $-\frac{1}{3} < k < 3$ C. $\frac{1}{2} < k < \frac{5}{3}$ E. $\frac{1}{3} < k < 3$

               D. $\frac{3}{2} < k \le \frac{8}{3}$ F. $-3 < k < \frac{1}{3}$

B. $-\frac{4}{9} < k < 4$

18. A circle with centre C(5,-3) passes through A(-2,1), and the point T lies on the tangent to the circle such that AT = 4. What is the length of the line CT?

A. 9   C. $\sqrt{95}$ E. $\sqrt{69}$

B. 18   D. $8\sqrt{2}$ F. 8

19. Evaluate $(6 \sin x)(3 \sin x) - (9 \cos x)(-2 \cos x)$.

A. 0

B. 0.5

C. 1

D. -1

E. 18

F. -18

20. In the figure to the right, all triangles are equilateral. What is the shaded area of the figure in terms of r?

A. $5r^2(2\sqrt{6} - 3\pi)$

B. $5r^2(5\sqrt{2} - 6\pi)$

C. $5r^2(3\sqrt{3} - \pi)$

D. $5r^2(4\sqrt{3} - 2\pi)$

E. $5r(2\sqrt{6 - 3\pi})$

F. $5r^2(5\sqrt{2} + 6\pi)$

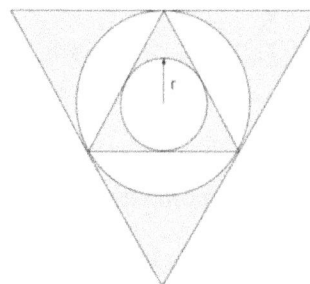

21. The binomial expansion is used to determine the value of $(3.12)^5$. What is the minimum number of terms that must be obtained in the expansion of $(3.12)^5$ in order to receive a result accurate to 1 decimal place?

A. 4   B. 5   C. 6   D. 7   E. 9   F. 8

22. For all θ, $(\sin(\theta) + \sin(-\theta))(\cos(\theta) + \cos(-\theta))$ is equal to which of the following expressions?

A. $2\sin\theta$

B. 0

C. 1

D. $4\sin\theta\cos\theta$

E. -1

F. $2\sin\theta\cos\theta$

23. What is the range of x values for which the inequality $|2x - 5| > 3|2x + 1|$ is valid?

A. $-2 < x < 4$

B. $x > 2, x > \dfrac{1}{4}$

C. $-2 < x < \dfrac{1}{4}$

D. $x > 2, x > 4$

E. $-2 < x < 4$

24. Which of the following is the equation of the circle whose diameter is the line segment connecting points (1,-4) and (3,6) and is reflected about the line $y = x$?

A. $(x + 4)^2 + (y - 1)^2 = 104$

B. $(x - 1)^2 + (y + 4)^2 = 104$

C. $(x - 1)^2 + (y - 2)^2 = 26$

D. $(x - 2)^2 + (y - 1)^2 = 2$

E. $(x+1)^2 + (y + 2)^2 = 26$

F. $(x + 2)^2 + (y + 1)^2 = 2$

25. A new computer does a calculations in b hours, and an old computer does c calculations in d minutes.

If the two computers work together, how many calculations can they perform in m minutes?

A. $60m\left(\dfrac{a}{b} + \dfrac{c}{d}\right)$

B. $m\left(\dfrac{60a}{b} + \dfrac{c}{d}\right)$

C. $m\left(\dfrac{a}{b} + \dfrac{c}{d}\right)$

D. $m\left(\dfrac{a}{60b} + \dfrac{c}{d}\right)$

E. $2m\left(\dfrac{30a}{b} + \dfrac{c}{d}\right)$

F. $m\left(\dfrac{a}{60b} + \dfrac{d}{c}\right)$

26. If -1 is a zero of the function $f(x) = 2x^3 + 3x^2 - 20x - 21$, then what are the other zeroes?

A. 1 and 3

B. -3 and 3

C. $-\frac{7}{2}$, 1 and 3

D. $-\frac{7}{2}$ and 3

E. -1 and 3

F. 1 and 7

END OF SECTION

Section 2

Question 1

A block is fired from point A with an initial velocity v at an angle θ. The plane BC is also at angle θ.

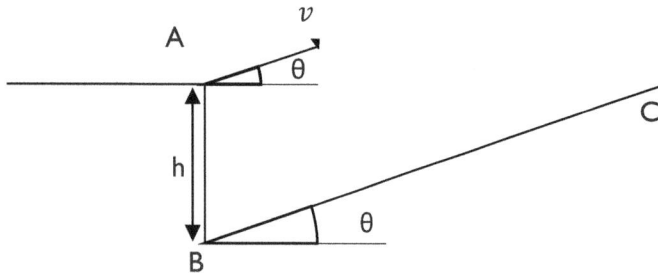

a) How long does it take the block to land on the plane?

A. $\sqrt{2h/g}$ C. $\sqrt{h/(g\cos\theta)}$ E. $h\sin\theta/v$

B. $\sqrt{2g\cos\theta/h}$ D. $\sqrt{h\tan\theta/g}$ F. $v\cot\theta/g$

b) How far up the plane from point B does the projectile land?

A. $h\sin\theta +$ $v\sqrt{2h/g}$ C. $v\sqrt{2h/g}$ E. $h\sin\theta + \dfrac{v^2}{g}$

B. $h\tan\theta$ D. $\sqrt{gh\sin\theta}/v$ F. $v\sqrt{h\cos\theta/g}$

c) When the block lands, its velocity perpendicular to the plane becomes 0. How much energy has been dissipated?

A. $mgh\sin\theta$ C. $mv^2\sin\theta/2$ E. $mgh\cos^2\theta$

B. $mv^2/4$ D. $mgh\cos\theta - mv^2\sin\theta.$ F. $mgh\tan\theta.$

d) What initial velocity is required for the projectile to land at the same height as when it was fired i.e. a distance h above point B?

A. $\sqrt{3gh}$ C. $\sqrt{gh\cos\theta}$ E. $gh\sqrt{2\sin\theta}$

B. $\sqrt{gh/\cos\theta}$ D. $3gh\cos\theta$ F. $\dfrac{1}{\sin\theta}\sqrt{gh/2}$

332

Question 2

Consider the circuit below. A, B and C are resistors. The power supply provides a voltage V.

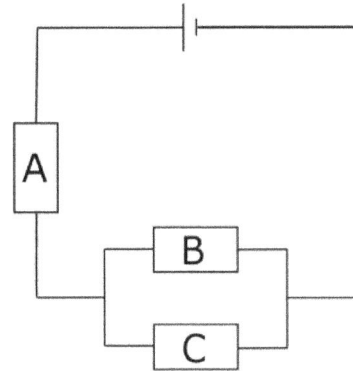

A. r/2 C. r√2 E. 3r

B. r/4 D. r/√2 F. 3r

a) Which of the following sets of units is a measure of power?

A. $V^2\Omega$ C. AV^{-1} E. $V\Omega$

B. $A^2\,\Omega$ D. VA^{-1} F. ΩV^{-2}

b) Let the resistances of A, B and C be 2R, R and R respectively. Which of the following statements is true?

A. The sum of the voltages across B and C equals the voltage across A.

B. The sum of the currents through B and C equals the current through A.

C. The current through A equals the current through B.

D. The voltage across A is equal to the voltage across C.

E. The power dissipated by A is equal to the power dissipated by B.

c) What is the total resistance of the circuit?

A. 4R   C. R   E. 5R/2

B. R/4   D. 2R/3   F. 3R

d) Assume all the resistors are cylinders with equal resistivity ρ and length. The radius of resistor B is r. The resistances of A and B are 2R and R respectively. What radius of C would be required for there to be a voltage of V/4 across resistor B?

Question 3

An object is initially stationary and then moves along one dimension. Its velocity, in m/s, has the following dependence on time, t, in seconds:

$v(t) = 10t^2 - t^3 - 24t$, for $t \geq 0$.

a) Apart from at t = 0, at which two times is the object stationary?

A. t = 3s, t = 8s   C. t = 3s, t = 7s   E. t = 2s, t = 12s

B. t = 4s, t = 6s   D. t = 2, t = -12s

b) What is the displacement, in m, of the object after 2 seconds?

A. 0   C. -15   E. -76/3

B. 81/2   D. -41/4   F. 17/5

c) At what time, in seconds, does the object experience its maximum acceleration?

A. 5   C. 10/3   E. 9/5

B. 3   D. 7/2   F. 11

d) At some point in time, the object collides with a wall and bounces off it at the same speed. Which of the following statements is true?

A. The object's momentum is the same before and after the collision.

B. The object transfers momentum to the wall during the collision.

C. The object loses energy in the collision which heats up the wall.

D. The momentum is not conserved in the collision.

E. Energy is not conserved during the collision.

Question 4

A block is sliding down a rough inclined plane with coefficient of friction μ.

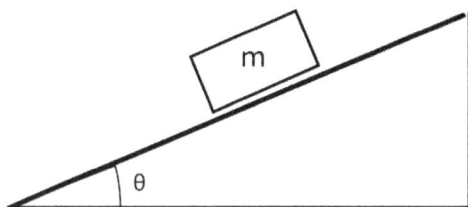

a) What is the normal force acting on the block?

A. $mg(\cos\theta - \mu\sin\theta)$ C. $mg\sin\theta$ E. $\mu mg(\sin\theta + \cos\theta)$

B. $\mu mg\tan\theta$ D. $mg(\sin\theta - \mu\cos\theta)$ F. $mg\cos\theta$

b) How much work is done by the frictional force as the block slides a distance d down the plane?

A. $\mu mgd\sin\theta$ C. $mg\cos\theta/d$ E. $mg\sin\theta/d$

B. $\mu mg\sin\theta/d$ D. $mgd\sin\theta$ F. $\mu mgd\cos\theta$

c) What distance d down the slope would the block have to travel to have a velocity v?

A. $g(\cos\theta + \mu\sin\theta)/v^2$ C. $v^2/(2g(\sin\theta - \mu\cos\theta))$ E. $g\sin\theta/v^2$

B. $v/(\mu g\sin\theta)$ D. v^2/g F. $v^2/(g(\cos\theta + \mu\sin\theta))$

END OF PAPER

ANSWERS

Paper A						Paper B					
Section 1A		Section 1B		Section 2		Section 1A		Section 1B		Section 2	
1	C	1	C	1	B	1	A	1	F	1	A
2	C	2	B	2	C	2	C	2	D	2	C
3	B	3	B	3	A	3	A	3	E	3	F
4	A	4	C	4	C	4	B	4	G	4	E
5	B	5	B	5	A	5	D	5	D	5	E
6	D	6	A	6	A	6	D	6	A	6	B
7	B	7	A	7	A	7	A	7	H	7	B
8	D	8	C	8	B	8	A	8	D	8	E
9	B	9	D	9	B	9	B	9	C	9	D
10	A	10	C	10	C	10	C	10	E	10	B
11	A	11	D	11	E	11	B	11	C	11	E
12	D	12	C	12	E	12	D	12	E	12	C
13	C	13	C	13	B	13	C	13	D	13	B
14	C	14	C	14	C	14	D	14	B	14	F
15	C	15	C	15	A	15	G	15	F	15	F
16	E	16	C	16	A	16	A	16	D	16	C
17	A	17	D			17	B	17	B		
18	C	18	F			18	C	18	A		
19	B	19	A			19	C	19	E		
20	C	20	E			20	A	20	C		
21	D	21	A			21	B	21	A		
22	C	22	C			22	E	22	B		
23	A	23	F			23	C	23	C		
24	A	24	B			24	E	24	C		
25	E	25	D			25	C	25	D		
26	B	26	A			26	E	26	D		
27	A					27	D				
28	B					28	A				

MOCK PAPER A ANSWERS

Section 1A

Question 1: C

This a constant acceleration problem so we need to use:

$$v^2 - u^2 = 2gs$$

We know everything apart from g so we can rearrange:

$$g = \frac{v^2 - u^2}{2s} = \frac{20^2 - 0^2}{2 \times 15} = \frac{400}{30} = 13.3\text{ms}^{-1}.$$

Question 2: C

We first calculate the useful output power using:

$$P_{out} = \frac{mgh}{t} = \frac{5000 \times 10 \times 3}{2} = 75\text{kW}.$$

We then now, using the efficiency that

$$IV \times 0.75 = 75 \Rightarrow IV = 100\text{kW}$$

Hence we can calculate the voltage

$$V = \frac{100,000}{50} = 2\text{kV}.$$

Question 3: B

The value of the comic in two years is given by:

$$25 \times (1.1)^2 = £30.25.$$

Question 4: A

Let us write two expressions, one for earth and one for the planet. Earth's expression for g_{earth} is given by:

$$g_{earth} = \frac{GM}{R^2} = \frac{G\rho\frac{4}{3}\pi R^3}{R^2} = \frac{4}{3}G\rho\pi R$$

For the planet, we can write:

$$g_{planet} = \frac{G'M'}{(R')^2} = \frac{G'\rho'\frac{4}{3}\pi(R')^3}{(R')^2} = \frac{4}{3}G'\rho'\pi R'$$

We also know the relationships between the constants in each universe, they are:

$$\implies G' = 2G, \rho' = 2\rho, R' = \frac{1}{2}R.$$

We can substitute these into the equation for the planet:

$$g_{planet} = \frac{4}{3}(2G)(2\rho)\pi\left(\frac{1}{2}R\right) = 2\left(\frac{4}{3}G\rho\pi R\right) = 2g_{earth}$$

Hence the ratio is given by:

$$\implies \frac{g_{planet}}{g_{earth}} = 2.$$

Question 5: B

We are told that the lamp is a full brightness – hence, it is operating at max power 60W. We can therefore calculate the current in the secondary coil $I_s = \frac{60}{12} = 5A$. We also know that the ratio of turns is equal to the ratio of voltage, using this we can write the following:

$$(0.75)I_pN_p = I_sN_s \implies I_p = \frac{I_sN_s}{(0.75)N_p}$$

We know that:

$$\frac{V_p}{V_s} = \frac{N_p}{N_s} \implies \frac{N_s}{N_p} = \frac{V_s}{V_p} = \frac{12}{200}$$

Therefore:

$$I_p = \frac{5\times12}{(0.75)\times200} = 0.4A.$$

Question 6: D

$$(5x - 1)^2 = y + 7$$

$$5x = 1 \pm \sqrt{y + 7}$$

$$\implies x = \frac{1}{5}\left(1 \pm \sqrt{y + 7}\right).$$

Question 7: B

The speed of a wave is given by $v = \lambda f$, where λ is the wavelength which has been given as 2 cm. We can work out the time period from the graph, which is 3s, hence the frequency is given by $f = \frac{1}{3}$, therefore the speed of the wave is:

$$\Rightarrow v = \lambda f = 2 \times \frac{1}{3} = 0.67 \text{ cms}^{-1}.$$

Question 8: D

The rate of change of momentum is a force and so it has the same units as weight, which is also a force.

Question 9: B

We can first rearrange to get: $3x^2 + 6x - 45 \leq 0$.

We divide by three and then factorise, treating it as an equation, and get: $(x + 5)(x - 3) \leq 0$.

Knowing that the quadratic passes the x-axis at the points $x = 3$ and $x = -5$ we can plot the graph and see where it goes below zero. This occurs at $x \geq -5$ and $x \leq 3$.

Question 10: A

The work done by the car's brakes must be equal to the kinetic energy of the car and the force must be applied in the opposite direction to the car's motion. Hence:

$$(-F)d = \frac{mu^2}{2} \Rightarrow F = \frac{-mu^2}{2d}.$$

Question 11: A

If we let a be the nucleon number and b be the proton number, we can see how they change with each emission.

	Original	After beta emission	After alpha emission	After second beta emission
Nucleon Number	a	a	a − 4	a − 4
Proton Number	b	b + 1	b - 1	b

We can see the final nuclide has the same proton number, but a different nucleon number than the original. Hence, it is an isotope of the original element.

Question 12: D

We know that the ratio B:C is equal to A:C × B:C. Therefore, we can set up a quadratic equation in x:

$$\frac{x+1}{2} \times \frac{x}{4} = \frac{3}{4}$$

$$4x(x+1) = 24$$

$$4x^2 + 4x - 24 = 0 \Longrightarrow x^2 + x - 6 = 0 \Longrightarrow (x+3)(x-2) = 0$$

Hence $x = 2, -3$.

Question 13: C

As x is inversely proportional to y^2, we can write the following relation, where k is the constant of proportionality.:

$$x = \frac{k}{y^2}$$

We can calculate k by substituting in the two known values:

$$k = xy^2 = 12 \times (2\sqrt{3})^2 = 12 \times 12 = 144$$

Hence, we can calculate y when $x = 9$:

$$9 = \frac{144}{y^2}$$

$$y^2 = \frac{144}{9} = 16 \Longrightarrow y = 4.$$

Question 14: C

We first need to calculate the amount of thermal energy generated by the heater within the kettle, we can do this as we know the power and the time,

$$E = Pt = 14,000 \times 120 = 1,680,000W$$

We can then use the thermal capacity equation to calculate the mass as we know the temperature change, as the boiling point of water is 100°C, and the heat capacity of water. The thermal energy Q is what we have calculated above:

$$Q = mc\Delta T \Longrightarrow m = \frac{Q}{c\Delta T}$$

$$m = \frac{1,680,000}{4200 \times 75} = 5 \text{ kg.}$$

Question 15: C

The total charge *is*: $Q = (7.5 \times 10^{15}) \times (1.6 \times 10^{-19}) = (1.2 \times 10^{-3})$.

We can then use the $Q = It$ equation to calculate the current

$$I = \frac{Q}{t} = \frac{(1.2 \times 10^{-3})}{40} = (3.0 \times 10^{-5})\, \text{A}.$$

Question 16: E

$$\left(\sqrt{5} - \sqrt{3}\right)^2 = \left(\sqrt{5} - \sqrt{3}\right)\left(\sqrt{5} - \sqrt{3}\right) = 5 - 2\sqrt{15} + 3 = 8 - 2\sqrt{15}.$$

Question 17: A

This can be calculated using the following tree diagram:

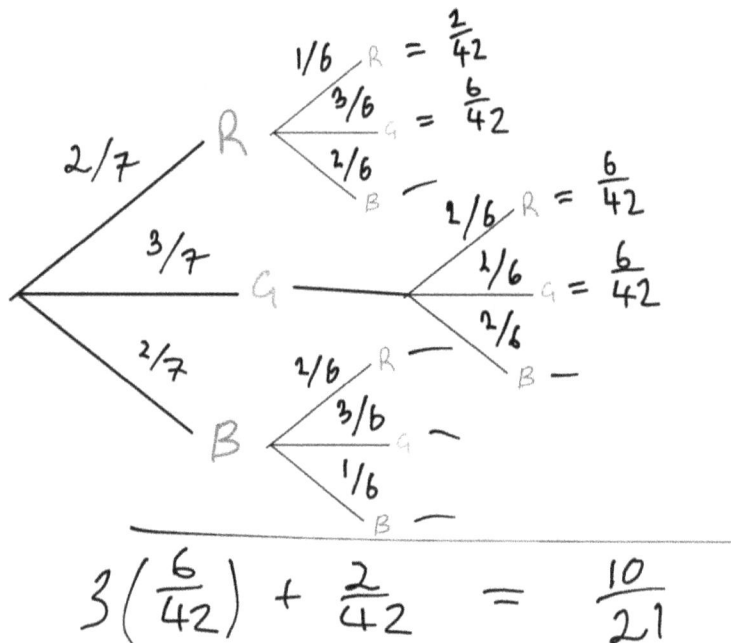

$$3\left(\frac{6}{42}\right) + \frac{2}{42} = \frac{10}{21}$$

Here we calculate all the probabilities associated with not picking a blue ball and then add them up.

Question 18: C

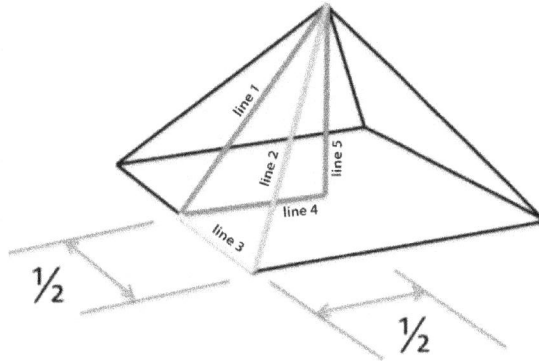

The question asks for the length of the red line and for that we first need the length of the green line. Hence, we can construct this triangle:

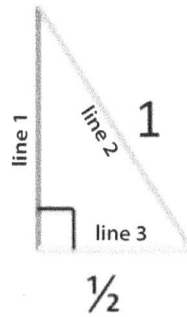

The length of the green line is given by Pythagoras' theorem:

$$1^2 = (\frac{1}{2})^2 + g^2 \Longrightarrow g = \sqrt{\frac{3}{4}} = \frac{\sqrt{3}}{2}$$

Then we can work out the height h using the red-green-blue triangle:

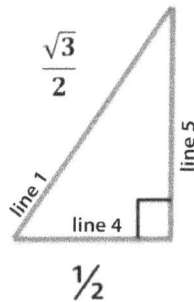

$(\frac{\sqrt{3}}{2})^2 = (\frac{1}{2})^2 + h^2 \Longrightarrow h = \frac{1}{\sqrt{2}}$.

Question 19: B

We know that the total resistance of a parallel combination is smaller than that the series combination of the same resistors, hence the parallel arrangement has a higher current than the series arrangement. This means the power provided to each bulb, which is equal to I^2R is higher in the parallel arrangement and hence the bulbs are brighter.

Question 20: C

This question is framed in a way that makes it seem much more complex than it actually is. If the elevator is descending at constant speed, then so is the mass. Constant speed tells us that the mass is not accelerating and hence there is no resultant force on the mass and hence the scale must still read mg.

Question 21: D

To calculate the equivalent expression, multiply out the brackets:

$$\left(2\sqrt{3} - 3\sqrt{2}\right)^2 = \left(2\sqrt{3}\right)^2 - 2 \times 2 \times 3 \times \sqrt{2 \times 3} + \left(-3\sqrt{2}\right)^2$$
$$= 12 - 12\sqrt{6} + 18$$
$$= 30 - 12\sqrt{6}.$$

Question 22: C

At terminal velocity, there is no resultant force and so the air resistance must equal the skydiver's weight:

$$kv^2 = mg$$
$$k \times 50^2 = 75 \times 10$$
$$k = \frac{750}{2500}$$
$$= 0.3.$$

Question 23: A

For every blue pen, there are $\frac{9}{4}$ red pens and for every red pen there are $\frac{10}{3}$ black pens. Therefore, for one blue pen, there are $\frac{9}{4} \times \frac{10}{3} = \frac{15}{2}$ black pens, giving a ratio of 2:15.

Question 24: A

The total resistance of two identical resistors in parallel is $\frac{R}{2}$. The resistance of the top line of the circuit is therefore $\frac{3R}{2}$. The total resistance from A to B is given by:

$$\frac{1}{R_{AB}} = \frac{1}{R} + \frac{1}{\left(\frac{3R}{2}\right)} = \frac{1}{R} + \frac{2}{3R} = \frac{5}{3R}$$

$$\therefore R_{AB} = \frac{3R}{5}.$$

Question 25: E

The sum of all 5 numbers is 120. After excluding one number, the mean is now 20 and the sum is 80. Therefore, the excluded number must be the difference between the two sums.

Question 26: B

A constant braking force means that the car has constant deceleration. The equation $v^2 - u^2 = 2as$ can therefore be used. Since the car comes to a stop, $v = 0$. Rearranging this equation and using Newton's second law of motion, where $a = \frac{F}{m}$, gives the solution:

$$-u^2 = 2\frac{F}{m} \times 2L \Rightarrow F = -\frac{mu^2}{4L}.$$

Question 27: A

The two proportionality relationships state that $A = k\sqrt{B}$ and $B = \frac{m}{C}$. Substituting in the values gives $k = \frac{6}{\sqrt{4}} = 3$ and $m = 8 \times 2 = 16$. Therefore, $A = 3\sqrt{B}$ and $B = \frac{16}{C}$, so:

$$A = 3\sqrt{\frac{16}{C}} = \frac{12}{\sqrt{C}}.$$

Question 28: B

The amount of lead increases over time, leaving only options B and C possible. The polonium decays to lead faster earlier on and approaches a horizontal asymptote as there is a finite amount of polonium to decay. This leaves option B as the correct answer.

END OF SECTION

Section 1B

Question 1. C

This question requires the knowledge of Archimedes' principle. This states that the upthrust on an object is equal and opposite in direction to the weight of the fluid displaced by the object.

In this case, the weight of the water that the ship displaces is equal to the upthrust on the ship. Therefore, statement 1 must be true.

This force must also be equal to the weight of the ship as the ship is stationary (it is floating, therefore not rising or sinking). This means that statement 4 must be true.

Therefore, both 1 and 4 are true, which is given by C.

Question 2. C

Firstly, the kinetic energy of the rubber ball must be determined. This is given by:

$$KE = \frac{1}{2}mv^2$$

In the question, it is stated that half of this energy is used to heat up the rubber ball. Therefore, the energy which will heat up the ball is given by:

$$Heat\ Energy = \frac{1}{2}KE = \frac{1}{4}mv^2$$

The equation for heat energy is $Q = mc\Delta T$. Therefore, the two expressions can be equated.

$$\Rightarrow mc\Delta T = \frac{1}{4}mv^2$$

$$\Rightarrow \Delta T = \frac{m}{4mc}v^2 = \frac{1}{4c}v^2.$$

Therefore, the answer is given by expression C.

Question 3. B

To determine the acceleration of the masses, the two individual masses and the whole system need to be considered one by one. The force diagram has been provided on the left.

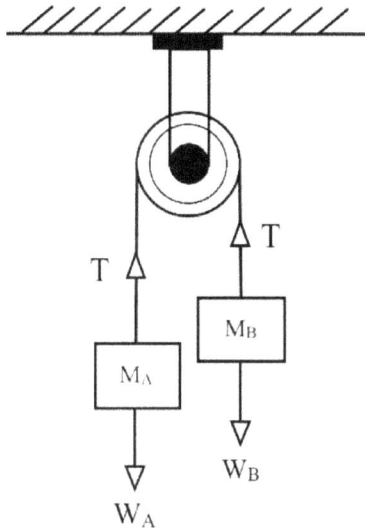

As mass A is larger, we expect the whole system to move towards it. So, let's define the downwards direction as positive. The equation of motion of the block A can be written using Newton's second law:

$$F = ma \Rightarrow M_A a = W_A - T = M_A g - T$$

For the second mass B, the following is obtained similarly:

$$F = ma \Rightarrow M_B a = T - W_B = T - M_B g$$

Tension is always constant through a rope or wire. Therefore, T is the same

in both of the above equations. T can be eliminated from both of the equations to obtain:

$$T = M_A g - M_A a, \quad T = M_B a + M_B g$$

$$\Rightarrow M_B a + M_B g = M_A g - M_A a$$

The acceleration is also the same for both masses (they're connected via the rope and therefore must move together.) Therefore, $a = \frac{(M_A - M_B)}{(M_A + M_B)}$ ms^{-2}, which is given by B.

Question 4. C

The equation for elastic potential energy is given by:

$$E = \frac{1}{2}kx^2$$

This is the energy stored within the spring when it is extended or contracted by a length x. Therefore, if we set the spring constant to be some value k, the energy stored in the left spring is equal to the expression above.

For the second spring, the extension is a fifth of the length x. As it's an identical spring, the spring constant is the same. Therefore, the elastic potential energy stored in the second spring is:

$$E_1 = \frac{1}{2}k\left(\frac{x}{5}\right)^2 = \frac{1}{2}k\frac{x^2}{25}.$$

Therefore, $E_1 = \frac{E}{25}$ which is given by C.

Question 5. B

The turns ratio is the ratio of the turns of wire on the primary coil to the number of wires on the second. In this case, it's 3:1, which means the primary coil has three times the number of coils of the second coil.

We know that, for transformers:

$$\frac{V_S}{V_P} = \frac{N_S}{N_P} = \frac{I_P}{I_S}$$

Therefore, the voltage in the second coil is one third of the voltage of the primary coil. Therefore, it will be 10V x (1/3) = 3.3V.

The resistance of the lamp is 5 ohms. Therefore, using V = IR, we obtain that the current is:

$$I = \frac{V}{R} = \frac{3.3}{5} \approx 0.7 \text{ A, which is given by B.}$$

Question 6. A

The period of the pendulum of length L is: $P = 2\pi\sqrt{\frac{L}{g}}$.

For a pendulum of length 2L, the period is given by:

$$P' = 2\pi\sqrt{\frac{2L}{g}} = 2\pi\sqrt{2}\sqrt{\frac{L}{g}}$$

In terms of the original pendulum, we obtain the ratio:

$$P' = \sqrt{2}\, 2\pi\sqrt{\frac{L}{g}} = \sqrt{2}\, P.$$

Therefore, the ratio of P to P' is $\sqrt{2}$.

Question 7. A

The electrostatic repulsion between the two charges makes them want to separate and get further apart. As q_B gets closer, it gets slower due to the repulsion between the two charges. Therefore, its speed will slowly diminish until it comes to a stop.

The work done by a force is equal to the force multiplied by the distance the force is applied over. The work done by the electrostatic force to move a charge q_B a distance r is given by:

$$\Rightarrow W = Fd = \left(\frac{q_A q_B}{4\pi\epsilon_0 r^2}\right) \cdot r = \frac{q_A q_B}{4\pi\epsilon_0 r}$$

When it stops, it is the closest it can get to the charge Q. This is known as the minimum distance of approach. This will occur when all of the kinetic energy of q_B has been converted into work done.

$$\frac{1}{2}mv^2 = \frac{q_A q_B}{4\pi\epsilon_0 r} \Rightarrow r = \frac{q_A q_B}{2\pi\epsilon_0 m v^2}.$$

This is given by A.

Question 8. C

The ideal gas assumptions simplify the behaviour of a real gas to make it easier to model. The assumptions are essentially that the intermolecular forces and particle sizes are neglected in an ideal gas and all collisions are elastic - however, the thermodynamic quantities, such as pressure, temperature and volume, do not need to be constant.

Therefore, statement C is not an assumption is the ideal gas model.

Question 9. D

This question may appear challenging – but it provides you with everything you need to answer it. Recall the relationship between the kinematic quantities velocity and acceleration:

$$v = \frac{dx}{dt}, \quad a = \frac{dv}{dt} = \frac{d}{dt}\left(\frac{dx}{dt}\right) = \frac{d^2x}{dt^2}$$

Therefore, we obtain (recall that the derivative of $cos(\alpha y)$ is $-\alpha sin(\alpha y)$):

$$s(x,t) = \frac{1}{5}cos\frac{2\pi}{\lambda}(x - vt)$$

$$v(x,t) = \frac{ds(x,t)}{dt} = \frac{1}{5}\frac{d}{dt}\left(cos\frac{2\pi}{\lambda}(x - vt)\right) = \frac{1}{5}\cdot\frac{2\pi}{\lambda}v\sin\frac{2\pi}{\lambda}(x - vt)$$

$$a(x,t) = \frac{d^2s(x,t)}{dt^2} = \frac{1}{5}\cdot\frac{2\pi}{\lambda}v\frac{d}{dt}\left(\sin\frac{2\pi}{\lambda}(x - vt)\right) = -\frac{1}{5}\cdot\left(\frac{2\pi}{\lambda}v\right)^2\cos\frac{2\pi}{\lambda}(x - vt)$$

We now have an expression for the acceleration. To find the maximum acceleration, we need the amplitude which is just the constant outside of the cosine expression (the amplitude of $A\cos(\alpha y)$ is A).

$$a_{max} = \left|\frac{1}{5}\left(\frac{2\pi v}{\lambda}\right)^2\right| = \left|\frac{4\pi^2 v^2}{5\lambda^2}\right| \text{ ms}^{-2}.$$

Therefore, the answer is given by the following expression which is equivalent to option D.

Question 10. C

The half-life of a radioactive substance is the time taken, in seconds, for the sample to decay to half its original size. Therefore, if the half-life of the substance is 45 seconds, that means that after 45 seconds, you will have 50% of the original sample remaining.

When 90 seconds have passed, the sample size will again half and therefore you will be left with only 25% of your original sample.

After another 45 seconds has passed (giving a total of 135 seconds), the sample will decay again and only 12.5% of your original sample size will remain.

As 110 seconds is between 90 and 135 seconds, you can estimate that the percentage of the sample remaining will be somewhere between 12.5% and 25%. The only option which satisfies this condition is C.

Question 11. D

For total internal reflection, the light ray needs to be incident at an angle greater than the critical angle. Statements B and E therefore are false as total internal reflection cannot occur in these cases.

In total internal reflection, all the light is reflected back inside the first medium and none is transmitted to the second medium. For any angle less than the critical angle, we do not get total internal reflection. Instead, there is some reflection back into the original medium and some refraction into the new medium. Therefore, A and C are both incorrect. The light ray partially reflects and refracts until the critical angle is reached.

Therefore, statement D is correct.

Question 12. C

The equation for intensity is given by: $Intensity = \frac{Power}{Area}$.

To power five bulbs with a power output of 60 W each, the total power output required is:

5 x 60 W = 300 W

However, as the solar panels have an efficiency of 20%, the power input to the solar panels needs to be much greater than this:

300/0.2 = 1500 W

Therefore, by substituting the value for the intensity and rearranging the equation above, we obtain:

$$\Rightarrow A = \frac{P}{I} = \frac{1500}{1360} \approx 1.102.$$

As each panel has an area of 0.25 m², this means that we require roughly 4.411 solar panels. By rounding up to the next integer value, the number of solar panels required is 5 which is given by C.

Question 13. D

The half-life is the time taken for half of the radioactive nuclei in a sample to undergo radioactive decay. The half-life is not related to the mass of the radioactive substance that is remaining to decay or limited to any specific type of radiation. It is used for all types of radiation.

Therefore, the correct definition is given by D.

Question 14. C

The state is initially a gas, the temperature will decrease until there is a change in state. From a gas, the next state must be a liquid - this process must therefore be condensation.

For further explanation of the graph, the temperature of the gas is steadily decreasing; the fact that it's kept in a container is therefore not an issue here as cooling down causes the molecules to get closer together. In heating, a constant volume is an issue as the gas cannot expand.

As it cools, the intermolecular bonds form and the gas turns into a liquid. This increases the potential energy of the molecules, instead of their kinetic energy. As temperature is the average kinetic energy of the molecules, we expect it to remain constant as the gas turns into a liquid. This process of the substance going from gas to liquid is known as condensation. Therefore, the answer is C.

Question 15. C

The object is accelerating in the positive x-direction. This is due to a constant force – so, we can deduce that the acceleration must also be constant (as F = ma). If we sketched the acceleration-time graph, it would be a straight horizontal line as the acceleration is not changing. Therefore, statement 2 is false. The acceleration cannot be linear as it is not varying.

Acceleration is the derivative of velocity with respect to time. Therefore, if the acceleration-time graph is constant, then the velocity-time graph must be linear (you can equally think about the SUVAT equation v = u + at). Therefore, statement 1 is also false.

Statement 3 may be true as we don't know which direction the object is travelling in. The acceleration is in the positive x-direction and, therefore, the object is speeding up in this direction, but the object may initially have a velocity in the negative x-direction. It will slow down due to the acceleration and then speed up in the positive x-direction due to the acceleration. Therefore, the answer is C.

Question 16. C

The object comes momentarily to rest when its velocity is zero. To determine its velocity, we must differentiate the displacement with respect to time.

$$v = \frac{ds}{dt} = 2t^2 - 7t + 3 = 0$$

We need to determine the solutions of this quadratic; these are easily obtained by factorising the equation:

$$\Rightarrow 2t^2 - 7t + 3 = 0 \Rightarrow (2t - 1)(t - 3) = 0$$

This has the solutions $t = 3, \frac{1}{2}$ seconds which correspond to the times at which the particle becomes stationary. Therefore, the earliest time at which the object is stationary is given by C.

Question 17. D

As the ball is thrown upwards, its velocity must be positive as that is defined to be the positive direction. The acceleration due to freefall always acts vertically downwards, so it must act in that direction. Therefore, it must be negative. Of course, once it reaches the top, the velocity will change. However, for its upwards motion, D provides the correct description.

Question 18: F

Intensity is proportional to the square of the amplitude. By doubling the amplitude, the intensity becomes four times its original value.

In this case, we want the reverse process – to go from intensity to amplitude. Therefore, for double the initial intensity:

$$I \propto A^2 \Rightarrow 2I \propto 2A^2 \propto \left(\sqrt{2}A\right)^2$$

Therefore, for double the intensity, the amplitude must be at $\sqrt{2}$ times its original value. This is given by F.

Question 19. A

It cannot be F as then the sound would simply be delayed but we cannot hear the sun at any time. It also cannot be E as diffraction is a wave phenomenon and so both light and sound waves undergo diffraction.

C and D are both incorrect as the wavelength and frequency are individual properties of light – some light waves have very low frequency and long wavelength, just like sound waves, and so, this is not a reasonable explanation. B is also incorrect as both light and sound waves are progressive. Therefore, the only suitable explanation is given by A. This is because longitudinal waves require a medium to travel; they are propagated by vibrations within the medium and cannot travel in free space. As sound is a longitudinal wave, it cannot travel in space. Light, however, is a transverse wave and these waves do not require a medium to propagate.

Question 20. E

This is a standard projectile motion question. Initially, we need to resolve the velocity into its horizontal and vertical components. In the horizontal direction, the velocity is $v_0 cos(30°) = \frac{\sqrt{3}}{2}v_0$ and in the vertical direction, the velocity is given by $v_0 sin(30°) = \frac{v_0}{2}$.

In the vertical direction, we need to use the SUVAT equations as there is a constant acceleration on the ball. The initial velocity is given by $\frac{v_0}{2}$., the displacement is 2 metres as this is the height of the wall, the acceleration is - g m/s² and the final velocity should be 0.

$$\Rightarrow \quad u = \frac{v_0}{2} \, ms^{-1}, \quad v = 0 \, ms^{-1}, \quad a = -g = -10 \, ms^{-2}, \quad s = 2m$$

Using $v^2 = u^2 + 2as$, as this is the SUVAT equation which includes all our quantities of interest, we can rearrange the equation for v_0.

$$v^2 = u^2 + 2as \Rightarrow 0 = \left(\frac{v_0}{2}\right)^2 - 2 \cdot 10 \cdot 2 \Rightarrow 160 = v_0^2$$

$$\therefore v_0 = \sqrt{160} = 4\sqrt{10} \, ms^{-1}.$$

This is given by option E.

Question 21. A

The total resistance of two resistors in parallel is given by $\frac{1}{R_T} = \frac{1}{R_1} + \frac{1}{R_2}$. The power dissipated is given by $P = \frac{V^2}{R_T}$, and thus:

$$P = V^2 \times \frac{1}{R_T}$$

$$\therefore P = V^2 \times \left(\frac{1}{R_1} + \frac{1}{R_2}\right)$$

Question 22. C

As the slide becomes less steep the magnitude of the child's acceleration decreases, as a smaller component of their weight is acting parallel to the slide. As the acceleration is still positive, the child's speed will continue to increase.

Question 23. F

The plank of wood pivots around the end in contact with the ground. This means that the force is applied at a distance of 4 m from the pivot and the weight of the block acts at a distance of 1 m from the pivot.

At the point at which the plank starts to lift, the moments must be in equilibrium:

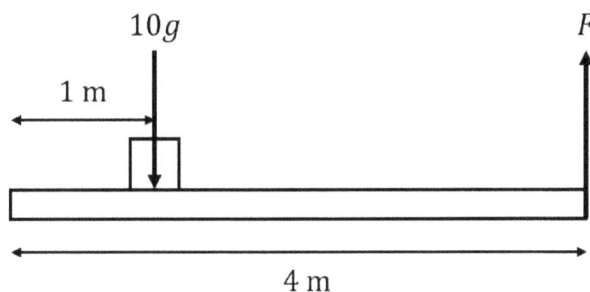

$$10 \text{ kg} \times 10 \text{ Nkg}^{-1} \times 1 \text{ m} = F \times 4 \text{ m}$$

$$F = 25 \text{ N}.$$

Question 24. B

Potential difference is defined as the work required to move a charge between two points, per unit of the charge. Potential difference is therefore calculated as:

$$\frac{\text{Work Done(J)}}{\text{Charge(C)}}$$

Thus, the units are JC⁻¹.

Question 25. D

Consider the motion of the skydiver independently to the parachute and define the upwards direction as positive:

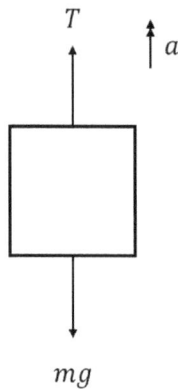

Apply Newton's second law:

$F = ma$

$T - mg = ma$

$T - 70 \times 10 = 70 \times 5$

$T = 1050 \, \text{N}.$

Question 26. A

The same battery is used in each case, so the potential difference across the battery will be the same in both circuits, thus B and D are incorrect. The current in the circuit is determined by Ohm's law: $V = IR$, where R is the total resistance of the circuit. Since the total resistance of the bulbs will be higher in the circuit with two bulbs and the potential difference, V, is the same, the current will be lower, thus C and E are incorrect, and A is the correct answer.

END OF SECTION

Section 2

Question 1

a) Answer: B

- Firstly, the time taken to reach the bottom plate must be determined. The force on the particle is $eE + mg$, providing an acceleration of $\frac{eE}{m} + g$. Apply the following SUVAT equation to the vertical direction:

$$s = ut + \frac{1}{2}at^2$$

Thus, as $u = 0$, the time taken to reach the bottom plate is:

$$\frac{d}{2} = \frac{1}{2}\left(g + \frac{eE}{m}\right)t^2$$

$$t^2 = \frac{d}{g + \frac{eE}{m}} \Rightarrow t = \sqrt{\frac{d}{g + \frac{eE}{m}}}.$$

We see this diverges for a finite negative value of E, since the denominator is 0 when $E = -\frac{mg}{e}$, so here we expect the time to approach infinity. The only graph with this feature is B.

b) Answer: C

- Apply the above formula, taking $E = -\frac{mg}{2e}$, and $d = 5$:

$$t = \sqrt{\frac{5}{g + e\left(-\frac{g}{2e}\right)}} \Rightarrow t = \sqrt{\frac{5}{g - \frac{g}{2}}}$$

$$\Rightarrow t = \sqrt{\frac{5}{10 - 5}} \Rightarrow t = \sqrt{\frac{5}{5}} = 1.$$

Now apply $s = ut + \frac{1}{2}at^2$ again, in the horizontal direction. Since there is no horizontal acceleration:

$$s = (5)(1) + \frac{1}{2}(0)(1)^2 = 5 \text{ m}.$$

c) Answer: A From the hint, we know eE is a force, and so has units $N = kgms^{-2}$. Furthermore, e is a charge so has units $C = As$. So the units of E are $\frac{N}{C} = \frac{kgms^{-2}}{sA} = kgms^{-3}A^{-1}$.

d) Answer: C With $E = -\frac{4gm}{e}$, the electric force is $4\ times$ greater than the gravitational force gm. The net force is thus upwards, so they move towards the top plate. In the limit, we have many particles in the top plate, so a strong positive charge on it, producing a net upward force attracting the negative particle.

e) Answer: A The force of the field on the two particles is the same, but in opposite directions, as they have equal but opposite charge. Since the particles have the same mass, the gravitational force is the same on each. We can slowly change E such that, for some value of E, the particles would hit different plates by ensuring the force due to the field is greater in magnitude than the weight of the particles.

Question 2

a) Answer: A Firstly, we must calculate the resistive forces. They clearly act up the slope, as the bus moves down the slope. Speed is constant, so there is no acceleration - we are in equilibrium. So, we know that forces down the slope are equal to the forces up slope. The only downwards force is the component of weight, which must equal the resistance.

$$F = W sin\theta = 5 \times 10^3 \text{ kg} \times 10 \text{ Nkg}^{-1} \times \frac{1}{40} = 1250 \text{ N}.$$

b) Answer: A We are now driving up the slope, so resistive forces act down the slope. Again, forces up slope must balance the forces down slope, as there is no acceleration. The forces down the slope are given by:

$$T = 500\text{N} + 5 \times 10^3 \times 10 \times \frac{1}{40}\text{N} = 1750 \text{ N}.$$

The first term gives resistive forces and the second gives the component of weight down the slope.

Power is given by force times velocity:

$$P = Fv = 1750 \text{ N} \times 12 \text{ kmh}^{-1} \times \frac{5 \text{ ms}^{-1}}{18 \text{ kmh}^{-1}} \approx 5.8 \text{ kW}.$$

The factor of $\frac{5}{18}$ converts kmh^{-1} to ms^{-1}.

c) Answer: C Our power is $10,000 \text{ Js}^{-1}$ over a period of 1000 seconds. Therefore, the energy required is:

E = Pt = $10000 \times 1000 = 1 \times 10^7$ J.

This is the energy per half tonne; so, per tonne, the truck produces 2×10^7 J.

d) Answer: B Up the slope, there is a driving force of 1000 N, and 500 N resistive force, so there is a net force of 500N.

$$F = ma \Rightarrow 500 = 5000a \Rightarrow a = \frac{1}{10}$$

Now using $s = ut + \frac{1}{2}at^2$, we can determine t:

$$50 = 0t + \frac{1}{2}\frac{1}{10}t^2 \Rightarrow t^2 = 1000 \Rightarrow t \approx 32 \text{ s}.$$

Question 3
a) Answer: C

Gamma waves and radiowaves are on the two opposing ends of the EM spectrum. Radiowaves have large wavelengths and low frequencies, and therefore carry lower every than gamma waves.

b) Answer: E

EM waves propagate at different speeds in different materials, EM waves are transverse and all waves, including EM waves, can diffract. They can also be created and destroyed in various ways – for example, gamma waves can be created by decay. Therefore, all of the statements are false.

c) Answer: E

The doppler shift applies, and we get a blueshift, as the peaks are getting compressed by the movement of the star towards us. Shorter wavelengths correspond to the "blue" end of the spectrum. Intensity remains the same, and so (4) and (5) are false. (1) is also false, as the speed of a distance object doesn't affect its apparent size if measured instantaneously.

Question 4
a) Answer: B

Recall voltage is proportional to resistance, and that voltage splits evenly in parallel. Also recall the formula for combining resistance in parallel.

1. Statement 1 is false, since the sum of currents across b, c and h sum to the current across a – current splits in parallel.
2. Statement 2 is true as the current is preserved in series, and a and f are in series.
3. Statement 3 is true - we find the resistance across (e, f) is:

$$\Rightarrow \frac{1}{R_{ef}} = \frac{1}{R} + \frac{1}{R} = \frac{2}{R} \rightarrow R_{ef} = \frac{R}{2}.$$

This is the same as that of the bulb and thus so is the voltage.

4. Statement 4 is false – we just calculated the resistance of the (e, f) section, and it is different to that at a, so the voltage is also different.

b) Answer: C

Since the resistors have equal resistance, the current across the cell must be $3I$, using the fact that currents add when the wires meet and current is conserved in series.

Using the resistance in parallel formula for each parallel section (as in (a)), the resistance across the whole circuit is just the sum of its components $R + \frac{R}{2} + \frac{R}{2} + \frac{R}{3} = \frac{7R}{3}$, where the terms are clockwise around the cell.

Using V = IR, we find that $V = \frac{7R}{3}(3I) = 7RI$.

c)   **Answer: A**

Power is given by $\frac{V^2}{R}$. First, we need to calculate the old power output. The voltage across the bulb is simply the voltage across the circuit, multiplied by the proportion of resistance of the component compared to the whole circuit. From above, this is

$$\frac{\frac{R}{2}}{\frac{7R}{3}} = \frac{3}{14}$$

Therefore, the voltage is $\frac{3V}{14}$ and so:

$$\Rightarrow P = \frac{\left(\frac{3V}{14}\right)^2}{\frac{R}{2}} = \frac{9V^2}{98R}.$$

When the resistance doubles, we now find that the resistance across the triple of resistors is:

$$\frac{1}{\frac{1}{R}+\frac{1}{R}+\frac{1}{2R}} = \frac{2R}{5}$$

The new total resistance is:

$$\Rightarrow R + \frac{R}{2} + \frac{R}{2} + \frac{2R}{5} = \frac{12R}{5}$$

The proportion of resistance is then:

$$\frac{\frac{R}{2}}{\frac{12R}{5}} = \frac{5}{24} \Rightarrow P = \frac{\left(\frac{5V}{24}\right)^2}{\frac{R}{2}} = \frac{25V^2}{288R}.$$

d)   **Answer: A**

By the resistivity formula, resistance is inversely proportional to the cross-sectional area. So, if the diameter is halved, the new resistance of the resistor h is $\frac{R}{4}$.

Now taking the cell to have voltage V, originally, the current is given by $\frac{V}{R}$ and, after the change, is given by $\frac{V}{R/4} = \frac{4V}{R}$, and so the current has increased by 300%.

END OF PAPER

MOCK PAPER B ANSWERS

Section 1A

Question 1: A

Firstly, we can rearrange the inequality to get $x^2 + 5x + 4 \geq 0$. We can then factorise the quadratic to get $(x + 1)(x + 4) \geq 0$. If this were an equation, we would solve it to get $x = -1$ and $x = -4$. We would then plot the equation knowing that it crosses the x-axis at the above-mentioned points. We can use our graph to see where it goes above zero, leading to our solutions $x \leq -4$ and $x \geq -1$.

Question 2: C

Using the equation $N_p I_p = N_s I_s$ we can calculate I_s for 100% efficiency.

$$I_s = \frac{N_p I_p}{N_s} = \frac{1200 \times 0.25}{500} = 0.6 \text{ A.}$$

We can now account for the efficiency which means the secondary current is actually $0.83 \times 0.6 = 0.5$ A.

Question 3: A

We know that $h = 2r$ hence we can equate the volumes and cancel terms:

$$\frac{\pi R^2 (2r)}{3} = \frac{4\pi r^3}{3} \implies R^2 = 2r^2$$

$$\therefore R = r\sqrt{2}.$$

Question 4: B

The gradient m_y of $y = 2x - 2$ is 2. Therefore, the gradient of the line that intersects it m_f must equal $-\frac{1}{2}$ as $m_y m_f = -1$ for the lines to be perpendicular. Both lines also pass through $x = 1$ and therefore the y-coordinate of the intersection point is $y = 2(1) - 2 = 0$.

As we have the gradient and a point we can calculate the equation of the line $f(x)$ using $y - y_1 = m_f(x - x_1)$ where $m_f = -\frac{1}{2}$ and $(x_1, y_1) = (1,0)$:

$$y - 0 = -\frac{1}{2}(x - 1)$$

$$\implies y = -\frac{1}{2}x + \frac{1}{2}.$$

Question 5: D

$$(y + 2)^2 = 7x - \frac{1}{2} \implies 7x = (y + 2)^2 + \frac{1}{2}$$

$$\therefore x = \frac{(y+2)^2}{7} + \frac{1}{14}.$$

359

Question 6: D

Using equations for power such as $P = Et$ and $P = Fv$ we can work by a process of elimination to arrive at D as our answer. Or, we can directly see that D is incorrect using the following:

$$\Rightarrow P = E/t = Fd/t$$

$$\therefore [P] = kgm^2s^{-2} /s = kg\ m^2\ s^{-3}.$$

This is clearly not equal to D – therefore, D is not a unit of power.

Question 7: A

We can first calculate the current flowing using $I = \frac{V}{R} = \frac{10}{500} = 0.02A$. This means that in 5 seconds a total charge of $Q = It = 0.02 \times 5 = 0.1C$ passes through the resistor. Finally, if we divide this by the charge of an electron we arrive at our answer:

$$N = \frac{0.1}{1.6 \times 10^{-19}} = 6.25 \times 10^{17}\ \text{electrons}.$$

Question 8: A

If the temperature increases, the resistance of the NTC thermistor will decrease. As the components are in series, this will decrease the total resistance of the circuit - hence increasing the current. The power dissipated by the fixed resistor is given by $P = I^2R$ and hence because R is staying the same and I is increasing, the power dissipated P must also increase.

Question 9: B

This is a constant acceleration problem hence we need to use the constant acceleration equation,

$$v^2 = u^2 + 2as$$

$$v^2 = 0 + 2 \times 10 \times 35 = 700$$

$$\Rightarrow v = \sqrt{700} = 10\sqrt{7}ms^{-1}.$$

Question 10: C

We know that the elastic energy is converted into kinetic energy and that an extension x produces a speed v hence we know that $\frac{1}{2}kx^2 = \frac{1}{2}mv^2$. Increasing the extension to $2x$ gives us,

$$\frac{1}{2}k(2x)^2 = \frac{1}{2}m(v')^2$$

We can rearrange this to

$$4(\frac{1}{2}kx^2) = \frac{1}{2}m(v')^2$$

We know what $\frac{1}{2}kx^2$ is and so we can substitute it in

$$4(\frac{1}{2}mv^2) = \frac{1}{2}m(v')^2$$

Cancelling terms, we obtain:

$$4v^2 = (v')^2 \Rightarrow v' = 2v.$$

Question 11: B

At 6.15, the hour hand is $\frac{1}{4}$ of the way between 6 and 7. The angle between 6 and 7 is 30° - hence one quarter of that is 7.5°. The minute hand is at 15 minutes past – therefore, the total angle is 90 + 7.5 =97.5°.

Question 12:D

The wavelength λ of the progressive is a minimum of $0.5 \times 8 = 4$m. Using the wave equation, we can then calculate the speed of the wave $v = \lambda f = 4 \times 10 = 40 \text{ ms}^{-1}$.

Question 13: C

The percentage efficiency of a motor is equal to the useful power output P_{out} divided by the input power P_{in}. Hence, we need to calculate the output power:

$$P = \frac{E}{t} = \frac{mgh}{t}$$

As the mass is gaining gravitational potential energy. We know that $\frac{h}{t}$ is the constant speed v given in the question – therefore, we get:

$$P = mgv = 10 \times 10 \times 0.5 = 50 \text{ W}$$

Hence, the percentage efficiency is:

$$\frac{50}{100} \times 100 = 50\%.$$

Question 14: B

The new value of the car after three years is given by,

$$£25000 \times (0.9)^3 = £18,225$$

Therefore, the reduction in value is:

$$£25,000 - £18,225 = £6,775.$$

Question 15: G

The ball is falling at terminal velocity hence its velocity is constant and therefore its kinetic energy cannot be increasing hence 1 is false. If the velocity is constant, then there is no net force on the ball hence 3 is true and from this we know that the drag force must be equal to the weight hence the drag force $D = mg = 0.1 \times 10 = 1$N hence 2 is true.

Question 16: A

The ratio A:C can be calculated by multiplying the ratios A:B and B:C as follows:

$$\frac{A}{C} = \frac{A}{B}\frac{B}{C} = \frac{4}{3}\frac{5}{6} = \frac{20}{18} = \frac{10}{9}.$$

Question 17: B

The area A of a triangle is given by $\frac{1}{2}bh$ where b is the base and h is the height, hence we get

$$A = \frac{1}{2}\left(5 + \sqrt{3}\right)\left(3 - \sqrt{3}\right) = \frac{1}{2}\left(12 - 2\sqrt{3}\right) = \left(6 - \sqrt{3}\right).$$

Question 18: C

The correct equation is C as we know that the atomic weight must balance. Therefore, by going systematically from left to right, we obtain:

$$1 + 235 = x + y + w(1) \implies x + y + w = 236 \implies x = 236 - (y + w).$$

Question 19: C

On the graph, acceleration is taking place when the cars velocity is increasing. To calculate the distance, we need to calculate the area under these two sections of the graph. The area A_1 under the first acceleration is the area of a triangle given by:

$$A_1 = \frac{1}{2}bh = \frac{1}{2} \times 2 \times 15 = 15$$

The area A_2 under the second acceleration is the area of a trapezium given by:

$$A_2 = \frac{a + b}{2}h = \frac{15 + 25}{2} \times 5 = 100$$

Therefore, the total distance travelled while accelerating is $A_1 + A_2 = 115$ m.

Question 20: A

Let v_α be the velocity of the alpha particle and v_N the velocity of the nucleus. Initially, the momentum is zero as the nucleus is stationary. After the emission, the total momentum is $mv_\alpha - (M - m)v_N$. This must equal zero and so $mv_\alpha = (M - m)v_N$. We can rearrange to get an expression for v_N,

$$v_N = \frac{mv_\alpha}{(M - m)}$$

We now need to express v_α in terms of the kinetic energy E.

$$E = \frac{1}{2}m(v_\alpha)^2 \implies v_\alpha = \sqrt{\frac{2E}{m}}$$

Therefore, we get:

$$v_N = \frac{m\sqrt{\frac{2E}{m}}}{(M-m)} = \frac{\sqrt{2mE}}{(M-m)}.$$

Question 21: B

The problem becomes a lot easier if the expression inside the brackets is simplified first.

$$\implies \left(\frac{8x^{\frac{5}{2}}y^{-\frac{1}{2}}}{2\sqrt{xy^{\frac{3}{2}}}}\right)^{\frac{1}{2}} = (4x^2y^{-2})^{\frac{1}{2}} = 2xy^{-1}.$$

Question 22: E
Newton's second law states that F = ma, where the F is the resultant force.

$T - Mg = Ma$
$T - 400 \times 10 = 400 \times 2$
$\therefore T = 4800$ N.

Question 23: C
There are 88 boys in total. 20 play tennis and 45 go swimming, so 23 must do athletics. This means that 57 girls do athletics and, since 15 go swimming, 40 are left to play tennis.

Question 24: E
After one half-life, the level of radiation falls to half the initial value. It is required to fall to $\frac{1}{64} = \frac{1}{2^6}$ of the original level, which takes 6 half-lives. The half-life of the material is 15 days and so the required number of days is $6 \times 15 = 90$.

Question 25: C
$3x^2 + 5x - 2 \geq 0$

$(3x - 1)(x + 2) \geq 0$

$\therefore x \geq \frac{1}{3}, x \leq -2$

Question 26: E
Output power is equal to the energy gained by the mass per second, which is equal to its increase in gravitational potential energy per second.

$P_{out} = Fv = mgv = 75$ W

Since the output power is half the input power, the efficiency is 37.5%.

Question 27: D
The volume of the pipe is given by

$V = \pi(1.5^2 - 1^2) \times 2 = 2.5\pi$.

And, therefore, as density is defined as mass/volume:

$\rho = \frac{5000}{2.5\pi} = \frac{2000}{\pi}$.

Question 28: A
The initial momentum of X is zero, as it is at rest. Since momentum is conserved, the momentum of Y must be equal and opposite to the alpha particle.

END OF SECTION

Section 1B

Question 1: F

The change in momentum of the ball is: change in momentum = mass × (final velocity − original velocity).

Thus, if we define the original direction of the ball as positive, the change of momentum is:

$$0.5 \times \left(6 - (-12)\right) = 9 \text{ kgms}^{-1}$$

$$\text{Average force} = \frac{\text{change in momentum}}{\text{time taken}}$$

$$F_{\text{average}} = \frac{9}{0.3} \Rightarrow F_{\text{average}} = 30 \text{ N}.$$

Question 2: D

Recall the equation for energy transfer:

$$\text{Energy transferred} = \text{Power} \times \text{Time} = \text{Voltage} \times \text{Current} \times \text{Time}$$

Therefore, the battery holds a total energy of:

$$12 \times 0.5 \times 0.5 = 3 \text{ Wh}.$$

The area of the solar panel is $0.2^2 = 0.04 \text{ m}^2$. Therefore, the solar panel charges the battery at a rate of:

$$0.04 \text{ m}^2 \times 1000 \text{Wm}^{-2} \times 0.1 = 4 \text{ W}.$$

The charging time is therefore:

$$t = \frac{3 \text{ Wh}}{4 \text{ W}} \Rightarrow t = \frac{3}{4} \text{ hours} \Rightarrow t = 45 \text{ mins}.$$

Question 3: E

If light passes between two mediums of different optical densities, it will be refracted. In this situation the glass block is more optically dense than air, therefore the light travels more slowly in the glass block than in the air, thus B is true, and E is false. As the light enters a denser medium it is expected to bend towards the normal (dotted line), therefore $\theta_2 > \theta_1$, and C is correct. As it leaves the same medium it bends away from the normal by the same amount and will be travelling in the same direction as before it entered the medium, thus $\theta_1 = \theta_3$ and A is correct. Refraction does not affect the frequency of light, only its direction, thus D is correct.

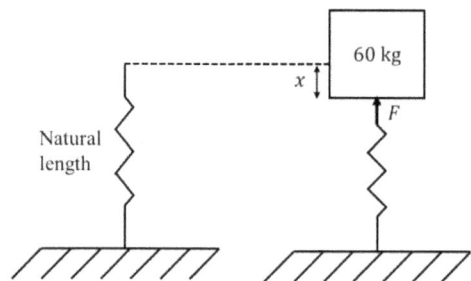

Question 4: G

The compression of the spring, x, is given using Hooke's law:

$F = kx \Rightarrow 60 \times 10 = 6000 \times x$

$\therefore x = 0.1$ m

The energy stored in the spring is given by:

$E = \frac{1}{2}kx^2 \Rightarrow E = \frac{1}{2} \times 6000 \times 0.1^2$

$\therefore E = 30$ J.

Question 5: D

Use the equation $v^2 = u^2 + 2as$. Taking downwards as positive, the rock has an acceleration $a = 10$ ms^{-2} and its displacement when it hits the ground is $s = 1$ m. Substitute these values into the equation:

$v^2 = (-4)^2 + (2 \times 10 \times 1)$

$v^2 = 16 + 20$

$v^2 = 36$

$\therefore v = 6$ ms^{-1}.

Question 6: A

The resistance of a wire is given by:

$R = \dfrac{\rho L}{A}$

Where ρ is resistivity and is constant for wires of the same material. For a second wire of twice the cross-sectional area and half the length, the resistance is:

$R_2 = \dfrac{\rho\left(\frac{L}{2}\right)}{2A} = \dfrac{\rho L}{4A} \quad \therefore R_2 = \dfrac{R}{4}$.

Question 7: H

The circuit described in the question is shown in the diagram below:

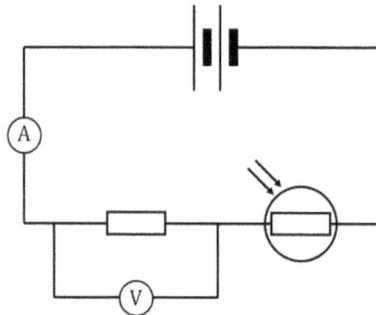

This is a potential divider circuit, where:

$$V = \frac{R_1}{R_1 + R_2}$$

The voltage is measured across the fixed resistor, so:

$$V = \frac{R_{fixed}}{R_{fixed} + R_{LDR}}$$

As light intensity decreases, the resistance of a light dependent resistor increases, so R_{LDR} increases and the voltage across the fixed resistor decreases according to the potential divider equation. This also means that the total circuit resistance increases and, for a constant applied voltage, Ohm's law shows that the current will decrease.

Question 8: D

Total internal reflection can only occur when light travels from a more optically dense medium into a less optically dense medium, excluding options A and B. Total internal reflection occurs when the angle of incidence is greater than the critical angle, which is equal to $\sin^{-1}(1/1.5) \approx 42$ degrees, which leaves option D as the correct answer.

Question 9: C

The frictional force on the 10kg block is given by $F_f = \mu R$, where R is the reaction force:

$$R = 10g = 100 \text{ N}$$

The resultant force on the block is therefore $mg - \mu R$, which can be substituted into Newton's second law using a desired acceleration of $a = 2$ ms^{-2}:

$$F = ma \Rightarrow mg - \mu R = 10a$$

$$10m - 0.8 \times 100 = 20 \Rightarrow 10m = 100 \Rightarrow m = 10 \text{ kg}.$$

Question 10: E

In both cases, the ball has only horizontal velocity as it leaves the slope, and so its vertical component of motion remains unchanged and it takes 3 s to land. As it is released from a higher position, the ball will be moving with a greater horizontal speed as it leaves the slope. In 3 seconds, it will therefore move further than 2 m before it lands.

Question 11: C

As air resistance is neglected, the only components of energy that the ball has are kinetic energy and gravitational potential energy. Conservation of energy states that:

$$(KE_{final} - KE_{initial}) + (GPE_{final} - GPE_{initial}) = 0$$

Take the initial state as the point at which the rock leaves the catapult and the final state as the rock's maximum height, when its speed is zero:

$$\left(0 - \frac{1}{2}mv^2\right) + (mg \times 85 - mg \times 5) = 0$$

The mass of the rock can now be cancelled out:

$$-\frac{1}{2}v^2 + g(85 - 5) = 0 \Rightarrow 80 \times 10 = \frac{1}{2}v^2$$

$$v^2 = 1600 \Rightarrow v = 40\,\text{ms}^{-1}.$$

Question 12: E

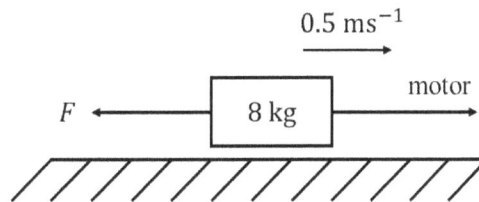

As the block is moving at a constant speed, the output force required from the motor is equal to the frictional force experienced by the block:

$$F = \mu mg \Rightarrow F = 0.6 \times 8 \times 10$$

$$\Rightarrow F = 48\,\text{N}$$

The power of the motor is given by force $\times$ velocity, and including the efficiency of 75%, gives the required input power as:

$$P = \frac{48 \times 0.5}{0.75} = 24 \times \frac{4}{3}$$

$$\Rightarrow P = 32\,\text{W}.$$

Question 13: D

The wavelength, λ, of a wave is given by:

$$\lambda = \frac{v}{f}$$

The frequency, f, is given by:

$$f = \frac{1}{T} = \frac{1}{1.5} = \frac{2}{3} \text{ Hz}$$

The speed, v, is given by:

$$v = \frac{60}{18} = \frac{10}{3} \text{ ms}^{-1}$$

Therefore, the wavelength is:

$$\Rightarrow \lambda = \frac{10}{3} \Big/ \frac{2}{3} = 5 \text{ m}.$$

Question 14: B

The total resistance of resistors in series is given by:

$$R_s = R_1 + R_2 + \cdots$$

The total resistance of resistors in parallel is by:

$$\frac{1}{R_p} = \frac{1}{R_1} + \frac{1}{R_2} + \cdots$$

Using the second formula for two identical resistors in parallel gives

$$\frac{1}{R_p} = \frac{1}{R} + \frac{1}{R} \Rightarrow R_p = \frac{R}{2}.$$

Using the formula for resistors in series shows that the resistance of each row is:

$$\Rightarrow R_{top} = R + \frac{R}{2} = \frac{3R}{2}, R_{middle} = R, R_{bottom} = 2R$$

The total resistance is therefore:

$$\frac{1}{R_{total}} = \frac{1}{R_{top}} + \frac{1}{R_{middle}} + \frac{1}{R_{bottom}} \Rightarrow \frac{1}{R_{total}} = \frac{2}{3R} + \frac{1}{R} + \frac{1}{2R}$$

$$\frac{1}{R_{total}} = \frac{4+6+3}{6R} \Rightarrow R_{total} = \frac{6R}{13}.$$

Question 15: F

First step is to find the height the ball travels upwards to the top of its arc (let this equal h_2):

Initial vertical component of velocity: $u_1 = 20\sin 30 = 10$.

Vertical velocity component at highest point: $v_1 = 0$. Acceleration: $a = -g$

Using the formula $v^2 = u^2 + 2as$:

$0^2 = (20\sin 30)^2 + (2 \times -g \times h_2) \Rightarrow 0 = 100 - 20h_2 \Rightarrow h_2 = 5$ m.

We know that it takes 3s for the ball to travel from its highest point down to the ground. At the highest point, the initial vertical velocity is 0 and it is accelerating towards the ground at g.

By calculating the total height the ball falls, s, we can calculate h using:

$\Rightarrow h = s - h_2 \Rightarrow s = ut + \frac{1}{2}at^2$

$s = 0 + \left(0.5 \times 10 \times (3^2)\right) \Rightarrow s = 45$ m.

$\Rightarrow h = 45 - 5 \rightarrow h = 40$ m.

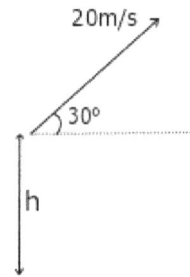

Question 16: D

We know that $(\sec x + \tan x)(\sec x - \tan x) = \sec^2 x - \tan^2 x$.

Using the trigonometric identity $\sec^2 x - \tan^2 x = 1$, as well as the information provided in the question, we know that:

$-5(\sec x + \tan x) = 1 \Rightarrow (\sec x + \tan x) = -\frac{1}{5}$.

By substitution, we know that $\sec x - \tan x + (\sec x + \tan x) = -5 + \left(-\frac{1}{5}\right)$:

$2\sec x = -5.2 \Rightarrow \sec x = -\frac{5.2}{2} = -2.6 = -\frac{13}{5}$.

Since $\sec x = \frac{1}{\cos x}$, $\cos x = \frac{1}{\sec x} = -\frac{5}{13}$.

Question 17: B

First, let us find the points along which any potential intersection between the line and the curve would take place, by setting the two equations equal to one another.

$x^2 + (3k - 4)x + 13 = 2x + k$
$x^2 + 3kx - 6x + 13 - k = 0$
$x^2 + 3(k - 2)x + 13 - k = 0$

Since the line and the curve do not intersect, we know that there must not be any real roots.

As such, by the discriminant condition, we know that $b^2 - 4ac < 0$.

$(3(k - 2))^2 - 4(13 - k) < 0$
$9(k^2 - 4k + 4) - 52 + 4k < 0$
$9k^2 - 32k - 16 < 0$
$(9k + 4)(k - 4)$

We know that the critical values therefore extend from $-\frac{4}{9} < k < 4$.

Question 18: A

The distance AC (equivalent to the radius of the circle) can be determined given the coordinates of A and C:

$A = (-2,1), C = (5,-3) \Rightarrow AC = \sqrt{(5 + 2)^2 + (1 + 3)^2} = \sqrt{65}$

To find the length of the line CT, we use Pythagoras' Theorem:

$CT^2 = AT^2 + AC^2 \Rightarrow CT^2 = 4^2 + 65$
$CT^2 = 81 \Rightarrow CT = 9$.

Question 19: E

$(6 \sin x)(3 \sin x) - (9 \cos x)(-2 \cos x) = 18 \sin^2 x + 18 \cos^2 x$

$\Rightarrow 18(\sin^2 x + \cos^2 x) = 18$.

Question 20: C

Define half the length of the inner equilateral triangle as *x*, and form a right-angled triangle by drawing a line from the centre of the inner circle to the inner triangle, defining the distance of that line as y.

$$\tan 30 = \frac{r}{x} \qquad\qquad\qquad\qquad \Rightarrow x = \frac{r}{\frac{1}{\sqrt{3}}} = \sqrt{3}r$$

$$\sin 30 = \frac{r}{y} \Rightarrow y = \frac{r}{1/2} = 2r$$

Using the formula for the area of a triangle, Area $= \frac{1}{2}ab \sin C$, in conjunction with the formula for area of a circle:

Area $= \pi r^2 \Rightarrow$ Area of the small circle $= \pi r^2$, Area of the big circle $= \pi(2r)^2 = 4\pi r^2$

Area of the small triangle $= \frac{1}{2}(2\sqrt{3}r)(2\sqrt{3}r)\sin 60 = 12\sqrt{3}\,r^2$

Therefore, the shaded area is: Shaded area $= (12\sqrt{3} - 4\pi + 3\sqrt{3} - \pi)r^2 = (15\sqrt{3} - 5\pi)r^2 = 5r^2(3\sqrt{3} - \pi)$.

Question 21: A

To binomially expand the expression given, we must rewrite it into the following form:

$$(3.12)^5 = (3 + 0.12)^5 = \left((3(1 + 0.04)\right)^5 = 3^5(1 + 0.04)^5$$

$$= 3^5(1 + 5(0.04) + 5\binom{4}{2}(0.04)^2 + \frac{5(4)(3)(0.04)^3}{6} + \cdots$$

$$= 3^5(1 + 0.20 + 0.016 + 0.00064) = 3^5 \times 0.00064 = 0.16.$$

Therefore, I must obtain 4 terms in the expansion.

Question 22: B

$(\sin(\theta) + \sin(-\theta))(\cos(\theta) + \cos(-\theta))$

$\Rightarrow (\sin\theta + -\sin\theta)(\cos\theta + \cos\theta) = 0(2\cos\theta) = 0$

Question 23: C

We can remove the modulus by squaring both sides:

$(2x - 5)^2 > \left(3(2x + 1)\right)^2$
$(2x - 5) = \pm 3(2x + 1)$

Solving this gives the critical values, the points of intersection, which are -2 and $\frac{1}{4}$. Therefore, for $(2x - 5)^2 > \left(3(2x + 1)\right)^2$, we require x to be within the range $-2 < x < \frac{1}{4}$.

Question 24: C

The midpoint of the two points given is (2,1), which is the centre of the circle. The distance between (1,-4) and (2,1) is $\sqrt{(2-1)^2 + \left(1-(-4)\right)^2} = \sqrt{26}$. This is the radius of the circle.

The equation of the circle pre-reflection, therefore, is $(x-2)^2 + (y-1)^2 = 26$.   Upon reflection in the line $y = x$, the x and y coordinates of the circle change places, but the radius remains the same. Thus, the equation of the circle becomes $(x-1)^2 + (y-2)^2 = 26$.

Question 25: D

Since the new computer does *a* calculation in *b* hours, it does $\frac{a}{60b}$ calculations in one minute. Simply add the individual rates together and multiply their sum by m minutes total to receive: $m\left(\frac{a}{60b} + \frac{c}{d}\right)$.

Question 26: D

Since -1 is a zero of the function, $(x+1)$ is a factor of the overall polynomial. By long division or synthetic division, we can determine that $\frac{2x^3+3x^2-20x-21}{x+1} = 2x^2 + x - 21$.

Factoring $2x^2 + x - 21 = 0$, we get: $(2x+7)(x-3) = 0$. Therefore, the roots are $x = -\frac{7}{2}$ or $x = 3$.

END OF SECTION

Section 2

Question 1

a) Answer: A

Resolve the velocity and acceleration parallel and perpendicular to the plane. Note, this is different to the regular x and y directions, so be careful with your algebra!

The perpendicular velocity is initially 0 and the acceleration is $g\cos\theta$ towards the plane. The distance to the plane is $h\cos\theta$. Apply $s = ut + \frac{1}{2}at^2$:

$$h\cos\theta = \frac{1}{2}g\cos\theta\, t^2 \ \Rightarrow\ t^2 = 2h/g \ \Rightarrow\ t = \sqrt{2h/g}.$$

b) Answer: A

Acceleration parallel to the plane is $g\sin\theta$ in the opposite direction to the initial velocity v. Therefore:

$$s = ut + \frac{1}{2}at^2 \ \Rightarrow\ s = vt - \frac{1}{2}g\sin\theta \times 2h/g \Rightarrow s = vt - h\sin\theta$$

However, the projectile starts a distance $h\sin\theta$ from B. Therefore, the projectile lands at a distance $v\sqrt{2h/g}$ from B.

c) Answer: F

For the projectile to land at a height h above the point B it must land at a distance of $h/\sin\theta$.

$$v = \frac{h}{\sin\theta}\sqrt{g/2h} \ \Rightarrow\ v = \frac{1}{\sin\theta}\sqrt{gh/2}.$$

d) Answer: E

The block takes $\sqrt{2h/g}$ to reach the plane. The acceleration perpendicular to the plane is $g\cos\theta$. Therefore, the velocity perpendicular to the plane is $\sqrt{2gh}\cos\theta$ when it lands, which is easily obtained using the SUVAT equation $v = u + at$.

The kinetic energies perpendicular and parallel to the plane are independent of each other, so the energy dissipated is simply $\frac{m}{2}\left(\sqrt{2gh}\cos\theta\right)^2$. This gives $mgh\cos^2\theta$.

Question 2

a) Answer: B

$P = I^2 R$. The units of current are amps, A, and units of resistance is ohms, Ω. Therefore $A^2\Omega$ is indeed a correct unit for power.

b) Answer: B

The sum of the currents through B and C is equal to the current through A – this is one of Kirchoff's laws. The potential difference is equal for the resistors in parallel, where as current is divided between the branches.

c) Answer: E   Resistors in parallel add in reciprocal.

$$\Rightarrow \frac{1}{R} + \frac{1}{R} = \frac{2}{R}$$

Therefore, the resistance of B and C together is R/2. The resistances in series then add to give 5R/2.

d) Answer: D   Resistance is given by: $R = \frac{\rho L}{A}$. The voltages across B and C are equal.

Require the voltage across the parallel resistors to be V/4, which is clear if you consider this problem as a potential divider. Require:

$$\frac{R_{BC}}{R_A + R_{BC}} = \frac{1}{4} \Rightarrow 3R_{BC} = R_A \Rightarrow R_{BC} = \frac{2R}{3}.$$

Adding resistors in parallel gives:

$$R_{BC} = \frac{R_B R_C}{R_B + R_C} \Rightarrow 2R/3 = \frac{R R_C}{R + R_C} \Rightarrow R_C = 2R.$$

Therefore, we need half the area to give twice the resistance, so the radius of C must be $r/\sqrt{2}$.

Question 3

a) Answer: B

$v(t) = 10t^2 - t^3 - 24t$. This factorises to $v(t) = -t(t-4)(t-6)$.

Therefore, the object is stationary at t = 4s and t = 6s after the initial t = 0 s.

b) Answer: E

Displacement is the integral of velocity with respect to time.

$$\Rightarrow s(t) = -\frac{t^4}{4} + \frac{10t^3}{3} - 12t^2$$

By inputting t = 2s into this equation, we obtain the position at that instant as -76/3.

c) Answer: C

Maximum positive acceleration occurs at a stationary point of the acceleration. Differentiate velocity for acceleration:

$$\Rightarrow a(t) = -3t^2 + 20t - 24$$

Now differentiate a(t) again to find stationary points.

$$\Rightarrow \frac{da}{dt} = -6t + 20$$

Therefore, maximum acceleration is at t = 10/3 s. It is clear that it is a maximum as the acceleration is a negative quadratic. You can also find the second derivative to verify the nature of the stationary point.

d) Answer: B

The objects momentum changes direction during collision so momentum must be transferred to the wall during the collision, so it is conserved. The ball does not lose energy as its speed is the same before and after.

Question 4

a) Answer: F

Resolve the weight of the block parallel and perpendicular to the slope. This gives $N = mg \cos \theta$, as the components perpendicular to the plane must be balanced.

b) Answer: F

The frictional force is given by μN. The work done by a force is given by $W = F \times d$. As such the work done by the frictional force is $\mu N \times d$, we can use the answer from part a to obtain:

$\Rightarrow W = \mu mgd \cos \theta$.

c) Answer: C

The force down the slope is given by: $mg \sin \theta - \mu mg \cos \theta$.

Therefore, the energy converted into kinetic energy is:

$\Rightarrow dmg(\sin \theta - \mu \cos \theta) = mv^2/2$

$\therefore d = v^2/(2g(\sin \theta - \mu \cos \theta))$.

END OF PAPER

FINAL ADVICE

Arrive well-rested, well-fed and well-hydrated

The ENGAA is an intensive test, so make sure you're ready for it. Ensure you get a good night's sleep before the exam (there is little point cramming) and don't miss breakfast. If you are taking water into the exam, make sure you've been to the toilet before, so you don't have to leave during the exam. Make sure you're well rested and fed in order to be at your best!

Move on

If you find yourself struggling on a particular question, move on. Every question has equal weighting and there is no negative marking. In the time it takes to answer one hard question, you could gain three times the marks by answering the easier ones. Be smart to score points - especially in Section 2 where some questions are far easier than others.

Afterword

Remember that the route to a high score is a methodical approach and consistent practice. Do not fall into the trap that "*you can't prepare for the ENGAA*"– this could not be further from the truth. With knowledge of the test, some useful time-saving techniques, and plenty of practice, you can dramatically boost your score.

Work hard, never give up and do yourself justice.

Good luck!

Acknowledgements

I would like to express my sincerest thanks to the many people who helped make this book possible, especially the 15 Oxbridge Tutors who shared their expertise in compiling the huge number of questions and answers.

Rohan

About Us

We currently publish over 85 titles across a range of subject areas – covering specialised admissions tests, examination techniques, personal statement guides, plus everything else you need to improve your chances of getting onto competitive courses such as medicine and law, as well as into universities such as Oxford and Cambridge.

This company was founded in 2013 by Dr Rohan Agarwal and Dr David Salt, both Cambridge Medical graduates with several years of tutoring experience. Since then, every year, hundreds of applicants and schools work with us on our programmes. Through the programmes we offer, we deliver expert tuition, exclusive course places, online courses, best-selling textbooks and much more.

With a team of over 1,000 Oxbridge tutors and a proven track record, UniAdmissions have quickly become the UK's number one admissions company.

Visit and engage with us at:

Website (UniAdmissions): www.uniadmissions.co.uk

Facebook: www.facebook.com/uniadmissionsuk

What's Next?

- ✓ Three books crammed into one!
- ✓ Includes complete Course profiles and College insights from Oxbridge admissions tutors and current students
- ✓ 100+ of examples of real personal statements written by Oxbridge Applicants
- ✓ Each statement analysed with feedback
- ✓ Covers every major subject at Oxford & Cambridge
- ✓ Work experience advice- how to arrange it and how to stand out from the crowd

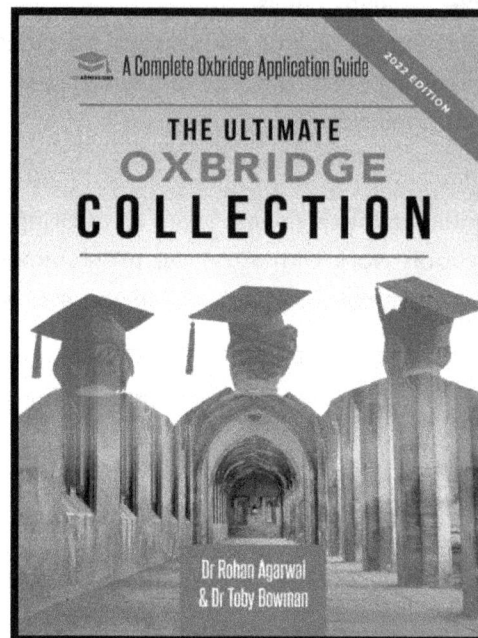

A Complete Oxbridge Application Guide
2022 EDITION
THE ULTIMATE
OXBRIDGE
COLLECTION
Dr Rohan Agarwal & Dr Toby Bowman

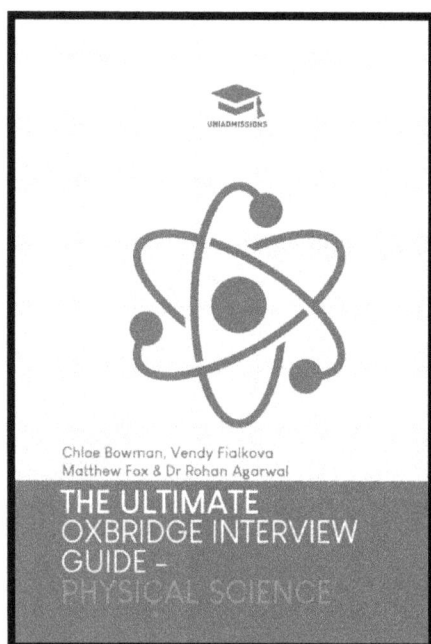

UNIADMISSIONS
Chloe Bowman, Vendy Fialkova
Matthew Fox & Dr Rohan Agarwal
THE ULTIMATE
OXBRIDGE INTERVIEW
GUIDE –
PHYSICAL SCIENCE

- ✓ 200+ Real Oxbridge interview Questions
- ✓ Covers Engineering, Natural Sciences, Chemistry, Material Sciences, Physics, Maths, Computer Sciences, Earth Sciences interview questions at Oxford and Cambridge
- ✓ Includes model answers for every question with commentary from Admissions Tutors
- ✓ Written by Cambridge & Oxford Science Tutors & Professors

YOUR FREE BOOK

Thanks for purchasing this Ultimate Collection Book. Readers like you have the power to make or break a book –hopefully you found this one useful and informative. *UniAdmissions* would love to hear about your experiences with this book. As thanks for your time, we'll send you another ebook from our Ultimate Guide series absolutely <u>FREE</u>!

★ ★ ★ ★ ★   ✓ Posted publicly as Amazon Customer | Edit

Write your review here

How to Redeem Your Free Ebook:

1) Either scan the QR code or find the book you have on your Amazon purchase history, or your email receipt, to help find the book on Amazon.

2) On the product page at the Customer Reviews area, click 'Write a customer review'. Write your review and post it! Copy the review page or take a screen shot of the review you have left.

3) Head over to www.uniadmissions.co.uk/free-book and select your chosen free ebook!

Your ebook will then be emailed to you – it's as simple as that!

Alternatively, you can buy all the titles at:

www.uniadmissions.co.uk